Knowledge For The Anthropocene

MULTIDISCIPLINARY MOVEMENTS IN RESEARCH

This high-quality series aims to publish reviews and original research in areas crossing the traditional boundaries between academic disciplines. As the value of multidisciplinary collaboration becomes clearer, the academic world is increasingly moving away from siloed research practices towards a more nuanced and open understanding of the challenges and complexity facing society today. Learning from, and recognizing the values and insights of multiple disciplines, can result in new and creative ways of thinking, and produce high-impact research to help address intractable problems.

| 8 | Localization and globalization of core adaptive knowledge
Alexander K. Lautensach | 107 |

PART III   ANTHROPOCENE ECONOMICS

9	The end of Holocene economics Richard Heinberg	120
10	Precursors of an economics for the Anthropocene Daniel Dahm and Günter Koch	132
11	Deep adaptation and collapsology Jason Monios and Gordon Wilmsmeier	145
12	Genuine savings and economics for the Anthropocene Eoin McLaughlin and Cristián Ducoing	157

PART IV   JUSTICE IN THE ANTHROPOCENE

13	Epistemic injustice Sabelo J. Ndlovu-Gatsheni	167
14	The urgency for epistemic and political climate justice Jacobo Ocharan, Velina Petrova and Irene Guijt	178
15	Towards global environmental governance Julia M. Puaschunder	194
16	Transition agendas: going beyond consumerism? Boris Manov and Asen Balabanov	204

PART V   KNOWLEDGE SYSTEMS FOR THE ANTHROPOCENE

17	Scientific knowledge for the Anthropocene Marc Zimmer	213
18	The sciences of knowledge Francisco Javier Carrillo	225
19	Knowledge as world capital: global knowledge Alexander Ruser	240
20	Adaptive value of traditional knowledge Michael Blakeney	249

Contents

List of figures	x
List of contributors	xi
Foreword: knowing what to know, what to do and how to do it in the Anthropocene	xiii
Noel Castree	
Preface	xvi
Acknowledgments	xxi

Introduction to *Knowledge For The Anthropocene*   1
Francisco Javier Carrillo

PART I   KNOWLEDGE AND THE PLANETARY EMERGENCY

1   A portable philosophy toolkit for the Anthropocene   11
     Carlos Jesús García-Meza

2   Existential challenges to knowledge   22
     Bertrand Guillaume

3   Social psychological drivers of climate change denial   30
     Irina Feygina

4   Media accountability before the climate crisis   42
     Gabriel Valerio-Ureña, Jorge Asprón and Nalleli Salazar

PART II   ANTHROPOCENE LITERACY

5   A terminology for the Anthropocene   54
     Ernesto Contreras

6   A directory of digital resources about the Anthropocene   76
     Paulo David Soasti-Bareta

7   Educating for the Anthropocene   98
     Audrey Groleau, Chantal Pouliot, Isabelle Arseneau

DEDICATION

We dedicate this volume to Paulo David Soasti-Bareta for his generous support in the early stages of the editorial process and his commitment to the WCI.
Francisco Javier Carrillo and Günter Koch

© Francisco Javier Carrillo and Günter Koch 2021

All rights reserved. No part of this publication may be reproduced, stored in a retrieval system or transmitted in any form or by any means, electronic, mechanical or photocopying, recording, or otherwise without the prior permission of the publisher.

Published by
Edward Elgar Publishing Limited
The Lypiatts
15 Lansdown Road
Cheltenham
Glos GL50 2JA
UK

Edward Elgar Publishing, Inc.
William Pratt House
9 Dewey Court
Northampton
Massachusetts 01060
USA

A catalogue record for this book
is available from the British Library

Library of Congress Control Number: 2021947585

This book is available electronically in the **Elgar**online
Geography, Planning and Tourism subject collection
http://dx.doi.org/10.4337/9781800884298

ISBN 978 1 80088 428 1 (cased)
ISBN 978 1 80088 429 8 (eBook)

Printed and bound in Great Britain by TJ Books Limited, Padstow, Cornwall

Knowledge For The Anthropocene
A Multidisciplinary Approach

Edited by

Francisco Javier Carrillo

The World Capital Institute and Emeritus Professor of Knowledge Based Development, Tecnológico de Monterrey, México

Günter Koch

Humboldt Cosmos Multiversity, Tenerife, Spain

MULTIDISCIPLINARY MOVEMENTS IN RESEARCH

Cheltenham, UK • Northampton, MA, USA

PART VI   IMAGINATION IN THE ANTHROPOCENE

21	Designing post-human futures *Raphaële Bidault-Waddington*	263
22	Integral ecology: reconnecting nature, culture, and knowledge *Sam Mickey*	276
23	Visuality conditions under the Anthropocene *Irmgard Emmelhainz*	284
24	The aesthesis of plastic capitalism *Amanda Boetzkes*	297

PART VII   CO-CREATING FUTURES

25	Democracy in the Anthropocene *David W. Orr*	307
26	Envisioning scenarios for the Anthropocene *David Arthur Sampson*	316
27	The farthest we can see *Anthony Hodgson*	328
28	Knowledge for the Anthropocene: an agenda *Francisco Javier Carrillo*	339

Conclusion to *Knowledge For The Anthropocene*   358
Günter Koch

Index   365

Figures

12.1	Global CO_2 social cost under two scenarios. CO_2 estimated at $131 and $1455	160
12.2	Genuine savings and GDP per capita in USA, UK, Australia and Chile. 1870–2000	162
14.1	The carbon inequality 'dinosaur' of emissions growth from 1990 to 2015	180
21.1	Toward Alien Cosmologies, critical future topology (V6)	268
24.1	Jim Shaw, Heap, 2009. McDonaldland toys, Styrofoam, metal rods, 64 × 24 × 77 IN. (162.6 × 61 × 195.6 CM)	298
24.2	Chris Jordan, Midway: message from the gyre (albatross), 2011	303
27.1	The multidimensionality of the present moment	330
27.2	Taking a long view in three horizons	333
27.3	The current displacement between old and new epochs	336

Contributors

FRONT MATTER

Francisco Javier Carrillo, The World Capital Institute and Tecnológico de Monterrey, México

Noel Castree, University of Technology Sydney, Australia and Manchester University, UK

Günter Koch, The World Capital Institute and Humboldt Cosmos Multiversity, Spain

CHAPTERS

Isabelle Arseneau, Université Laval, Canada

Jorge Asprón, Tecnológico de Monterrey, México

Asen Balabanov, South West University "Neofit Rilski", Bulgaria

Raphaële Bidault-Waddington, LIID Future Lab, France

Michael Blakeney, The University of Western Australia, Australia

Amanda Boetzkes, University of Guelph, Canada

Francisco Javier Carrillo, The World Capital Institute and Tecnológico de Monterrey, México

Ernesto Contreras, Tecnológico de Monterrey, México

Daniel Dahm, World Future Council & United Sustainability Group, Germany

Cristián Ducoing, Lund University, Sweden

Irmgard Emmelhainz, Independent writer, researcher, and lecturer, Mexico City, Mexico

Irina Feygina, Behavioral Science Consultant for Climate and Clean Energy, New York, USA

Carlos Jesús García-Meza, Tecnológico de Monterrey, México

Audrey Groleau, Université du Québec à Trois-Rivières, Canada

Irene Guijt, Oxfam, UK

Bertrand Guillaume, University of Technology of Troyes, France

Richard Heinberg, Post Carbon Institute, USA

Anthony Hodgson, Trustee, H3Uni, UK

Günter Koch, The World Capital Institute and Humboldt Cosmos Multiversity, Spain

Alexander K. Lautensach, University of Northern British Columbia, Canada

Boris Manov, South West University "Neofit Rilski", Bulgaria

Eoin McLaughlin, University College of Cork, Ireland

Sam Mickey, University of San Francisco, USA

Jason Monios, Kedge Business School, France

Sabelo J. Ndlovu-Gatsheni, University of Bayreuth, Germany

Jacobo Ocharan, Oxfam, Spain

David W. Orr, Oberlin College, USA

Velina Petrova, Oxfam, USA

Chantal Pouliot, Université Laval, Canada

Julia M. Puaschunder, The New School, USA

Alexander Ruser, University of Agder, Norway

Nalleli Salazar, Tecnológico de Monterrey, México

David Arthur Sampson, Arizona State University, USA

Paulo David Soasti-Bareta, Tecnológico de Monterrey, México

Gabriel Valerio-Ureña, Tecnológico de Monterrey, México

Gordon Wilmsmeier, Universidad de Los Andes, Colombia

Marc Zimmer, Connecticut College, USA

Foreword: knowing what to know, what to do and how to do it in the Anthropocene

Noel Castree

Since around 2010 the term 'Anthropocene' has become a buzzword in many academic circles. It names a new, all-encompassing and threatening reality that demands a global and highly sustained response. A concept initially coined and given substance by geoscientists, in recent years numerous social scientists and humanists have sought to understand the causes, effects and implications of the Anthropocene. For researchers across the disciplines, the Anthropocene is breaking down old analytical and topical barriers, even as it reveals sharp differences in cognitive and normative perspectives. Many in the arts have also joined the fray, within and beyond the universities. By the 2030s, the term is likely to have become a societal keyword across the globe. People in all walks of life will come to realise just how profound human impacts on the atmosphere, biosphere, cryosphere, hydrosphere and lithosphere actually are – in large part because various 'epistemic communities' in universities and government agencies will continue to provide robust explanations, arresting evidence and worrying forecasts pertaining to Earth System change. If 'the Anthropocene' signifies anthropogenic impacts on pretty much *everything* in the physical environment, it also implies the need to also change *everything* in the ways of thinking and acting that have squeezed the Earth out of its Holocene envelope. And make no mistake: changes to thought and action need to happen *now* because the future effects of business-as-usual will be very significant indeed for the generations to come, and for non-human species, processes and phenomena. It will also require coordinated action: a few nations acting alone will not suffice. But the 'wickedness' of the Anthropocene problem, the variety of possible means and ends required to 'fix' it, and the inertia caused by those favouring the status quo will make action very slow, very fragmented and highly contested. Where does 'knowledge' feature in this incipient socio-ecological transformation? What kinds of knowledge are required to steer a path forward? Whose knowledge should exert influence in the world to come? What areas of ignorance do we need to shine a light on with new or repurposed knowledge? How can different bodies of knowledge be brought into a productive dialogue? And can knowledge be deployed effectively and quickly?

This book seeks to answer these large and very important questions. Traditionally, at least in the West, knowledge meant 'justified true belief'. It was predominantly associated with the natural sciences, the validity of whose knowledge was demonstrated practically through various technologies ('if it works it must be true!'). But, for all their benefits, the knowledge and technologies posed certain risks to human health and the physical environment. In many cases, science and technology became vectors of violence, as with nuclear weapons and the agricultural machinery that remade the ecology of New South Wales, where I currently reside. Despite its supposed autonomy and objectivity, as far back as the 1960s science was shown to be entangled in deeply value-laden social relationships, institutional practices and political agendas cross-cut by struggles for power and influence. As its size and scope increased (in the realms of higher education, government and business), it became ever more contested, throwing-up large ethical questions about who and what benefits and loses in the name of 'scientific progress'. This is not to say that we need to relinquish science and technology in order to create a 'good Anthropocene'. But we do need to ensure that they operate in a wider network of knowledge and practice that will ensure they're governed in ways that include diverse perspectives on how best to rise to the 'Anthropocene challenge'.

In many cases, these perspectives challenge older assumptions about what counts as 'knowledge'. Knowledge is not always about representing 'the truth' because 'justified belief' can relate to ethics, norms and values as much as to facts, objects, or material processes. In addition, 'truth seeking' always occurs in the context of contingent decisions about what's worth knowing and what's not. Knowledge can also be tacit and expressed through learnt practices rather than being made explicit and enduring in research articles, books or manuals. 'Know what' and 'know how' can therefore come in many forms, and be efficacious in very diverse contexts. But the reality of modern life is that certain forms of knowledge have come to dominate thought and action in key domains. For instance, in universities, businesses and state treasuries, particular concepts and methods of economics long ago became normalised globally – one adverse effect of this was the global financial crisis of 2008–9. Looking ahead, can we displace 'hegemonic knowledge' when responding to the Anthropocene challenge? If not, what sort of new hegemonic knowledge do we need to institute? And can we elasticate what counts as knowledge in order to let other voices be heard, so as to mitigate at least some of the epistemic exclusions that hegemony necessarily creates?

Here we face some very significant challenges. On the one hand, geoscientists are sounding the alarm, with many declaring a 'planetary emergency'.

But on the other hand, our economic and social context is not conducive to constructive dialogue about that emergency. We see deep political divisions within and between the most powerful countries (e.g. the USA, China and Russia); at the time of writing we have a global pandemic that's dominating human affairs to the exclusion of almost everything else; we have a coterie of business leaders who have an awful lot to lose by any switch towards 'green growth', let alone 'de-growth'; we have a fragmented news media landscape, along with ascendant social media, and together they are eroding the basis for shared understanding while fanning the flames of discord; and we have a proliferation of 'alternative facts' and 'fake news' whereby knowledge in its various forms is being substituted by opinion, conviction and wishful thinking. In short, ours is a fractured and very unhappy world at a time when common purpose based on shared values and knowledge is imperative. If the global response to anthropogenic climate change teaches us anything, it's that denial and conflict will precede any meaningful response to the very real threat of a 'bad Anthropocene'. Meanwhile, another kind of response – equally concerning – looms on the horizon: that of science and technology offering us 'emergency solutions' to the Anthropocene challenge that are enacted quickly in 10 or 20 years from now (without proper consultation) and which produce a set of irreversible and undesirable knock-on effects at various spatio-temporal scales.

The chapters of this book cannot provide definitive answers to the profound issues they explore. But they can help us on the journey towards a knowledgeable response to the epic changes human activity has unintentionally instigated. The journey will be incredibly arduous. Unspeakable harm will be inflicted; naïve hope will be trampled under the weight of too many unwelcome events. But the journey may also bring the very best out of people. We desperately need a kinder, more cooperative, more reflective society to emerge from the rubble of neoliberal globalisation. Acute income inequality, the forces of economic competition, the dominance of secular rationality, the prevalence of consumerism, the doctrine of endless growth, the lingering effects of colonialism, religious intolerance, cultural enmity, the hollowing-out of 'public goods', and the persistence of aggressive nationalisms: somehow we need knowledge that can help us tackle all these things while also preventing runaway change to our precious planet. The knowledge will stretch far beyond the realms of science and technology. But for knowledge to matter we also need thought leadership, courageous people willing to speak out, the right institutions to connect all the stakeholders, and the requisite resources to enhance our collective literacy about 'the human planet' that is ours to protect.

Preface

CONTEXT

This book and its twin volume, *City Preparedness for the Climate Crisis*, are part of current work at the World Capital Institute (WCI) in order to evolve the fields of Knowledge-Based Development and Knowledge Cities – the scientific and technical sides of the WCI work – into those of *Knowledge for the Anthropocene* (Volume 1) and *City Preparedness for the Climate Emergency* (Volume 2). The choice faced by the WCI is the same as many individuals and organizations face in their daily experience: coming to terms with Anthropocene realities as the single most pressing issue of our age and the need to re-invent our lives accordingly.

Both volumes work as mutual companions and together add to a conceptual and practical approach towards the Anthropocene. These two volumes carry a formal symmetry, while the contents are quite specific to each. The original selection of topics for both volumes was based on prior work that led to the ongoing transformation of the WCI work in the second decade of our century.

The World Capital Institute is an independent international think- and do-tank "whose purpose is to further the understanding and application of knowledge as the most powerful leverage of development" (WCI webpage at www.worldcapitalinstitute.org). For that purpose, it has created and internationally propelled the discipline of Knowledge-Based Development (KBD) and its application to urban development under the better-known category of the Knowledge City, which stands for a best model representation for a socio-economic and infrastructural pattern. KBD and the Knowledge City aim at a balance of all collective capital forms (both tangible and intangible) in the evolution of a human activity system. Under this perspective, the main societal performance benchmark is not monetary mass increase as expressed by the GDP (Gross Domestic Product), but the improvement in the everyday and long-term experience of the majority of citizens on the basis of knowledge creation and distributed capitalization. The bulk of the literature on Knowledge Cities has been led by WCI associates and the leading journal in the field – the *International Journal of Knowledge Based Development* (IJKBD), one of the offsprings of WCI, was created and continues to be edited by prominent WCI members. Additionally, the WCI conveys the main annual conference on KBD and K-Cities: The Knowledge Cities World Summit (KCWS) as well as the annual Most Admired Knowledge City Awards (MAKCi).

In public discourse, the Knowledge City is often equated with the Smart City, which we believe is an inadmissible simplification. The term Smart City was originally coined from the side of technologists and from industry and is still understood by the majority today as a city that is organized for the benefit of its citizens, in particular by means of information technology. In our sense, the Knowledge City seeks a communal value balance, which brings us to the topic of this first volume, namely how a public policy can be shaped using all its potentials and in all dimensions of social existence (Carrillo, 2006; Carrillo et al., 2014; Koch, 2016).

Knowledge-based Development (KBD) evolved from the fields of Knowledge and Intellectual Capital Management at the organizational level to the wider realm of societal value and knowledge base. KBD focused on the intangible (or 'knowledge-based') collective value, that is, all those forms of capital beyond the traditional financial and physical assets that can leverage social development. Hence, KBD has been defined as "the collective identification and enhancement of the value set whose dynamic balance furthers the viability and transcendence of a given community" (Carrillo, 2014: 416).

The promise of KBD has been to contribute to the understanding and design of human coexistence in knowledge-intensive societies. Such understanding involved a major challenge for it implied coming to terms with knowledge as a natural and behavioural phenomenon as well as with its economic implications. Advancing on this endeavour helped to set up an agenda for knowledge societies beyond economic productivity, rather aiming at a qualitative evolution in the human condition.

But if KBD is about net future social value and related human preferences, then the realities of the Anthropocene, as Naomi Klein would have it, put our whole value systems in perspective (Klein, 2015). Therefore, the WCI has increasingly shifted its focus over the last years towards the knowledge implications of and for the Anthropocene. In particular, it has paid attention to two key concerns that provide continuity to its developmental and urban traditions: (a) what is the social role of knowledge and what new shape might it take as the realities of the Anthropocene unfold? And (b) how can cities be best prepared to deal with the upcoming challenges of the Anthropocene?

The move from Knowledge-Based Development and Knowledge Cities towards Knowledge for the Anthropocene and City Preparedness for the Climate Crisis is highlighted by the recent 'Anthropocene Turn' as described by an editorial in the *International Journal of Knowledge-Based Development* (Carrillo, 2019). This book now is right at the core of a movement that is bound to attract increasing interest as we enter the Anthropocene epoch.

BOOK PLAN

This book conveys a wide spectrum of scientists, philosophers, activists, decision-makers, artists and humanists to tackle these questions from a trans-disciplinary perspective. The content structure is aimed at a general academic audience, rather than the specialists of each particular field. Even for specialists, the interest and value of this work lies in the interconnections and cross references that are made possible by the mosaic of data, theories, cases, models, methods and experiences collected in this book. Those interconnections will be underscored by a general conclusion, emphasizing some of the links as they emerge from the joint reading of all chapters. Individually, each chapter serves as reference to a specific topic. The aim is not to provide a consistent or complete picture of all the challenges we are confronted with in the Anthropocene, but rather to collect a multitude of approaches and arguments that need to be considered when discussing such a complex topic.

Even with a wider academic audience in mind, an emphasis has been given to readability and clarity. Each relatively brief chapter is intended to contribute a contemporary perspective on a selected topic. The philosophical and political standpoints, theoretical and methodological approaches – although generally convergent – allow for a naturally broad range of points of view. Writing styles also vary from a colloquial essay to the empirical analysis, from the journalistic to the strictly academic, from the monograph to the artistic rendering. The underlying belief is that the reader will benefit from an adequate set of realizations and analyses integrated under the common challenges of Knowledge for the Anthropocene and hopefully will use it as a source for further elaborations.

The book is organized in seven parts, going from the more fundamental ethical, epistemic, ontological and aesthetic foundations to the political, scientific, managerial and technical.

Part I introduces the reader to the meaning of Anthropogenic Existential Risks, their philosophical implications and the behavioural and public opinion challenges it poses to Knowledge Systems. While reference is constantly being made to physical science evidence for climate change, its relative presence is minor given that Earth, Atmospheric, Environmental, and Life Sciences are sufficiently well documented elsewhere.

Instead, Part II serves a reference purpose, with Chapter 5 providing a glossary of terms balancing the physical sciences terms with those from the social and behavioural science and humanities. Also, Chapter 6 provides a directory of organizations and movements concerned with the Anthropocene thus serving as a source of information for further investigations – a subject of great concern for the WCI.

The bulk of chapters deal with behavioural, economic, social and cultural aspects including economic aspects of the Anthropocene and alternative economic cultures (Part III), and issues of ethics and justice (Part IV). Part V looks into the challenges to knowledge systems, including formal and informal knowledge. The final chapters consider some of the conditions from economics, politics and art determining potential outcomes, including explorations of the limits of human imagination (Part VI). The final part (VII) looks at possible futures and our means to envisage them. The book closes with a set of conclusions by Günter Koch, offering his interpretation of the common emerging threads, including main lessons, new questions and further research.

The intended target for this book is the academic public at large, including researchers and graduate students as well as science communicators. It can also be of interest for general graduate courses on climate change. To specialists on social issues of the Anthropocene and Environmental Sciences, it provides a novel look at the central matter of what new significance the socio-economic and cultural knowledge base can acquire before the Anthropocene. We also would hope that it may serve as a reference to journalists and the media, as a competent source for elaborations on the future of the planet and societies in this new age of Anthropocene.

This would be the first volume to directly deal with the subject of Knowledge for the Anthropocene. A growing number of climate scientists, philosophers, environmental activists, decision-makers, politicians, social scientists and the public at large are beginning to realize the inadequacy of the current scientific and education establishment to meet the likely scenarios of the climate emergency. If the COVID-19 pandemic, which governed during the period when this book was compiled, is any indication, the lack of health systems preparedness, sufficient infrastructure and medical systems and services personnel, alert networks for adequate timing, scientific advice integration, global crisis governance, R&D coordination efforts, proper social communication tools, economic and social response capacities, to name but a few, experienced even by some of the most capital and technology-intensive societies, is a major reason for concern. It is a cause of no small wonder that even the most powerful country in the world is also the one where the pandemic toll in both confirmed cases and deaths in absolute numbers has been the highest at the time when this book was compiled. Whatever the final global pandemic cost might be in human lives, suffering and economic stagnation, it might pale in comparison with the foreseeable scenarios posed by Anthropogenic Global Existential Risks. New climate research tends to increasingly suggest that such scenarios are looking worse and closer by the day.

However, it must also be acknowledged that new forms of scientific and diplomatic cooperation during the pandemic led to the huge accomplishment of developing alternative vaccines in record time through several international cooperation mechanisms. To the extent knowledge is still regarded as a global common good, the potential for international synergies exist.

We now turn firsthand to the research community in the hope that this collective work opens up much needed collaborations across disciplines to consolidate the emerging field of Knowledge for the Anthropocene. If it contributes to raise new relevant and productive questions and suggest effective ways to tackle them, its purpose would have been accomplished.

Javier Carrillo and Günter Koch

REFERENCES

Carrillo, F. (Ed.) (2006). *Knowledge cities*. Elsevier.
Carrillo, F. J. (2014). What 'knowledge-based' stands for? A position paper. *International Journal of Knowledge-Based Development*, 5(4), 402–421.
Carrillo, F. J. (2019). The Anthropocene Turn in Knowledge Based Development. *International Journal of Knowledge-Based Development*, 10(4), 293.
Carrillo, F. J., Yigitcanlar, T., García, B., & Lönnqvist, A. (2014). *Knowledge and the city: Concepts, applications and trends of knowledge-based urban development*. Routledge.
Klein, N. (2015). *This changes everything: Capitalism vs. the climate*. Simon & Schuster.
Koch, G. (2016). Knowledge Cities vs. Smart Cities – A discussion triggered on the case of Timisoara, Romania. In: Conference Proceedings of the 14th NETTIES Conference on "Smart Learning". Polytechnical University of Timisoara, Romania. Published by IAFeS, Vienna.

Acknowledgments

This book was conceived as COVID-19 emerged, configured as it spread globally, carried out through the pandemic and finalized as it peaked. Every stage of the editorial process became far more complex and required so much more effort than it would normally take. Like in almost all fields of human activity, the direct and indirect effects of the pandemic presented an obstacle at times seemingly unsurmountable. In the end, the collective energy and positive spirit was able to complete the project in due time and form.

It was only thanks to that common sense of purpose and determination that this project was completed. We want to acknowledge each and every one of the women and men who have vested their best intellectual and emotional capacities into finalizing this work.

We want to acknowledge the collective support provided by the WCI team. As the twin volume project (this book, *Knowledge For The Anthropocene* together with its twin volume *City Preparedness for the Climate Crisis*, both as part of Edward Elgar's *Multidisciplinary Movements in Research* series) emerged from strategic planning at WCI, this initiative should be seen as the fruit of that communitarian endeavour. The whole team took part in the sessions leading to the strategic focus shift towards Knowledge for the Anthropocene and City Preparedness for the Climate Crisis focus. Several of them contributed substantial chapters on either volume. Above all, we want to thank David Soasti for the key role he played in the early stages. One of the biggest challenges of the editorial plan was to gather an international team of specialists in a broad range of specialized topics. David generously provided the patience and dedication to take on the correspondence for scouting such a diverse set of authors. Thanks to his steadfastness and cheerful devotion to the task, that challenging stage was dully accomplished. As a token of appreciation this book is dedicated to him.

A total of 38 authors of 17 different nationalities have taken part in this work. Each of them has a story to tell as to the obstacles the pandemic presented to complete their contributions. From health and family crises through job disruption, environmental hazards and country migration, all have been somehow affected by the pandemic, just as everyone else was the world over. The vast majority of them were able to complete their contributions and endure the review process and subsequent draft iterations. To each of them we acknowledge their professionalism and kind disposition.

A special thanks is due to Noel Castree, who has written the preface to this book. As we feel honoured by his generous acceptance, we acknowledge his unique contribution in the direction this book is pointing towards. A critical and interdisciplinary stand by the social sciences and the humanities is required before the pressing environmental issues of our times, as much as a bridge and transdisciplinary dialogue with the natural sciences of the Anthropocene. He has opened several productive avenues in that regard throughout his research career. Thank you, Noel.

Finally, we want to acknowledge the high standards and friendly attitude from our colleagues at Edward Elgar. Stephanie Hartley provided an effective and kind-hearted guidance throughout the whole editorial process. The suggestion to allocate both volumes within the "Multidisciplinary Movements in Research" series was particularly wise. The whole team taking care of the copyediting and proofreading tasks was both highly professional and kind.

This book is therefore the outcome of a fortunate addition of talent and goodwill. We are thankful to all who made it possible.

Introduction to *Knowledge For The Anthropocene*

Francisco Javier Carrillo

Why 'Knowledge for the Anthropocene'? How is this title relevant? Basically, it refers to how these two terms relate to each other. How we understand each of them on their own and then how a new relationship can be constructed between them. Such relationship can in turn set up new inter-determinacies, co-evolving in their subsequent understanding.

The term Anthropocene was introduced to name the proposed geological epoch following the Holocene (Crutzen & Stoermer, 2000). It is characterized by an overwhelming mark of human activity on Earth so as to leave a geo-stratigraphical record (Zalasiewicz et al., 2010; Zalasiewicz et al., 2019). While there is ongoing debate regarding the starting date of the Anthropocene, a favoured interpretation is to associate it to the 'Great Acceleration': the surge of human impact on Earth since the mid-20th century (Steffen, Broadgate et al., 2015; McNeill & Engelke, 2016). Given the enormous significance of these facts and the observed and potential consequences for the Biosphere, the term has also been adopted to comprise its wider social, economic and cultural implications (Castree, 2017; Cohen & Colebrook, 2017; Malabou, 2017; Clark & Szerszynski, 2021). This book adopts the wider use insofar as it conveys an essentially transdisciplinary 'Anthropocenic turn' in contemporary culture (Oldfield et al., 2014; Hamilton et al., 2015; Carrillo, 2019; Dürbeck & Hüpkes, 2020; Krogh, 2020).

Anthropogenic environmental impacts are heading to disrupt our way of life in deeper ways and at a wider scale than anything previously experienced by mankind. The public imaginary and the media often reduce the Anthropocene challenges to both Climate Change and Global Warming. Within this narrative, even when there is an acceptance that these are caused by human action, the tone is of some climate discomfort and eventual weather hazards that can be dealt with by greater resilience and appropriate technology. These are such diluted and politicized terms that both are being replaced by the most compelling 'Climate Crisis' or 'Climate Emergency' and 'Global Heating' respectively (Carrington, 2019; Ripple et al., 2019). Local discourses also tend to replace anodyne labels with more compelling ones, such as 'from warming to warning' (Manning, 2020).

Actually, the Climate Emergency is only one of nine anthropogenic vectors of Earth System disruption, each having the potential to severely lessen the human habitability of our planet. These are: Climate Change, Novel Entities (anthropogenic objects, materials and bio-actants), Stratospheric Ozone, Atmospheric Aerosol Loading (anthropogenic particles in the Atmosphere), Ocean Acidification, Biogeochemical Flows (Nitrogen and Phosphorus cycles), Freshwater Use, Change in Land Use and Biodiversity Loss (Extinction rate). Each of these 'Planetary Boundaries' is sustained by a delicate balance required to keep a 'Safe Operating Space' for Humanity (Rockström et al., 2009). Hence, trespassing each of these Planetary Boundaries conveys an Existential Risk to Humanity (Barnosky et al., 2012; Steffen, Richardson et al., 2015).

While Anthropogenic Global Existential Risks challenge the conditions of the Biosphere that prevailed through the Holocene, the existing social knowledge base seems inadequate to help us cope with such realities (Castree et al., 2014; Clark & Szerszynski, 2021). The idea of knowledge as a global transformative force shaped the concept of 'the Noosphere' in the early 1920s, thanks to the convergence of independent work by geologist Vladimir Vernadsky, philosopher Pierre Theilard de Chardin and mathematician Édouard Le Roy. They shared the idea of an evolutionary layer or global developmental stage based on reason, science and technology. Vernadsky, with his scientific background in biogeochemistry, was able to ground the concept in natural science.

The relationship between the Anthropocene as a geological epoch and human knowledge was noticed early (Crutzen & Stoermer, 2000, p. 17). However, the relationship between the Anthropocene and human knowledge goes well beyond an aspirational stage of evolution. Reflexively, modern Western technoscientific and economic culture has also been regarded as a major underlying driver of the Anthropocene, requiring to 'recognize the more radical normative implications' (Castree, 2017, p. 64). Hence, the knowledge-based culture accompanying the Great Acceleration has been regarded both as cause and potential solution (O'Brien, 2016; Castree, 2017; Cohen and Colebrook, 2017; Malabou, 2017; Few et al., 2021), a corollary of the basic tenet in Science and Technology Studies regarding the inter-determinacy between knowledge and society (Felt el al., 2016). On the one hand, we need to understand how our current knowledge practices and institutions have brought us here; such understanding requiring a new critical perspective in itself. On the other hand, we need to reinvent the knowledge systems that might still enable us to adapt to and mitigate anthropogenic impacts and, above all, to redefine the terms of relation between humans and planet Earth.

Chapters in this book address different aspects of knowledge, the Anthropocene, and the relation between both. While a diversity of approaches and perspectives has been welcomed and actively encouraged, the concept of knowledge system (KS) as an integrative framework resonates through several chapters. According to the concept developed at the Center for Knowledge Systems,[1] a KS is 'the alignment of value dimensions between an object, an agent and a context of knowledge' (see, e.g., Carrillo, 1998; Carrillo et al., 2019). While such concept allowed a very effective characterization of knowledge management processes over a quarter of a century, each of the core categories and perhaps the very idea of a knowledge system is in need of a major overhaul. Knowledge objects multiply and diversify, in a flat ontology realm such as the democracy of the beings proposed by Object-Oriented Ontology (Harman, 2018). Suddenly, objects and materials are not regarded anymore as inert entities, but come alive, as it were, in an entangled network of actors where texts, instruments, symbols, concepts, and so on stop being passive recipients of transitive action and engage themselves as agents.

Accordingly, knowledge agents morph into the rhizomatic universe of the others, a re-encounter with alterity and an openness towards the more-than-human world. The human is demoted from the centre of history and diluted into a geological scale. The casting of the knowledge theater explodes in a diaspora of characters that tends to correspond with the diversity of the world. The voices of the local, feminist, native, traditional, domestic, pragmatic, and other majorities at large formerly silenced or ignored by the anthropo- and euro-centric, male, white, colonial, capitalist worldview anchored in the knowledge establishment reclaim their stand, giving way to a societal and global choir (Braidotti, 2017). Beyond that, the more-than-human voices of the living and non-living allow the creatures and the critters, the mountain and the river, the galaxy and the molecule, to regain agency and co-design the world (Haraway, 2016).

Above all, knowledge contexts are harder to make sense of because the value systems providing semantic significance and economic worth have been falling apart. One by one, the foundational categories of modernity are demolished by the deconstructive fever of the postmodern. In parallel, the globalized capitalist establishment implodes in an ever increasing and paradoxical concentration and nullification of wealth. Politics is polarized and atomized into the party of each individual, only for identity to be further disaggregated into quantum events.

Hence, the known knowns of object, agent and context became known unknowns. The post-truth society challenges not just the public awareness of the climate emergency but the very possibility of public knowledge. A more robust foundation of a pragmatic common axiological base (capital system) is required. A 'genuine climate change' entails a new materiality: a basic common understanding on 'the things that matter' (Cohen & Colebrook, 2017, p. 135). Certainly, the Climate Emergency forces us to reconsider our common value base or capital system (Aeron-Thomas et al., 2004; Gleeson-White, 2015; Dietz et al., 2020; Gough, 2020). Seeking knowledge for the Anthropocene largely becomes learning to manage what matters to people: Managing Knowledge-based Value Systems (Carrillo, 1998; O'Brien & Wolf, 2010; Albizua & Zografos, 2014).

The existential challenges of the Anthropocene thus confront us with a double task: to redefine posthuman knowledge and to do so before the unprecedented and highly complex circumstances of the unfolding futures. Knowledge systems for the Anthropocene are cognitive hyperobjects (Morton, 2013) distancing from the past and evading predetermined futures.

Hence, if there is going to be a Knowledge for the Anthropocene, it has yet to be imagined. Not only the foundational questions of knowledge will need to be reset. The primordial categories will need to be defined afresh. What is to be known? Who will know? Who will benefit? For what purpose? What will be relevant dimension for aligning objects, agents and contexts in Anthropocenic knowledge systems? Knowledge discourses will have to be advanced upon new axes. Whose will be the narrative voice in a more-than-human history at once being challenged with survival and subsumption into a cosmic consciousness?

The monumental task of reinventing a knowledge for the Anthropocene is only rivalled by the challenge of articulating the global human experience, so that the urgent supranational policies required to enact viable futures are agreed and enforced. Or perhaps they are interdependent: 'Could it be that new histories of mentalities, which would bring together the geological, biological, and cultural current dimensions of historical (non) awareness, may open a new chapter of Anthropocenic study?' (Malabou, 2017, p. 52). The circumstance of the Anthropocene is at once an existential threat and an ontological option: 'Humanity is an effect of being unified by a critical condition' (Cohen & Colebrook, 2017, p. 130).

Hence, the set of chapters constituting this book aim at shedding some light upon these questions. Rather than individual chapters providing separate answers, it is their collective transdisciplinarity that becomes meaningful. Even in their plurality and intersectionality, the set of contributions compounds a mosaic that might help to articulate the 'new chapter of Anthropocenic study'.

The time is ripe for a publication like this book. Certainly, there is a substantial set of precursors from different disciplines to different aspects of understanding knowledge in the Anthropocene, as several of the contributions to this book show. Amongst the now abundant literature on Anthropocene-related issues, a large portion of works will somehow deal with aspects of knowledge. Actually, a substantial collection of works that do not mention knowledge for the Anthropocene per se, provide some of the seminal contributions to chapters covered by this book.

Yet, Knowledge in the Anthropocene as an analytic category on its own, has only partially been addressed in different platforms. The 2017 Special Issue of the *South Atlantic Quarterly* on 'Climate Change and the Production of Knowledge' (edited by Baucom & Omelsky, 2017) has been one of the most substantial precursors, while the abovementioned 2019 Editorial of the *International Journal of Knowledge Based Development* (Carrillo, 2019) incorporated this topic into the IJKBD editorial scope. Within scientific periodicals probably the special issue of *Current Opinion in Environmental Sustainability* on 'Advancing the science of actionable knowledge for sustainability' (Arnott et al., 2020) constitutes the most substantial recent contribution.

A more consistent channel of focused contributions has been that of individual papers in scientific journals (e.g., Baghel, 2012; Homsy & Warner, 2013; Seidl et al., 2013; Perret, 2014; Wark & Jandrić, 2016; Delanty & Mota, 2017; Edwards, 2017; Turnhout, 2018; Nakagawa & Payne, 2019; Featherstone, 2020; Jensen, 2020: Schemmel, 2020). The recently published study by Fazey et al. (2020) directly addresses issues related to the design of Knowledge Systems for the Anthropocene. Also, some research centres or think tanks besides the WCI case described above, are vesting dedicated efforts specifically to issues of knowledge in the Anthropocene, such as the Responses to Environmental and Societal Challenges for our Unstable Earth (RESCUE) Program running from 2009 through 2011 (Jäger et al., 2011) or the ongoing Changing Ecologies of Knowledge (CEKA) at the University of Oxford Institute for Science Innovation and Society.

In terms of books the contributions have so far been rather marginal. While the subheading to Section 1 of Dürbeck & Hüpkes' book (2020) is 'Creating Knowledge for the Anthropocene', chapters included deal with historical, metaphysical and cultural aspects of the Anthropocene. These are certainly important in the context of interdisciplinary approaches to the Climate Emergency but cover little in terms of knowledge systems and processes. Similarly, Renn's (2020) evolutionary account of the history of science from a cognitive structuralist perspective includes a chapter (16th of 17) on Knowledge for the Anthropocene that works as an epilogue and 'further research' section. While his monumental reconstruction of the history of science and technology is a landmark contribution, his incursion on our core topic remained incidental.

This book constitutes the first volume wholly and directly concerned with issues related to Knowledge in the Anthropocene. If knowledge is the most important leverage to human action and the leading force to civilization, how can our knowledge societies come to terms with the foreseeable realities of the climate crisis and face the emergency in the best possible way? What relevant knowledge will become critical to cope with deteriorating environmental conditions? How can science, technology, innovation and education be radically transformed for adequately responding? How can governments continue to serve their purpose? How need our mind frames evolve? These and other related questions are addressed in a direct and accessible manner by a set of international specialists. The common purpose is to provide a general perspective on the role that knowledge – the most important leverage to human action, may need to play in our not-so-distant future.

NOTE

1. The Center for Knowledge Systems operated at Tecnológico de Monterrey, México, from 1989 through 2014. Throughout its 25 years of existence, it was fully self-sufficient due to its combination of consultancy with research and teaching. The CKS mission statement was "To leverage the value creation capacity of individuals, organizations and societies through the research, design, implementation and learning on knowledge systems." Since 2013 it became the basis for the Strategic Research Group on Knowledge Societies.

REFERENCES

Aeron-Thomas, D., Nicholls, J., Forster, S., & Westall, A. (2004). *Social return on investment: Valuing what matters.* New Economics Foundation.

Albizua, A., & Zografos, C. (2014). A values-based approach to vulnerability and adaptation to climate change. Applying Q methodology in the Ebro Delta, Spain. *Environmental Policy and Governance, 24*(6), 405–422.

Arnott, J.C., Mach, K.J., & Wong-Parodi, G. (eds) (2020). Special Issue on advancing the science of actionable knowledge for sustainability. *Current Opinion in Environmental Sustainability, 42* (February), A1–A6, 1–82.

Baghel, R. (2012). Knowledge, power and the environment: Epistemologies of the Anthropocene. *Transciencie, 3*(1), 1–6.

Barnosky, A.D., Hadly, E.A., Bascompte, J., Berlow, E.L., Brown, J.H., Fortelius, M., … & Martinez, N.D. (2012). Approaching a state shift in Earth's biosphere. *Nature, 486*(7401), 52–58.

Baucom, I., & Omelsky, M. (2017). Knowledge in the age of climate change. *South Atlantic Quarterly, 116*(1), 1–18.

Braidotti, R. (2017). Critical posthuman knowledges. *South Atlantic Quarterly, 116*(1), 83–96.

Carrillo, F.J. (1998). Managing knowledge-based value systems. *Journal of Knowledge Management, 1*(4), 280–286.

Carrillo, F.J. (2019). The Anthropocene turn in knowledge based development. *International Journal of Knowledge-Based Development, 10*(4), 293.

Carrillo, F.J., Edvardsson, B., Reynoso, J., & Maravillo, E. (2019). Alignment of resources, actors and contexts for value creation. *International Journal of Quality and Service Sciences*, 1–31. doi.org/10.1108/IJQSS-08-2018-0077.

Carrington, D. (2019, 17 May). Why the *Guardian* is changing the language it uses about the environment. Last retrieved on 25 December 2020, from https://www

.theguardian.com/environment/2019/may/17/why-the-guardian-is-changing-the-language-it-uses-about-the-environment.

Castree, N. (2017). Global change research and the 'people disciplines': Toward a new dispensation. *South Atlantic Quarterly, 116*(1), 55–67.

Castree, N., Adams, W. M., Barry, J., Brockington, D., Büscher, B., Corbera, E., ... & Newell, P. (2014). Changing the intellectual climate. *Nature Climate Change, 4*(9), 763–768.

Clark, N., & Szerszynski, B. (2021). *Planetary social thought: The Anthropocene challenge to the social sciences*. John Wiley & Sons.

Cohen, T., & Colebrook, C. (2017). Vortices: On 'critical climate change' as a project. *South Atlantic Quarterly, 116*(1), 129–143.

Crutzen, P.J. and Stoermer, E.F. (2000). The Anthropocene. *IGBP Newsletter 41*. Royal Swedish Academy of Sciences.

Delanty, G., & Mota, A. (2017). Governing the Anthropocene: Agency, governance, knowledge. *European Journal of Social Theory, 20*(1), 9–38.

Dietz, T., Shwom, R.L., & Whitley, C.T. (2020). Climate change and society. *Annual Review of Sociology, 46*, 5.1–5.24.

Dürbeck, G., & Hüpkes, P. (eds) (2020). *The Anthropocenic Turn: The Interplay Between Disciplinary and Interdisciplinary Responses to a New Age*. Routledge.

Edwards, P. N. (2017). Knowledge infrastructures for the Anthropocene. *The Anthropocene Review, 4*(1), 34–43.

Fazey, I., Schäpke, N., Caniglia, G., Hodgson, A., Kendrick, I., Lyon, C., ... & Verveen, S. (2020). Transforming knowledge systems for life on Earth: Visions of future systems and how to get there. *Energy Research & Social Science, 70*, 101724.

Featherstone, M. (2020). Stiegler's ecological thought: The politics of knowledge in the anthropocene. *Educational Philosophy and Theory, 52*(4), 409–419.

Felt, U. et al. (2016). *The handbook of science and technology studies*, The MIT Press, 253.

Few, R., Spear, D., Singh, C., Tebboth, M.G., Davies, J.E., & Thompson-Hall, M.C. (2021). Culture as a mediator of climate change adaptation: Neither static nor unidirectional. *Wiley Interdisciplinary Reviews: Climate Change, 12*(1), e687.

Gleeson-White, J. (2015). *Six capitals, or can accountants save the planet?: Rethinking capitalism for the twenty-first century*. WW Norton & Company.

Gough, I. (2020). In times of climate breakdown, how do we value what matters? Open Democracy. Last retrieved on 24 February 2020, from https://www.opendemocracy.net/en/oureconomy/times-climate-breakdown-how-do-we-value-what-matters/.

Hamilton, C., Gemenne, F., & Bonneuil, C. (eds) (2015). *The Anthropocene and the global environmental crisis: Rethinking modernity in a new epoch*. Routledge.

Haraway, D.J. (2016). *Staying with the trouble: Making kin in the Chthulucene*. Duke University Press.

Harman, G. (2018). *Object-oriented ontology: A new theory of everything*. Penguin UK.

Homsy, G.C., & Warner, M.E. (2013). Climate change and the co-production of knowledge and policy in rural USA communities. *Sociologia Ruralis, 53*(3), 291–310.

Jäger, J., Pálsson, G., Goodsite, M., Pahl-Wostl, C., O'Brien, K., Hordijk, L., & Avril, B. (2011). Responses to environmental and societal challenges for our unstable earth (RESCUE), ESF Forward Look–ESF-COST 'Frontier of Science' joint initiative. *RESCUE report*.

Jensen, C.B. (2020). The Anthropocene eel: Emergent knowledge, ontological politics and new propositions for an age of extinctions. *Anthropocenes – Human, Inhuman, Posthuman*, *1*(1).

Krogh, M. (ed.) (2020). *Connectedness. An incomplete encyclopedia of the Anthropocene*. Strandberg Publishing.

Malabou, C. (2017). The brain of history, or, the mentality of the Anthropocene. *South Atlantic Quarterly*, *116*(1), 39–53.

Manning, P. (2020). *Body count – how climate change is killing us*. Simon & Schuster Australia.

McNeill, J.R., & Engelke, P. (2016). *The great acceleration: An environmental history of the Anthropocene since 1945*. Harvard University Press.

Morton, T. (2013). *Hyperobjects: Philosophy and ecology after the end of the world*. University of Minnesota Press.

Nakagawa, Y., & Payne, P.G. (2019). Postcritical knowledge ecology in the Anthropocene. *Educational Philosophy and Theory*, *51*(6), 559–571.

O'Brien, K. (2016). Climate change adaptation and social transformation. In Richardson, D. et al., *International Encyclopedia of Geography: People, the Earth, Environment and Technology*, Wiley-Blackwell, 1–8.

O'Brien, K.L., & Wolf, J. (2010). A values-based approach to vulnerability and adaptation to climate change. *Wiley Interdisciplinary Reviews: Climate Change*, *1*(2), 232–242.

Oldfield, F., Barnosky, A.D., Dearing, J., Fischer-Kowalski, M., McNeill, J., Steffen, W., & Zalasiewicz, J. (2014). The Anthropocene review: Its significance, implications and the rationale for a new transdisciplinary journal. *The Anthropocene Review*, *1*(1) 3–7.

Perret, M. (2014). Amphibians, affect and agency: On the production of scientific knowledge in the Anthropocene. *Berkeley Undergraduate Journal*, *27*(2), 132–159.

Renn, J. (2020). *The evolution of knowledge: Rethinking science for the Anthropocene*. Princeton University Press.

Ripple, W., Wolf, C., Newsome, T., Barnard, P., Moomaw, W., & Grandcolas, P. (2019). World scientists' warning of a climate emergency. BioScience. Last retrieved on 24 February 2020, from: https://hal.archives-ouvertes.fr/hal-02397151.

Rockström, J., Steffen, W., Noone, K., Persson, Å., Chapin III, F.S., Lambin, E., ... & Nykvist, B. (2009). Planetary boundaries: Exploring the safe operating space for humanity. *Ecology and Society*, *14*(2), 32–64.

Schemmel, M. (2020). Global history of science as a knowledge resource for the Anthropocene. *Global Sustainability 3*(22), 1–8.

Seidl, R., Brand, F.S., Stauffacher, M., Krütli, P., Le, Q. B., Spörri, A., ... & Scholz, R.W. (2013). Science with society in the Anthropocene. *Ambio*, *42*(1), 5–12.

Steffen, W., Broadgate, W., Deutsch, L., Gaffney, O., & Ludwig, C. (2015). The trajectory of the Anthropocene: The great acceleration. *The Anthropocene Review*, *2*(1), 81–98.

Steffen, W., Richardson, K., Rockström, J., Cornell, S.E., Fetzer, I., Bennett, E. M., ... & Folke, C. (2015). Planetary boundaries: Guiding human development on a changing planet. *Science*, *347*(6223).

Turnhout, E. (2018). The politics of environmental knowledge. *Conservation and Society*, *16*(3), 363–371.

Wark, M., & Jandrić, P. (2016). New knowledge for a new planet: Critical pedagogy for the Anthropocene. *Open Review of Educational Research*, *3*(1), 148–178.

Zalasiewicz, J., Waters, C.N., Williams, M., & Summerhayes, C.P. (eds) (2019). *The Anthropocene as a geological time unit: A guide to the scientific evidence and current debate*. Cambridge University Press.

Zalasiewicz, J., Williams, M., Steffen, W., & Crutzen, P. (2010). The new world of the Anthropocene. *Environmental Science and Technology, 44,* 2228–2231.

PART I

Knowledge and the planetary emergency

1. A portable philosophy toolkit for the Anthropocene
Carlos Jesús García-Meza

INTRODUCTION: PHILOSOPHY AND THE ANTHROPOCENE

The incontestable evidence of the Anthropocene and its portentous horizon of risk and uncertainty has led some thinkers in the social sciences and the humanities to explore and challenge the main philosophical assumptions of modernity (still affecting our contemporaneity), that is, how we think about ourselves as a living species and how we define our relation to the natural world and the world at large. Importantly, these are not just theoretical questions, but they crucially determine how we live and act (and therefore they also have to do with our future). These kinds of questions are formally dealt with by three branches of philosophy: ontology, epistemology, and ethics. Ontology is the study of existence, reality, and becoming, and tries to give answers to the primordial question of what the world is made of. Epistemology studies the nature, sources and validity of knowledge, and is oriented to the question of how reality or the world can be known (related to this, methodology deals with the methods that give access to or constitute that knowledge). Finally, ethics studies the good, the right and the valuable.

In mainstream philosophy – and thus in the social sciences and the humanities – ontology, epistemology and ethics share in two fundamental assumptions, both with a historical origin in the Enlightenment: *anthropocentrism* and *human exceptionalism*. First, anthropocentrism is the privileging of the human being against other living beings (and more so against the many entities that populate the world, be they organic, inorganic, or inert). The idea is that human beings are the most important entities in the world, so everything is interpreted and justified in terms of human values and experience. This (self-attributed) supremacy of the human establishes a neat separation between humanity and the rest of the world; in other words, a first binary distinction (humans/nonhumans) is created, out of which many other dual dichotomies have arisen: society/nature, mind/body (the classical Cartesian dualism), subject/object, and the like. Again, this basic distinction between human and nonhuman (or nature) rests on an *ontological* assumption: the world is composed of humans and nature, but humans are different, superior, better, than nature.

Second, if the prevalent ontology underpins a dualistic outlook to the world in which we humans see ourselves as separate *from* nature, then it may also lead us to assume that we are entitled to do things *to* nature (e.g., extracting resources and destroying ecosystems, polluting air and water, warming the atmosphere), without necessarily having any ethical sense of responsibility or guilt about it. In other words, the first ontological assumption (anthropocentrism) is accompanied by another one, human exceptionalism, in which the (human) prerogative of *agency*, defined as the capacity to act and effect changes on the world, is considered an attribute pertaining only to humans.

If these two ontological assumptions are taken for granted (consciously or not) then it follows that the concomitant epistemological and ethical thinking and actions will maintain the same logic of anthropocentrism and human exceptionalism. But what the Anthropocene has made very clear is that, paradoxically, the practical consequences of these two assumptions have had increasingly disastrous effects not only on other living beings (and nonliving entities) but also on *us*. What is needed is a different trifecta of ontology, epistemology and ethics that instead of holding itself to a logic of ontological separation and superiority (and the associated violence to other beings and entities and even to ourselves), may recognize the interdependence of the human with the rest of the world. Let us call this new ontological reconfiguration a *"more-than-human" world* (Whatmore, 2006), an ontological shift that may well allow us to acknowledge our ontological entanglement(s) in more-than-human realities that can relocate us in new, symbiotic, biophilic relationships with the planet, instead of the usual ones of separation, parasitism and destruction.

Before we enter into the philosophical terrain of *the new materialisms*, a general denomination to the ontology of a more-than-human world, a brief note on the title of this chapter. In contemporary academic spheres, theories and concepts are usually understood as "tools," that is, things that may help us to get a better understanding of a given topic. In this sense, this chapter offers a minimal set of tools, that is, a "portable toolkit" for readers who wish to take a first, non-professional neomaterialist philosophical look at the Anthropocene. For didactic sake, philosophical "tools" are *italicized* words in this chapter.

THE NEW MATERIALISMS

New materialisms are a variegated set of theories representing a branch of the so-called *ontological turn* that has been happening in theorization and empirical studies in contemporary social sciences and humanities since the mid-1990s. Inaugurating this philosophical perspective, Manuel DeLanda defined a neomaterialist approach as

> a philosophical stance which rejects ideas of progress not only in human history but in natural history as well. Living creatures, according to this stance, are in no way 'better' than rocks. Indeed, in a nonlinear world in which the same basic processes of self-organization take place in the mineral, organic and cultural spheres, perhaps rocks hold some of the keys to understand sedimentary humanity, igneous humanity and all their mixtures. (DeLanda, 1996)

After DeLanda, in 2000 Rosi Braidotti used the term of "new materialism" as well (Braidotti, 2000, p. 160).

The new materialisms comprise a transdisciplinary group of authors and theories which include, amongst the most noteworthy, Bruno Latour (*actor-network theory*, or ANT), Manuel DeLanda (*assemblage theory*), Karen Barad (*agential realism*), Jane Bennett (*vitalist materialism*), Rosi Braidotti (*posthumanism*), Graham Harman (*object-oriented ontology*), and Timothy Morton (*hyperobjects* theory). New materialisms are also associated with other theoretical perspectives such as process philosophy, affect theories, and non-representational theories (Coole and Frost, 2010; Depelteau, 2018; Fox and Alldred, 2017). All these theories and perspectives are very much indebted to feminist thinking, which infuses new materialisms' orientation to embodiment (human and nonhuman bodies), materiality, affect and emotions, as well as the acknowledgment of the importance of situatedness and relationality.

Each neomaterialist author has developed her or his own approach to the literature of the new materialisms, so relative divergences exist between them, but nonetheless all of them share in some basic tenets, related to the following themes of materiality, agency, and relationality.

THREE TENETS OF THE NEW MATERIALISMS

Materiality

Taking an ontological *turn to matter*, new materialisms put emphasis on the *materiality* of the world. Living and nonliving beings, material stuff of any kind, places and spaces (urban, architectural, natural, etc.), material forces (of water, or air, etc.), and even non-physical things (memories, ideas, thoughts, feelings, or imaginary and fantastic elaborations), are all considered as *matter*, in the sense of having the capacity to produce effects or changes in the world (therefore a capacity for *worlding*, i.e., the creation of new "worlds").

Traditionally, philosophy and the natural sciences have treated matter as something that is dumb, static, and inert: a passive stuff subjected to an external and active human dominion and control. Opposing this view, new materialisms claim that the whole field of matter (existence itself, including human and nonhuman entities) is "alive" and full of creativity, and has the capacity to produce effects and form new and complex entities. Microplastics, e-waste, pollutants, carbon molecules, electrical batteries, or "smart" textiles all have a real and highly visible efficacy that goes well beyond their descriptions (ontology) as scientific or technological products, commodities, or discourse (i.e., referents to pieces of information or of laws and regulations, or parts of our conversations). Instead of reducing the materiality of the world to culture and discourse, an analytical mode that conceals and disregards that very materiality rendering it trivial, new materialisms force us to think beyond discursive representations and ask us to recognize the very materiality of the world.

Contrary to what the natural sciences claim, matter for new materialisms is rather boundless, unpredictable and uncontrollable. Thus, for example, new materialisms reject the notion of nature as the background for human endeavors, in which "natural" matter is seen as passively waiting for humans to grant it meaning (for example, as a reserve of "natural" resources for exploitation, or as a business target for geo-engineering projects). New materialisms strongly reject this "bifurcation of nature" (Whitehead, 1929), the foundational division of the world into reciprocally exclusive categories (general/particular, mind/matter, science/philosophy, etc.), each struggling for winning over the other. Whitehead warns us to consider the loss we may experience if we follow this "bifurcation of nature." As he says, it is nonsensical attempting to ontologically determine what belongs to nature and what belongs merely to our (human) experience: "For us the red glow of the sunset should be as much a part of nature as are the molecules and electric waves by which men of science would explain the phenomenon" (Whitehead, 1929, p. 29). Thus, Whitehead rejects the dualistic thinking of decoupling nature (living entities, landscapes, things) from human culture (society, communication, language). Neomaterialist thinkers assert that this dualism is false, being a distinction that results, for instance, from analytical or research practices that work by tracing arbitrary boundaries between what counts (by analysts or researchers) as cultural and what as natural. What we actually have is, for instance, in Donna Haraway's coinage, *natureculture*, in which "all the actors become who they are in the dance of relating, not from scratch, not ex nihilo, but full of the patterns of their sometimes-joined, sometimes-separate heritages both before and lateral to this encounter" (Haraway, 2008, p. 25).

Regarding matter as active and productive leads to a *"flat ontology"* (DeLanda, 2006), which means that all entities in the world have the same ontological status. This implies that there is no ontological reason to defend and retain the many binary distinctions that frame and determine our accustomed outlook to the world: culture/nature, human/nonhuman, urban/rural, structure/agency, global/local, animate/inanimate, organic/inorganic, sentient/non-sentient, alive/inert, life/death, mind/matter, subject/object, language/matter, original/copy, abstract/concrete, representation/reality, base/superstructure, micro/macro, and so on. Given the materiality of the world, new materialisms operate outside of or beyond these dualisms. This does not mean that the components of each pair of opposites do not exist, are useless, or are not taken into account anymore. It is simply that the way of considering them is more in terms of inclusion and mutual implication, not of an a priori separation, autonomy, and the privileging of one term in a pair. Besides, the problem with this binary logic is that the stuff making up reality is considered *either* natural *or* social, active *or* passive, agent *or* acted upon, and so on, without allowing for the possibility to see reality as populated by, in Bruno Latour's term, *hybrids*, entities that do not quite readily belong to one or the other side of a given binary distinction. Examples of hybrids are corporations, the ozone layer, frozen embryos, digital surveillance systems, genetically modified organisms, or the COVID-19 pandemic.

One of the most interesting and original articulations of (neo)materiality is Timothy Morton's concept of *hyperobjects* (Morton, 2013). A hyperobject is a vast and complex entity dispersed in space and in time. Global warming, radioactive particles, Styrofoam consumables, radioactive particles, plastic articles, and fine dust, are examples of hyperobjects. According to Morton (2013, pp. 27–95), hyperobjects have five properties. (1) Hyperobjects are *viscous*: they stick to us despite our efforts to isolate and categorize them. The more one tries to study and know them, the stranger they become (they are, Morton says, *strange strangers*), and the more one tries to externalize and classify them, the more one finds oneself entangled in them. (2) Hyperobjects are *non-local*: they are indeterminately dispersed. (3) Hyperobjects are *temporally undulated*: they encompass multiple temporalities and rhythms, and they will far outlive us. (4) Hyperobjects are *phased*: one can only ever see a portion of them, as it is impossible for us to see them as a whole. (5) Hyperobjects are *interobjective*: they form a *mesh*, a heterogeneous assemblage of humans and nonhumans. Morton's characterization of hyperobjects as "strange strangers" means that they are completely out of reach to our limited intelligibility of the world in terms of bounded entities such as "nature" or "society."

Morton's concept of hyperobjects derives from *object-oriented ontology*, or OOO (Harman, 2017), another philosophical strand of the ontological turn that posits *objects* (instead of actors or actants) as the only components of the world. Basically, OOO deals with two philosophical concepts: *finitude* and *withdrawal*. Finitude is our failure to know or access objects and reality itself, since an inescapable ontological gap between things and appearances pervades the world. Withdrawal means that objects cannot be grasped by any means at all, since they always withdraw or stay separated from other objects (as humans, for instance), even from themselves. OOO assumes a *realist* view of the world (i.e., the world exists outside our minds), so objects (or hyperobjects) can only be known metaphorically, by allusive or vicarious means (Harman 2016, p. 17). This represents a great challenge to the endeavor of knowledge in the Anthropocene, that is, the epistemic efforts we have to make in order to come to grips with the radically new hyperobjects with which we cohabit now and in the future.

Agency

If the concept of *agency* is an exclusive human trait for mainstream social sciences and humanities, new materialisms challenge this assumption and extend the capacity to act to all the entities of the world, so it is no longer the case that only humans act upon nonhuman matter, which then it passively reacts. Plants, microbes, winds, rocks, rivers, or wearable apps do not simply react to or obey our actions; they have their own ways of acting. Nonhuman things can actually make us act much in the same way as they can respond to our actions; they have, as we do, a capacity to affect and be affected. Or as Jane Bennett (2010) puts it, things have *thing-power*. Entities have an agentic capability much like the bottle's label in *Alice in Wonderland* saying, "*Drink me*." For Bennett, matter is vibrant, that is, it is always changing and in a constant process of fluidity, and it also has the capacity of acting in the world (i.e., agency). Bennett defines *vitality* as "the capacity of things – edibles, commodities, storms, metals – not only to impede or block the will and designs of humans but also to act as quasi agents or forces with trajectories, propensities, or tendencies of their own" (Bennett, 2010, p. viii). In this same vein, and refusing anthropocentrism too, DeLanda (2006) claims that matter has its own morphogenetic capacities, rather than being just passive stuff at the mercy of humans. An interesting empirical case of this is Bennett's assemblage analysis of how an electricity blackout in the United States in 2003 was produced by many heterogeneous intertwined actants, including "coal, sweat, electromagnetic fields, computer programs, electron streams, profit motives, heat, lifestyles, nuclear fuel, plastic, fantasies of mastery, static, legislation, water, economic theory, wire and wood" (Bennett 2010, p. 25).

Defining herself as a "compostist," Donna Haraway uses the compost metaphor to describe the vast variety of human and nonhuman entities that happen to gather together in collectives and their capacity to generate vitalities (Haraway, 2016). This adds a second layer of meaning to the expression of "flat ontology": agency is distributed among all entities, so the world is continuously emerging and created in the form of entangled events.

Now, the concept of (human) agency is usually associated with (human) intentionality, so the category of "actor" or actant is reserved to human beings.

As a way to avert this traditional association, new materialisms have developed an alternate conceptualization. For instance, actor-network theory uses the term *actant*, which is also used by Bennett (2010); an actant is something that acts, brings about events, or generates effects in the world. Street bumps, chemical compounds, surplus behavioral data, CO_2, coltan, or African slave kids are all actants. The important point is that instead of thinking of each actant as a particular agent that can act on its own (i.e., having agency in itself), new materialisms claim that agency is not a property of individual actants, but is located in the interrelationships of actants. Assuming that nonhuman entities are agentic consequently causes the dissolution of the association of agency with intentionality. Even if we may still believe that we have and can exert our intentions, the myriad of actions of nonhuman actants will interfere with our own actions, so the final result of all those actions ("ours" and "theirs") cannot be easily determined in advance, nor can authorships be readily assigned to particular actants.

For ANT, the list of who or what constitutes an actor is, in the first place, undefined, and can include "human beings, machines, animals, 'nature', ideas, organizations, inequalities, scale and size" (Law, 2009, p. 141). Latour says that an actant "can literally be anything provided it is granted to be the source of action" (Latour, 1996, p. 373). More technically, the term *actant* (or *actor*) is any entity that can make other entities act, and thus "does modify a state of affairs by making a difference" (Latour, 2005, p. 71), or as Callon (1986; 1991) puts it, an actant produces associations between heterogeneous entities, be they tools, ideas, waste, policies, human skills or buildings. These associations can exist not only on a "local" level but can also extend in time and space (Latour, 2005).

As agency is distributed, individual actors never "act" by themselves, since action (agency) is enacted by the differences and connections made between actors. In methodological terms, this is called the *principle of symmetry*, which prescribes an equal (i.e., ontological) treatment to all actors. To avoid misunderstandings, a point of clarification is given by actor-network theorist John Law: "to say that there is no fundamental difference between people and objects is an analytical stance, not an ethical position" (Law, 1992, p. 383). The symmetrical treatment of human and nonhuman actors is then used to analyze how actor-networks are formed, how connections are stabilized, how actors enter, exit or develop, and how their power-relations shift therein.

Another implication of the "flat ontology" is that human anthropocentrism and exceptionalism are radically challenged. As mentioned, traditional ontology posits humans as exceptional. Whereas (common) matter is depicted as passive, predictable, and controllable, humans instead are imagined as active, rational, unpredictable and possessing agency. Presumably, humans have the capacity to effect changes in the world according to their will and intentions.

But if, as new materialisms claim, the materiality of the world is not passive, and both humans and nonhuman actions co-mingle with rather unknowable and uncontrollable outcomes, then the issue of "who" is in control is indeterminate. The Anthropocene is just one but crucial case for posing a *posthuman agency*. The serious and unintended outcomes of our actions on "nature" (anthropogenic events, such as extreme weather, floodings, species extinctions, global warming, glaciers melting, etc.) easily subvert any pretense of human dominion of the planet. This does not mean that it is precisely because of our actions that the world is in peril; it is just that anthropocentrism and human exceptionalism have to be abandoned, now that many biospheric processes are on the course of, perhaps, unprecedented irreversible and disastrous changes.

Disavowing human agency also has ethical and political implications in terms of advancing an "antihumanist" or posthuman stand in the social sciences and the humanities, as it can provide a ground for (a) a critical assessment of humans' negative impact on the environment, and (b) at the same time serve as an entry point to a more biophilic and entangled relationship with the planet itself. For example, the anthropocentric legal system that grants property rights to individuals and corporations could be challenged as it provides the means to the commodification of nature (i.e., culture/nature dualism) and the promotion of a global market-based logic for both human and nonhuman interactions, with evident and increasingly pernicious effects on all of "us": human society and the whole of the biosphere. On the other hand, the concept of posthuman agency can also help us to explore nonhumans' capacities and moral and legal rights previously set aside for humans only. Interesting cases in this direction are, for instance, these posthuman inquires: "Do glaciers listen?" (Cruikshank, 2005), and "How forests think?" (Kohn, 2013), which have a foundational origin in Christopher Stone's book of 1972, *Should Trees Have Standing?* (see Stone, 2010).

Relationality

By refusing any dualisms, new materialisms open up the prospect to explore how human and nonhuman entities affect each other and make things happen. A neomaterialist flat ontology allows the rejection of social macrostructures (neoliberalism, sociopolitical hierarchies, patriarchy, the State) as explanatory devices for society and even for nonhuman situations (e.g., natural events). For Latour, researchers "have confused what they should explain with the explanation" (Latour, 2005, p. 8), meaning that the "social" is not a kind of glue that holds social things together but rather is "what is glued together by many other types of connectors" (ibid, p. 5). Society, as nature, then, is not a "specific realm ... but only ... a very peculiar movement of reassociation and reassembling ... a trail of associations between heterogeneous elements" (ibid, p. 5).

Instead of appealing to systems or structures "out there," new materialisms propose "actor-networks," "phenomena," "naturecultures," or "assemblages" as the real collective enmeshments of human and nonhuman entities that constitute the world. The new materialisms research explores the agential relationships and reciprocities contained within entanglements, and this is done in a transdisciplinary fashion, that is, considering the ad hoc biological, social, geological, historical, aesthetic, physical, and so on, elements at hand, so as to study and depict not the event as a "thing" or state of things but as *becoming*, looking for the iterative processes of materialization in which the world (the event, phenomenon, etc.) comes about.

Actor-networks, assemblages and the like are emergent materializations which are fluid, unstable, and uncertain. At every moment, these assemblages are endlessly configured and reconfigured by their contacts with other impermanent phenomena. In the course of these events or becomings, the frontiers (their "identities," so to speak) of the participating entities (human and nonhuman) are established and kept for a while; nonetheless, the "separated" entities are transversally connected (Bennett, 2010). Instead of saying that entities have their own causal effects, new materialisms claim that entities should be perceived as *becoming with* one another.

A corollary of the neomaterialist ontology of relationality is that the world does not depend on *transcendent* things such as gods, capitalism, spirits, destiny, Gaia, or macrostructures, that is, pre-existing things which surpass, exceed or go beyond down-to-earth, empirical limits; the ontology of new materialisms is rather of *finitude* and *immanence*, that is, it exists or remains within the world or reality of materiality. In the language of the agential realism of Karen Barad, since entities are always inside the phenomenon, it is preferable to say that entities encounter each other as *intra-action*, instead of the customary, dualistic expression of inter-action (Barad, 2007).

CONCLUSION

The ontological turn, or more broadly, the philosophical shift of the new materialisms invites us to abandon the dualistic and transcendent thinking that permeates the current anthropocentric inquiries in the social sciences and the humanities, by radically rethinking the themes of materiality, agency and relationality from a posthuman framework of finitude and immanence and for a human and nonhuman eco-cohabitation. What implications might exist if we shift our philosophical outlook into a neomaterialist one? First, one directly appropriate to the Anthropocene, insofar as the new materialisms displace the attention to anthropocentrism and human exceptionalism towards posthuman existence in a more-than-human world, thus promoting a more symbiotic and biophilic entanglement with "nature" and the whole of world materiality.

Second, since for the new materialisms the world is not made of a priori structures and systems, this opens up the possibility for a really creative and productive engaging in events, becomings, and assemblages. And third, a good part of the new materialisms deal with the question of which politics and ethics are best fitted to do research that may help us to better understand the world and how to change it or intervene in it for the better. As Judith Butler says:

> We are all, in the very act of social transformation, lay philosophers, presupposing a vision of the world, of what is right, of what is just, of what is abhorrent, of what human action is and can be, of what constitutes the necessary and sufficient conditions of life. (…) But perhaps there is, prior to these questions, all of which are important questions, another question: the question of survival itself. (Butler, 2004, p. 205)

Biological and civilizational survival indeed is in fact what is at stake due to the human existential risk posed by the Anthropocene. New materialisms represent perhaps the adequate philosophical response to this predicament, being its *newness* that the Anthropocene precisely asks us for, insofar as current and past ontologies of the Holocene (e.g., totemism, animism, analogism and naturalism; see Descola, 2013) are "rendered obsolete" (Hamilton, 2020, p. 116) by the new geological age of the Earth System. New materialisms might as well be a "fifth ontology" (Hamilton, 2020), but, as Hamilton claims, it should necessarily imply an emotional disposition of *attunement* (see also Morton, 2013) to the dread of the Anthropocene. Of course, attunement, as important and fundamental as it is, is not the only emotion required by the Anthropocene, as the groundbreaking work of Glenn Albrecht on "psychoterratic emotions" has shown (see Albrecht, 2019). A pending research work is the theoretical and empirical exploration of the intermingling of the new materialism's tenets of materiality, agency and relationality with the novel field of psychoterratic emotions if we are to relatively successfully navigate the uncharted waters of the Anthropocene.

REFERENCES

Albrecht, Glenn. (2019). *Earth emotions: new words for a new world*. Cornell University Press.
Barad, K. (2007). *Meeting the universe halfway: quantum physics and the entanglement of matter and meaning*. Duke University Press.
Bennett, J. (2010). *Vibrant matter: a political ecology of things*. Duke University Press.
Braidotti, R. (2000). Teratologies. In: Ian Buchanan and Claire Colebrook (eds.), *Deleuze and feminist theory*. Edinburgh University Press, pp. 156–172.
Butler, J. (2004). *Undoing gender*. Routledge.

Callon, M. (1986). Some elements of a sociology of translation: domestication of the scallops and the fishermen of St Brieuc Bay. In: J. Law (ed.), *Power, action and belief: a new sociology of knowledge?* Routledge, pp. 196–223.

Callon, M. (1991). Techno-economic networks and irreversibility. In: J. Law (ed.), *A sociology of monsters*. Routledge, pp. 132–161.

Coole, D. H. and Frost, S. (2010). *New materialisms: ontology, agency, and politics.* Duke University Press.

Cruikshank, J. (2005). *Do glaciers listen? Local knowledge, colonial encounters, and social imagination.* University of British Columbia Press.

DeLanda, M. (1996). *The geology of morals. A neo-materialist interpretation.* Retrieved from http://www.t0.or.at/delanda/geology.htm.

DeLanda, M. (2006). *A new philosophy of society: assemblage theory and social complexity*. Continuum.

Depelteau, F. (ed.) (2018). *The Palgrave handbook of relational sociology*. Palgrave.

Descola, P. (2013). *Beyond nature and culture*. University of Chicago Press.

Fox, N. J. and Alldred, P. (2017). *Sociology and the new materialism: theory, research, action.* Sage.

Hamilton, C. (2020). Towards a fifth ontology for the Anthropocene. *Angelaki*, 25(4), 110–119.

Haraway, D. J. (2008). *When species meet*. University of Minnesota Press.

Haraway, D. (2016). *Staying with the trouble: making kin in the Chthulucene*. Duke University Press.

Harman, G. (2016). *Immaterialism: objects and social theory*. Polity.

Harman, G. (2017). *Object oriented ontology: a new theory of everything*. Penguin Random House.

Kohn, E. (2013). *How forests think: toward an anthropology beyond the human*. University of California Press.

Latour, B. (1996). On actor-network theory: a few clarifications. *Soziale Welt*, 47, 369–381.

Latour, B. (2005). *Reassembling the social: an introduction to actor-network-theory*. Oxford University Press.

Law, J. (1992). Notes on the theory of the actor-network: ordering, strategy and heterogeneity. *Systems Practice*, 5, 379–393.

Law, J. (2009). Actor network theory and material semiotics. In: B. Turner (ed.), *The new Blackwell companion to social theory*. Wiley-Blackwell, pp. 141–158.

Morton, T. (2013). *Hyperobjects: philosophy and ecology after the end of the world*. University of Minnesota Press.

Stone, C. D. (2010). *Should trees have standing? Law, morality, and the environment*. Oxford University Press.

Whatmore, S. (2006). Materialist returns: practising cultural geography in and for a more-than-human world. *Cultural Geographies*, 13(4), 600–609.

Whitehead, A. N. (1929). Theories of the bifurcation of nature. In: *The concept of nature: The Tanner Lectures delivered in Trinity College, November 1919*. Cambridge University Press, pp. 26–48.

2. Existential challenges to knowledge
Bertrand Guillaume

INTRODUCTION

Over generations, impacts of human activity on the planet have escalated, now affecting great biogeochemical cycles (e.g., water, carbon, nitrogen, phosphorus), the functioning of terrestrial and oceanic ecosystems, and fundamental regulating mechanisms of the Earth system (e.g., homeostasis feedbacks). Since the Industrial Revolution, and even more since the Great Acceleration after WWII (Steffen et al., 2015a), our pressures on the lithosphere, the hydrosphere, the cryosphere, the atmosphere and the biosphere have turned huge in scale and scope (Steffen et al., 2004), causing serious ecological threats to both humans and the nonhuman world (additional greenhouse effect, ozone depletion, biodiversity loss).

Such dramatic planetary changes have been driving us out of the Holocene—the period of the last ten to twelve millennia of relative climate stability which gave rise to human civilizations—with likely irreversible consequences towards a *terra incognita* of a new type: the Anthropocene (Crutzen, 2002). However, we are ourselves part of the biosphere and, as Renn (2018) recalls, the Earth is a dynamic system, so that our interventions introduce second-order changes, including "intangible" ones (Guillaume & Neuteleers, 2015), raising more specifically challenges in terms of knowledge, considered as a dynamic evolution as well, related to our understanding of the planet and our interrelations with it.

THE ANTHROPOCENE AND KNOWLEDGE

As already suggested by a key interdisciplinary symposium held at Clark University back in 1987, human action no longer *modifies* the Earth, as George Perkins Marsh could foresee when his masterwork "Man and Nature" was first revised (Marsh, 1874). It happens that human action now *transforms* the Earth, with a magnitude and to a pace never seen before, an idea embodied in the title of the subsequent volume to the conference (Turner et al., 1990). This unprecedented situation is both a consequence of knowledge evolution and a driver for the future of knowledge.

Planet Under Pressure

What is a really genuine novelty in our relations with the environment in the Anthropocene is the extent of potential ecological damage in scale and scope. As stated by Steffen et al. (2004) in a major International Geosphere Biosphere Programme (IGBP) report synthesis:

> For virtually all of human existence on the planet, interaction with the environment has taken place at the local, or at most the regional scale. The environment at the scale of the Earth as a whole (…) has been something within which people have had to operate, subject only to the great forces of nature and the occasional perturbations of extraterrestrial origin. (p. 2)

This is the true fundamental significance of the Anthropocene, the state of a "Planet under Pressure," as emphasized by the title of a major science conference organized by the global change research programs of the International Council for Science (ICSU) in London some ten years ago (Brito & Stafford-Smith, 2012). Over the past decades evidence has mounted that we now have to deal with planetary-scale effects of human activities on the complex series of interacting biogeochemical processes that transport and transform matter and energy. Because of this human-induced perturbation of the Earth System, which provides the conditions for life on the planet, there is now "something new under the sun" in environmental history (McNeill, 2000). Of course, the biblical reference to the Ecclesiastes and its cyclical conception of time suggests a rupture (Hamilton, 2016), a leap in our connection to nature: with the Anthropocene, the idea that "A generation goes, and a generation comes, yet the earth remains as ever" (Seow, 2008, p. 100) does not hold anymore.

Back in 1933, American philosopher Alfred North Whitehead stressed how from Ancient Greece to the end of the 19th century, the shared assumption was that "each generation will substantially live amid the conditions governing the lives of its fathers and will transmit those conditions to mould with equal force the lives of its children" (Whitehead, 1933, pp. 92–93). Not only is this assumption now false regarding the evolution of technology and culture, as Whitehead precisely argued when looking at the length of the timespan of important change compared to that of a single human life, but it has even become false regarding the evolution of climate, a huge novelty regarding the conditions of life that Whitehead still believed was a matter of several thousand years.

INTERDEPENDENCE AND FEEDBACKS: THE TECHNOCENE

What could not be conceived at the time, namely that mankind would be facing as well a novelty of *ecological* conditions in the timespan of one generation or so, is in fact the result of the accelerating evolution of knowledge and society (Whitehead, 1933, p. 91) established in history, from ancient technologies to the scientific age, then the Industrial Revolution, the "Dawn of the Century" and modern technology of his time. Yet the one clearly stems from the other: the unprecedented pressure on global ecology, through population growth, transformative agency and economic globalization, is the result of the successful accumulation and dissemination of knowledge embodied in human technology and material culture.

The development of both science and technology over centuries, and the implementation of their powerful alliance in contemporary society, have fueled the drivers of the Anthropocene. Thus, the Anthropocene is also a Technocene, namely a tale of the contingent history of knowledge evolution and its side-effects for humanity and the planet, say from the first fire age of celestial meteors and telluric lava to the later fire ages of fossil fuels and nuclear power (Braje, 2015). Incidentally, the awareness of global environmental change (and actually all what we now know about the state of the planet) is also the result of our successful evolution among life on Earth, exemplified by both a new planetary metabolic component (say: the "technosphere") and epistemic or cultural learning in human society, maybe echoing the Vernadskian and Teilhardian idea of a "noosphere" (Levit, 2000; Guillaume, 2014).

In particular, knowledge on the anthropogenic drift of the Earth System has been driven by the International Geophysical Year of 1957–58 and subsequent scientific programs such as the above-mentioned International Geosphere Biosphere Programme in the period 1987–2015 (now "Future Earth"). Politically, the history of geoengineering exemplifies for instance how strategic defense and scientific knowledge were closely intertwined during the Cold War (Fleming, 2010; Hamblin, 2013). Epistemologically, we could speak of a co-evolution between knowledge and the Anthropocene. Now, just as science and technology have enabled global environmental change, entering the Anthropocene conversely confronts us with fundamental challenges regarding knowledge (Renn, 2018).

The Future of Knowledge

Acknowledging that cultural and epistemic evolution has been pushing human global society to the limits of the planet (Rockström et al., 2009; Steffen et al.,

2015b) calls for new ways of thinking in order to achieve sustainability. It is about nothing else than avoiding the collapse of the biophysical basis of human civilization and its social complexity (Tainter, 1988; Diamond, 2005), but also about saving values and knowledge within culture, and maybe even their very possibility in the future (Guillaume & Neuteleers, 2015).

THE FUTURE OF KNOWLEDGE: EXISTENTIAL CHALLENGES?

Thinking processes are allowed by fluid hierarchies of cognitive structures, the complex interdependencies of which Jürgen Renn (2018) calls the "architecture of knowledge." Both individual and collective, knowledge is obviously not limited to the activity of the mind of individuals but has also social and material dimensions. It can be stored, shared, and circulated in space and time through symbolic systems and material culture of societies, resulting in generation, transformation, and evolution (accumulation or loss) of knowledge within dynamic clusters of thinking (Renn, 2018). In such a view, science is just a specific form of co-evolving knowledge implicating both mental theories and cultural practices. In what follows, I briefly discuss three challenges for the future of knowledge to cope with after the Holocene: on the one hand the integration of knowledge, on the other hand the implementation of knowledge, and finally the appropriation of knowledge.

Knowledge Integration: Epistemology

In the Anthropocene, a first challenge to existing systems of knowledge is what German physicist Hans Joachim Schellnhuber (1999) calls the "second Copernican revolution." Referring to the well-known transformation of scientific knowledge some 500 years ago, which put the Earth in its correct astrophysical context, Schellnhuber (1999, p. C19) describes the recent emergence of a new epistemic system likely to "enable us to look back on our planet to perceive one single, complex, dissipative, dynamic entity, far from thermodynamic equilibrium—the 'Earth system'." He further explains that the latter "scientific revolution" is in a way the opposite of the former one, for what we need in the future are "macroscope" instruments (de Rosnay, 1975), rather than magnification techniques in the past, in order to understand our planet as a whole, encompassing at all scales the interrelated dynamic processes of its various components (including the global technosphere and society).

Finally, as Schellnhuber (1999) suggests, one might consider three different options towards such a holistic integration of knowledge, which would most probably have to be combined to produce better representations of the world. The first aims at acquiring information and obtaining a panoramic view of the Earth from above (e.g., using remote sensing); the second is about data generation and the digital simulation of the planetary machinery (e.g., through computer modelling); the third involves actual mimicry of the biosphere at smaller scales (e.g., the Biosphere 2 experiment). Of course, because epistemic (r)evolutions are not limited to innovation in thinking nor technology only, but also involve a transformation of social networks and dynamics, such a focus on knowledge integration in the Anthropocene depends on rearrangements of scientific communities (Renn, 2018).

Knowledge Implementation: Politics and Ethics

A second challenge for the future of knowledge concerns its implementation. This issue refers to the application, in due time, of adequate concepts and practices of global environmental politics and ethics in society, based on new knowledge (e.g., the mental and cognitive basis of the above-mentioned "integrated knowledge" epistemology), so as to warrant regulatory mechanisms likely to avoid civilizational collapse, or even self-destruction (Jonas, 1984; Rees, 2013).

In an innovative piece, American historians of science Naomi Oreskes and Erik M. Conway (2013, 2014) offer an example of what could unfold instead from the point of view of knowledge. In their essay, blending the two genres of science-fiction and history in a quite original way, they imagine a future historian looking back on a past that is our present. Doing so, they picture a period of ecological collapse by the end of the 21st century brought about by our failure to act based on robust scientific knowledge on climate change from the late 1980s on, a dark age they label "the Period of Penumbra" in an ironical reference to the legacy of the Enlightenment.

American philosopher Dale Jamieson (2014) also seems to suggest he knows how the story ends, when he claims in the subtitle of his book that "the struggle against climate change failed," twenty-five years after the UN General Assembly adopted (back in 1988) a resolution declaring climate change to be a priority concern of mankind. He also refers to Enlightenment and "reason in dark time" when it comes to the explanation of the failures of climate policy: scientific ignorance; science-policy gap; organized denial and political partisanship; inadequate ethics; weak political institutions.

Of course plenty of approaches have been suggested for decades, such as supporting low-carbon initiatives, pricing emissions, regulating technologies, changing lifestyles, or increasing natural sinks. Yet without proper and strong knowledge implementation in the near future, our civilization might collapse and even disappear, or at least such an existential prospect would be left to chance.

Knowledge Appropriation: Anthropology

A third challenge for the future is knowledge appropriation, which is how people make sense of knowledge in terms of culture. From this point of view, it appears that global environmental change is a dramatic example of what Timothy Morton (2010, 2013) calls "hyperobjects," namely entities of so vast temporal and spatial dimensions that they challenge the idea of knowledge relative to what things are in the first place. Likewise, an associated limit to knowledge appropriation in the Anthropocene is, following German sociologist Ulrich Beck (1999), the passage from modern risks to contemporary threats, that is, when the "language" of quantification we have been using to think and act comes to fail to account for the uncertain consequences we put into the world to such a point that knowledge becomes meaningless. Philosophically, this echoes, in the framework of cultural and epistemic evolution, the idea once put forward by German philosopher Günther Anders (1956) of a "Promethean gap," namely a gap between, on the one hand, what knowledge allows us to do with technology and, on the other hand, the limited capacity we have for ultimately imagining it, and emotionally coping with it. Knowledge, however, should not cause such a feeling of disjointedness with reality, but instead make us fit with it.

Thus, knowledge after the Holocene should not only unite the sciences—and rearrange epistemic communities alone—as suggested above in the framework of "knowledge integration," but also include methods so as to unite them with the humanities, and bring changes within society and culture as well. Some twenty years ago, American biologist Edward O. Wilson (1998) already described such a synthesis of knowledge from different horizons in order to create a common ground of explicative relationship to the world, coining the term "consilience." A decade later, Dutch geologist and geophysiologist Peter Westbroek (2009) stressed the need of new realistic myths giving interest and meaning to human existence and values, scientific research being a specific source of mythic thought. Referring notably to the "Gaïa Hypothesis" (Lovelock, 1979) as conceived of in the heart of science—or "global ecology"—he discussed four concepts with mythical echoes in relation with global environmental change and human existence, namely: deep time; emergence (e.g., the origin of life); the cumulative development of the Earth; our solidarity with life on Earth and nature.

CONCLUSION

Early technology and material culture (e.g., tools, artefacts) had once been at the heart of the slow transformation of the biological evolution of our species into cultural evolution. In the last centuries or so, scientific theories and material practices (modern science and technology) have been playing the same role in the transition towards epistemic evolution for mankind across the globe. In the Anthropocene, our growing and unprecedented pressure on the Earth system involves both global environmental change and new knowledge generation, now co-evolving in a new timespan, the former being even likely to threaten the very idea of the latter.

A truly interdisciplinary inquiry—including both the natural sciences and the humanities—into the functioning of this co-evolution complexity might be a condition of the sustainable existence of human civilization on Earth. The quest of such a second-order learning loop in epistemic evolution should not only integrate biophysical measures and human dimensions and values, but also include cultural narratives likely to adjust social behavior, typically under the form of science-based myths for our future life on what, in the first place, is our home planet.

REFERENCES

Anders, G. (1956). *Die Antiquiertheit des Menschen. Über die Seele im Zeitalter der zweiten industriellen Revolution*. C.H. Beck.

Beck, U. (1999). *World Risk Society*. Polity Press.

Braje, T. J. (2015). Earth Systems, Human Agency, and the Anthropocene: Planet Earth in the Human Age. *Journal of Archeological Research, 23*(4), 369–396. https://doi.org/10.1007/s10814-015-9087-y

Brito, L. & Stafford-Smith, M. (2012). *State of the Planet Declaration*. Conference Planet Under Pressure, New Knowledge Towards Solutions conference, 2012.

Crutzen, P. (2002). Geology of mankind. *Nature, 415*, 23. https://doi.org/10.1038/415023a.

de Rosnay, J. (1975). *Le macroscope. Vers une vision globale*. Seuil.

Diamond, J. (2005). *Collapse: How Societies Choose to Fail or Survive*. Penguin Books.

Fleming, J. R. (2010). *Fixing the Sky: The Checkered History of Weather and Climate Control*. Columbia University Press.

Guillaume, B. (2014). Vernadsky's philosophical legacy: a perspective from the Anthropocene. *The Anthropocene Review, 1*(2), 137–146. https://doi.org/10.1177/2053019614530874

Guillaume, B. & Neuteleers, S. (2015). The intangibles of climate adaptation: philosophy, ethics and values, in K. O'Brien & E. Selboe (Eds.), *The Adaptive Challenge of Climate Change* (pp. 24–40). Cambridge University Press.

Hamblin, J. D. (2013). *Arming Mother Nature: The Birth of Catastrophic Environmentalism*. Oxford University Press.

Hamilton, C. (2016). The Anthropocene as Rupture. *The Anthropocene Review, 2*(1), 59–72. https://doi.org/10.1177/2053019616634741.

Jamieson, D. (2014). *Reason in a Dark Time: Why the Struggle Against Climate Change Failed—and What It Means for Our Future*. Oxford University Press.

Jonas, H. (1984). *The Imperative of Responsibility: In Search of Ethics for the Technological Age*. University of Chicago Press.

Levit, G. (2000). The Biosphere and the Noosphere. Theories of V. I. Vernadsky and P. Teilhard de Chardin: A Methodological Essay. *Archives Internationales d'Histoire des Sciences, 50*(144), 160–177.
Lovelock, J. (1979). *Gaia: A New Look at Life on Earth*. Oxford University Press.
Marsh, G. P. (1874). *The Earth as Modified by Human Action*. Scribner, Armstrong & Co.
Morton, T. (2010). *The Ecological Thought*. Harvard University Press.
Morton, T. (2013). *Hyperobjects. Philosophy and Ecology after the End of the World*. University of Minnesota Press.
McNeill, J. (2000). *Something New under the Sun: An Environmental History of the Twentieth-Century World*. W. W. Norton & Company.
Oreskes, N. & Conway, E. M. (2013). The Collapse of Western Civilization: A View from the Future. *Daedalus, 142*(1), 40–58. https://doi.org/10.1162/DAED_a_00184.
Oreskes, N. & Conway, E. M. (2014). *The Collapse of Western Civilization: A View from the Future*. Columbia University Press.
Rees, M. (2013). *Our Final Century: Will the Human Race Survive the Twenty-first Century?* Heinemann.
Renn, J. (2018). The Evolution of Knowledge: Rethinking Science in the Anthropocene. *Journal of History of Science and Technology, 12*, 1–22.
Rockström, J., Steffen, W., Noone, K., Persson, Å., Chapin, III, F. S., Lambin, E., Lenton, T. M., Scheffer, M., Folke, C., Schellnhuber, H., Nykvist, B., De Wit, C. A., Hughes, T., van der Leeuw, S., Rodhe, H., Sörlin, S., Snyder, P. K., Costanza, R., Svedin, U., Falkenmark, M., Karlberg, L., Corell, R. W., Fabry, V. J., Hansen, J., Walker, B., Liverman, D., Richardson, K., Crutzen, P., & Foley, J. (2009). Planetary Boundaries: Exploring the Safe Operating Space for Humanity. *Ecology and Society, 14*(2), 32. https://doi.org/10.5751/ES-03180-140232.
Schellnhuber, H. J. (1999). 'Earth System' Analysis and the Second Copernican Revolution. *Nature*, 402, C19–C23. https://doi.org/10.1038/35011515.
Seow, C. L. (2008). *Ecclesiastes: A New Translation with Introduction and Commentary*. Yale University Press.
Steffen, W., Sanderson, R. A., Tyson, P. D., Jäger, J., Matson, P. A., Moore III, B., Oldfield, F., Richardson, K., Schellnhuber, H. J., Turner, B. L., & Wasson, R. J. (2004). *Global Change and the Earth System: A Planet Under Pressure*. Springer.
Steffen, W., Broadgate, W., Deutsch, L., Gaffney, O., & Ludwig, C. (2015a). The trajectory of the Anthropocene: The Great Acceleration. *The Anthropocene Review, 2*(1), 81–98. https://doi.org/10.1177/2053019614564785.
Steffen, W., Richardson, K., Rockström, J., Cornell, S. E., Fetzer, I., Bennett, E. M., Biggs, R., Carpenter, S. R., de Vries, W., de Wit, C. A., Folke, C., Gerten, D., Heinke, J., Mace, G. M., Persson, L. M., Ramanathan, V., Reyers, B., & Sörlin, S. (2015b). Planetary Boundaries: Guiding Human Development on a Changing Planet, *Science, 347*(6223), 736. https://doi.org/10.1126/science.1259855.
Tainter, J. A. (1988). *The Collapse of Complex Societies*. Cambridge University Press.
Turner, B. L., Clark, W. C., Kates, R. W., Richards, J. F., Mathews, J. T., & Meyer, W. B. (Eds.) (1990). *The Earth as Transformed by Human Action: Global and Regional Changes in the Biosphere over the Past 300 Years*. Cambridge University Press.
Westbroek, P. (2009). *Terre ! Des menaces globales à l'espoir planétaire*. Seuil.
Whitehead, A. N. (1933). *Adventures of Ideas*. The Free Press.
Wilson, E. O. (1998). *Consilience: The Unity of Knowledge*. Alfred A. Knopf.

3. Social psychological drivers of climate change denial
Irina Feygina

INTRODUCTION

Despite an ever-growing body of knowledge about the anthropogenic causes of climate change, its relentless progression, and its rapidly escalating impacts, there continues to be a rift in public perceptions of climate change, primarily in Western developed countries (Capstick et al., 2015). Increasing scientific certainty and clarity about the causes and consequences of climate change has been paralleled by a growing and calcifying polarization into rigid groups that either accept the science and are concerned about the implications of climate change, or disregard climate science, are skeptical of human contributions to climate change, and are failing to support action to ameliorate it (Björnberg et al., 2017). How can we understand this divide?

Skepticism and denial of climate change, and resistance to engaging in solutions, is a particularly stark case of people failing to address risks posed to them personally as well as to their families, communities, and society at large. Behind such seemingly irrational and contradictory behavior are powerful social and psychological drivers which must be understood and worked with if we are to stand a chance of responding effectively and rapidly enough to the climate crisis (Wong-Parodi & Feygina, 2020).

THE SOCIAL PSYCHOLOGICAL ROOTS OF CLIMATE CHANGE DENIAL

There are a number of powerful drivers of climate change inaction (Gifford, 2011; Swim et al., 2011; Weber, 2016); this review is focused on the psychological and ideological dynamics that drive climate skepticism and denial. It is important to note that these processes are taking place in the context of a well-funded, politically supported effort to undermine trust in climate science, cast doubt on the anthropogenic causes of climate change, and impede the will for action – in an effort to prevent regulation and a shift away from fossil fuels

(Brulle, 2018; Farrell, 2016; McCright et al., 2016; Supran & Oreskes, 2017). We are, therefore, observing an interaction between individual, societal, and political processes of incredible complexity and psychological charge.

Another important point is that the drivers reviewed here, though distinct, are in many ways interrelated. Most stem from deeply seated needs for social belonging, whether through connection, inclusion, affiliation, status, and the upholding to established ideas, practices, and moralities (Feygina, 2013). As such, they underline the importance of considering the social and moral context in which any communication about and response to climate change takes place (Santos & Feygina, 2017).

Social Identity

People's most inexorable need is to belong, and it manifests in a powerful tendency to align our beliefs, attitudes, and behaviors with the groups we are part of, in order to ensure inclusion and acceptance, and feel valued and respected (Abrams & Hogg, 1990). These social identity processes exert a powerful impact on all facets of people's lives, including responses to climate change (Fielding & Hornsey, 2016). With many politically charged topics, people's reactions are based not on fact or personal preference, but on whether their group or political party endorses or rejects a certain position. In a striking example, presenting the same policy as either put forward by one's political party or the opposing party led to opposing reactions (Cohen, 2003). Similarly, when their political identity was made salient to a group of conservatives in Australia, they reported less acceptance of anthropogenic climate change and less willingness to support climate policies than when their identity was not directly evoked (Unsworth & Fielding, 2014). In a parallel approach, cultural cognition theory illustrates that people align their beliefs about climate change and perceptions of risk with that of their groups – rather than base them on scientific information – resulting in polarization (Kahan et al., 2011).

An important implication of the identity basis of people's reactions to climate change is how much we are influenced by descriptive social norms – the ways in which the majority of a group's members do things, as well as injunctive norms – what is considered correct and desired by the group (Cialdini et al., 1990). People align with actual and perceived behaviors of their group members in order to feel accepted, and to avoid disapproval or rejection (van Kleef et al., 2015). A great deal of social cohesion and learning takes place through social normative processes. In the environmental domain, social norm approaches have been successful in getting people to reuse towels in hotels and thereby reduce water and energy waste (Goldstein et al., 2008), to reduce littering (Cialdini et al., 1990), and to increase recycling behaviors (Schwab et al., 2014), among many other examples. Importantly, social norms have also been successful in helping people reduce their energy use at home (Jachimowicz et al., 2018) and in public settings (Dwyer et al., 2015). Similarly, and importantly, evidence shows that social influence processes underlie support for renewable energy: the choice to install solar panels is predicted by the likelihood of one's neighbors doing so (Bollinger & Gillingham, 2012).

Not only are we influenced by implicit social processes, such as norms, but also by explicit communication about the positions, attitudes, and beliefs that our group members hold – and this is particularly true of those in positions of authority and those we trust. For example, communicating about the impacts of and scientific consensus on climate change through trusted messengers (local weathercasters; Feygina et al., 2020), as well as through interpersonal conversations (Goldberg et al., 2019), led to greater acceptance of the anthropogenic sources of climate change. This was true across the ideological spectrum – even for people who identify as conservative, for whom climate acceptance does not align with their group identity. Similarly, children who talked to their parents about climate change were able to shift their attitudes toward acceptance of the science, especially for male and conservative parents (Lawson et al., 2019).

It is important that social norms which support climate acceptance and action are communicated clearly and misperceptions are corrected. For example, people tend to falsely believe that most people in their group are not concerned about climate change, and therefore stop themselves from discussing the issue. In fact, concern is much greater and more widespread than people realize, and updating this perception is important for supporting climate conversations (Geiger & Swim, 2016).

System Justification

People have a deeply seated, fundamental motive to perceive not only themselves and their groups, but also the larger systems and ideologies that they inhabit, as fair, legitimate, just, beneficial, and stable. Meeting this need serves many important functions, including feeling safe and secure, and maintaining a sense of connection to others in the system. While all people experience it, system justification is related to greater conservatism. This motive can come at a cost: When aspects of the system are unjust, unsuccessful, or unsustainable, the need for system justification can lead people to dismiss or reject the truth in favor of maintaining a positive view of the system. As a result, people fail to see problems and commit to taking actions to ameliorate them (Jost et al., 2004).

Climate change is a poignant case of system justification (Feygina et al., 2010; Fritsche et al., 2012; Jylhä & Akrami, 2015). The Western capitalist socio-economic system, and the many aspects that make it powerful and successful, are the cause of climate change. This includes the use of fossil fuels for development and economic advancement, the externalization of ecological costs for the sake for growth and increasing consumption needed to drive economic growth, and the reliance on ideologies of progress and technological development that subjugate many groups of people, as well as nature. To embrace climate change is to acknowledge these detrimental dynamics and the ways in which the overarching system is not just, beneficial, or sustainable. This is a profound threat to established beliefs and worldviews, and the status quo as a whole, which some people respond to by dismissing the problem in order to maintain a positive view of the system and grabbing onto information that casts doubt on the evidence (Feygina et al., 2010). The result is skepticism and denial, and an unwillingness to support climate solutions.

Personality Tendencies

There are a number of personality-level differences that are associated with greater resistance to acknowledging and addressing climate change. One is Belief in a Just World (BJW) – a faith in the world as a just and secure place where every person gets what they deserve. BJW can lead to blaming people for misfortunes outside of their control (Lerner & Miller, 1978). As with system justification, climate change threatens one's faith in the fairness and benevolence of the world, and those with greater BJW respond to messages about climate change with defensive skepticism and resistance to action (Feinberg & Willer, 2011).

Another personality tendency is Social Dominance Orientation (SDO) – a need for one's group to be superior and have control over other groups, a preference for rigid power hierarchies, and consequent support for inequality and prejudice (Pratto et al., 1994). Those greater in SDO are stronger supporters of human domination of nature (Milfont et al., 2013), more skeptical of climate change and less willing to engage in personal action (Stanley et al., 2017), and less concerned about environmental injustice resulting from economic profit gained by their group (Jackson et al., 2013). A closely related tendency is Right Wing Authoritarianism (RWA) – a preference for obeying established authorities and social norms, and active disapproval of those who fail to do so (Altemeyer, 1981). RWA contributes to support for inequality and resistance to change, to a preference for economic and industrial growth at any environmental cost, and unwillingness to support sustainability or environmental protection (Peterson et al., 1993; Schultz & Stone, 1994).

Political Orientation and Ideology

The most striking disparity between climate acceptance and skepticism falls along political party and ideological lines. This disparity has been observed across most Western countries (Capstick et al., 2015), but is the most pronounced in the United States, where liberals and Democrats tend to accept the science of climate change and express support for climate action, while conservatives and Republicans tend to doubt the science, fail to acknowledge anthropogenic causes of climate change, and resist solutions (McCright & Dunlap, 2011; McCright et al., 2016). This disparity is, in part, explained by the mechanisms described above. Specifically, politically conservative individuals report stronger tendencies toward system justification, and this contributes to their climate skepticism (Feygina et al., 2010). Similar relationships hold for belief in a just world, social dominance orientation and right-wing authoritarianism (Jost et al., 2004; Pratto et al., 1994; Schultz & Stone, 1994). Moreover, studies suggest that conservatives are opposed to proposed solutions to climate change (Campbell & Kay, 2014), which entail regulation, taxation, and a transformation of the energy system toward renewables, and conflict with conservative ideological positions (Jacques & Knox, 2016). These positions are reinforced by political leadership and social authority figures, who spread misinformation about climate science and advocate for denial and inaction. Through the forces of social identity and cultural cognition (Fielding & Hornsey, 2016; Kahan et al., 2011), these positions become a matter of group belonging, and therefore of personal dignity, respect, and worth, giving rise to ever-growing polarization.

The ideological divide also stems from differences in the moral foundations on which people rely when interpreting and responding to climate change and its solutions. Communication about climate change has emphasized its risks and harms, our responsibility for mitigation, as well as social and climate injustice – which align more directly with a liberal concern for care and prevention of harm, as well as fairness (Dickinson et al., 2016). Conservatives, on the other hand, place greater value on loyalty to their groups, respect for authority, and purity (Graham et al., 2009) – which are not typically evoked in calls to ameliorate climate change. Since all these dimensions are implicated in climate impacts and solutions, placing an emphasis on them may be more effective at engaging conservatives (Hurst & Stern, 2020), though the evidence is mixed (Day et al., 2014).

THE MECHANISM OF CLIMATE CHANGE DENIAL: MOTIVATED COGNITION

The above review of several key psychological processes underlying climate denial points to the mechanism by which an ever-growing body of knowledge and evidence about climate change does not result in public acceptance and action (Hornsey et al., 2016). The powerful needs to conform to the beliefs espoused by one's group, to see the system in a positive light, and to align with and uphold extant ideologies results in motivated reasoning – a biased means of perceiving, processing, and remembering of information in line with one's beliefs and goals (Kunda, 1990). Motivated reasoning allows a person to maintain and protect their worldviews, preferences, and perceptions by dismissing, misunderstanding, or failing to recall evidence that contradicts and undermines these (Dunning, 1999; Lewandowsky & Oberauer, 2016; Taber & Lodge, 2006).

There is extensive evidence of motivated reasoning in response to climate information. People who are more motivated to protect the system, when exposed to climate information, have poorer recall and perceive less risk (Hennes et al., 2016), or avoid that information altogether (Shepherd & Kay, 2012). Exposure to scientific news sources results in lower support for climate mitigation among conservatives compared to liberals (Hart et al., 2015), and overall higher levels of education and literacy are associated with greater rejection of scientific knowledge among those motivated by partisanship (Lewandowsky & Oberauer, 2016) and by group affiliation (Kahan et al., 2012).

These findings suggest not only that presenting people with scientific evidence for climate change is not likely to impact their beliefs if it conflicts with their motives and identities, but that education and exposure to science does not undo, and at times may even exacerbate, the detrimental effects of motivated cognition.

USING PSYCHOLOGICAL INSIGHTS TO FOSTER CLIMATE ENGAGEMENT

Given the powerful drivers of climate denial, it is important to create engagement approaches that address the roots of skepticism and draw on psychological knowledge to improve effectiveness. Rather than try to push through the resistance created by socially and ideologically based motives, which is unlikely to be effective, it is important to work with underlying needs to protect and uphold the groups and systems we belong to – which are, in fact, facing a lot of risk due to climate change. Addressing the need to protect the worldviews, moralities, ways of life, and institutions people feel connected to and dependent on, framing action in terms of identity, patriotism and the strength of one's society and groups, and emphasizing the alignment between economic and ecological well-being are likely to ameliorate some of the social and ideological threat people feel in the face of climate change, and support engagement.

Communication efforts need to refocus from a knowledge transfer approach that aims to convey science-based information about climate change, which is unlikely to be successful due to motivated cognition processes, toward communicating locally relevant information through community-based trusted messengers in a way that speaks to people's needs and realities (Feygina et al., 2020). Moreover, communication efforts need to prioritize conversation, bi-directional information exchange, listening, feedback, and relationship building.

The findings reviewed here also make clear that threatening communication about the problem can often backfire, whereby people cope by denying or minimizing the threat. Instead, a shift toward actionable, meaningful solutions can be empowering and increase efficacy. Solutions that align with conservative values, such as support for the free-market, patriotism, or passing traditions on to future generations, are likely to meet with greater engagement (Campbell & Kay, 2014, Feygina et al., 2010). Communicating in ways that connect with conservative moral proclivities, by emphasizing loyalty, authority, and purity, may improve support for climate action (Wolsko et al., 2016). Moreover, shifting the perceived social norm around climate mitigation, especially for conservatives, is imperative for breaking down the identity-based polarization and legitimizing engagement in solutions (Ehret et al., 2018).

CONCLUSION

In sum, people's responses to climate change are rooted in deeply seated social belonging processes. Polarization and denial stem from perceived threats to the identities, groups, as well as systems and ideologies people inhabit and depend on. Such threats lead to biased processing of climate information, perpetuating doubt and disengagement. Efforts to foster support for climate solutions need to acknowledge and address these psychological barriers, communicate in ways relevant to people's experiences, needs and realities, refocus on actionable solutions, and honor the larger social, economic, and ideological context in which responses to climate change unfold.

REFERENCES

Abrams, D., & Hogg, M. A. (Eds.) (1990). *Social identity theory: Constructive and critical advances.* Springer-Verlag.

Altemeyer, B. (1981). *Right-wing authoritarianism.* University of Manitoba Press.

Björnberg, K. E., Karlsson, M., Gilek, M., & Hansson, S. O. (2017). Climate and environmental science denial: A review of the scientific literature published in 1990–2015. *Journal of Cleaner Production, 167,* 229–241. https://doi.org/10.1016/j.jclepro.2017.08.066

Bollinger, B., & Gillingham, K. (2012). Peer effects in the diffusion of solar photovoltaic panels. *Marketing Science, 31*(6), 900–912. https://doi.org/10.1287/mksc.1120.0727

Brulle, R. J. (2018). The climate lobby: A sectoral analysis of lobbying spending on climate change in the USA, 2000 to 2016. *Climatic Change, 149*(3–4), 289–303. https://doi.org/10.1007/s10584-018-2241-z

Campbell, T. H., & Kay, A. C. (2014). Solution aversion: On the relation between ideology and motivated disbelief. *Journal of Personality and Social Psychology, 107*(5), 809–824. https://doi.org/10.1037/a0037963

Capstick, S., Whitmarsh, L., Poortinga, W., Pidgeon, N., & Upham, P. (2015). International trends in public perceptions of climate change over the past quarter century: International trends in public perceptions of climate change. *Wiley Interdisciplinary Reviews: Climate Change, 6*(1), 35–61. https://doi.org/10.1002/wcc.321

Cialdini, R. B., Reno, R. R., & Kallgren, C. A. (1990). A focus theory of normative conduct: Recycling the concept of norms to reduce littering in public places. *Journal of Personality and Social Psychology, 58*(6), 1015–1026. https://doi.org/10.1037/0022-3514.58.6.1015

Cohen, G. L. (2003). Party over policy: The dominating impact of group influence on political beliefs. *Journal of Personality and Social Psychology, 85*(5), 808–822. https://doi.org/10.1037/0022-3514.85.5.808

Day, M. V., Fiske, S. T., Downing, E. L., & Trail, T. E. (2014). Shifting liberal and conservative attitudes using moral foundations theory. *Personality and Social Psychology Bulletin, 40*(12), 1559–1573. https://doi.org/10.1177/0146167214551152

Dickinson, J. L., McLeod, P., Bloomfield, R., & Allred, S. (2016). Which moral foundations predict willingness to make lifestyle changes to avert climate change in the USA? *PLOS ONE, 11*(10), e0163852. https://doi.org/10.1371/journal.pone.0163852

Dunning, D. (1999). A newer look: Motivated social cognition and the schematic representation of social concepts. *Psychological Inquiry, 10*(1), 1–11.

Dwyer, P. C., Maki, A., & Rothman, A. J. (2015). Promoting energy conservation behavior in public settings: The influence of social norms and personal responsibility. *Journal of Environmental Psychology, 41,* 30–34. https://doi.org/10.1016/j.jenvp.2014.11.002

Ehret, P. J., Van Boven, L., & Sherman, D. K. (2018). Partisan barriers to bipartisanship: Understanding climate policy polarization. *Social Psychological and Personality Science, 9*(3), 308–318. https://doi.org/10.1177/1948550618758709

Farrell, J. (2016). Corporate funding and ideological polarization about climate change. *Proceedings of the National Academy of Sciences, 113*(1), 92–97. https://doi.org/10.1073/pnas.1509433112

Feinberg, M., & Willer, R. (2011). Apocalypse soon? Dire messages reduce belief in global warming by contradicting just-world beliefs. *Psychological Science*, *22*(1), 34–38. https://doi.org/10.1177/0956797610391911

Feygina, I. (2013). Social justice and the human–environment relationship: Common systemic, ideological, and psychological roots and processes. *Social Justice Research*, *26*(3), 363–381. https://doi.org/10.1007/s11211-013-0189-8

Feygina, I., Jost, J. T., & Goldsmith, R. E. (2010). System justification, the denial of global warming, and the possibility of "system-sanctioned change." *Personality and Social Psychology Bulletin*, *36*(3), 326–338. https://doi.org/10.1177/0146167209351435

Feygina, I., Myers, T., Placky, B., Sublette, S., Souza, T., Toohey-Morales, J., & Maibach, E. (2020). Localized climate reporting by TV weathercasters enhances public understanding of climate change as a local problem: Evidence from a randomized controlled experiment. *Bulletin of the American Meteorological Society*, *101*(7), E1092–E1100. https://doi.org/10.1175/BAMS-D-19-0079.1

Fielding, K. S., & Hornsey, M. J. (2016). A social identity analysis of climate change and environmental attitudes and behaviors: Insights and opportunities. *Frontiers in Psychology*, *7*, 1–12. https://doi.org/10.3389/fpsyg.2016.00121

Fritsche, I., Cohrs, J. C., Kessler, T., & Bauer, J. (2012). Global warming is breeding social conflict: The subtle impact of climate change threat on authoritarian tendencies. *Journal of Environmental Psychology*, *32*(1), 1–10. https://doi.org/10.1016/j.jenvp.2011.10.002

Geiger, N., & Swim, J. K. (2016). Climate of silence: Pluralistic ignorance as a barrier to climate change discussion. *Journal of Environmental Psychology*, *47*, 79–90. https://doi.org/10.1016/j.jenvp.2016.05.002

Gifford, R. (2011). The dragons of inaction: Psychological barriers that limit climate change mitigation and adaptation. *American Psychologist*, *66*(4), 290–302. https://doi.org/10.1037/a0023566

Goldberg, M. H., van der Linden, S., Maibach, E., & Leiserowitz, A. (2019). Discussing global warming leads to greater acceptance of climate science. *Proceedings of the National Academy of Sciences*, *116*(30), 14804–14805. https://doi.org/10.1073/pnas.1906589116

Goldstein, N. J., Cialdini, R. B., & Griskevicius, V. (2008). A room with a viewpoint: Using social norms to motivate environmental conservation in hotels. *Journal of Consumer Research*, *35*(3), 472–482. https://doi.org/10.1086/586910

Graham, J., Haidt, J., & Nosek, B. A. (2009). Liberals and conservatives rely on different sets of moral foundations. *Journal of Personality and Social Psychology*, *96*(5), 1029–1046. https://doi.org/10.1037/a0015141

Hart, P. S., Nisbet, E. C., & Myers, T. A. (2015). Public attention to science and political news and support for climate change mitigation. *Nature Climate Change*, *5*(6), 541–545. https://doi.org/10.1038/nclimate2577

Hennes, E. P., Ruisch, B. C., Feygina, I., Monteiro, C. A., & Jost, J. T. (2016). Motivated recall in the service of the economic system: The case of anthropogenic climate change. *Journal of Experimental Psychology: General*, *145*(6), 755–771. https://doi.org/10.1037/xge0000148

Hornsey, M. J., Harris, E. A., Bain, P. G., & Fielding, K. S. (2016). Meta-analyses of the determinants and outcomes of belief in climate change. *Nature Climate Change*, *6*(6), 622–626. https://doi.org/10.1038/nclimate2943

Hurst, K., & Stern, M. J. (2020). Messaging for environmental action: The role of moral framing and message source. *Journal of Environmental Psychology*, *68*, 101394. https://doi.org/10.1016/j.jenvp.2020.101394

Jachimowicz, J. M., Hauser, O. P., O'Brien, J. D., Sherman, E., & Galinsky, A. D. (2018). The critical role of second-order normative beliefs in predicting energy conservation. *Nature Human Behaviour*, *2*(10), 757–764. https://doi.org/10.1038/s41562-018-0434-0

Jackson, L. M., Bitacola, L. M., Janes, L. M., & Esses, V. M. (2013). Intergroup ideology and environmental inequality: Intergroup ideology. *Analyses of Social Issues and Public Policy*, *13*(1), 327–346. https://doi.org/10.1111/asap.12035

Jacques, P. J., & Knox, C. C. (2016). Hurricanes and hegemony: A qualitative analysis of micro-level climate change denial discourses. *Environmental Politics*, *25*(5), 831–852. https://doi.org/10.1080/09644016.2016.1189233

Jost, J. T., Banaji, M. R., & Nosek, B. A. (2004). A decade of system justification theory: Accumulated evidence of conscious and unconscious bolstering of the status quo. *Political Psychology*, *25*(6), 881–919. https://doi.org/10.1111/j.1467-9221.2004.00402.x

Jylhä, K. M., & Akrami, N. (2015). Social dominance orientation and climate change denial: The role of dominance and system justification. *Personality and Individual Differences*, *86*, 108–111. https://doi.org/10.1016/j.paid.2015.05.041

Kahan, D. M., Jenkins-Smith, H., & Braman, D. (2011). Cultural cognition of scientific consensus. *Journal of Risk Research*, *14*(2), 147–174.

Kahan, D. M., Peters, E., Wittlin, M., Slovic, P., Ouellette, L. L., Braman, D., & Mandel, G. (2012). The polarizing impact of science literacy and numeracy on perceived climate change risks. *Nature Climate Change*, *2*(10), 732–735. https://doi.org/10.1038/nclimate1547

Kunda, Z. (1990). The case for motivated reasoning. *Psychological Bulletin*, *108*(3), 480–498. https://doi.org/10.1037/0033-2909.108.3.480

Lawson, D. F., Stevenson, K. T., Peterson, M. N., Carrier, S. J., Strnad, R., & Seekamp, E. (2019). Children can foster climate change concern among their parents. *Nature Climate Change*, *9*(6), 458–462. https://doi.org/10.1038/s41558-019-0463-3

Lerner, M. J., & Miller, D. T. (1978). Just world research and the attribution process: Looking back and ahead. *Psychological Bulletin*, *85*(5), 1030–1051. https://doi.org/10.1037/0033-2909.85.5.1030

Lewandowsky, S., & Oberauer, K. (2016). Motivated rejection of science. *Current Directions in Psychological Science*, *25*(4), 217–222. https://doi.org/10.1177/0963721416654436

McCright, A. M., & Dunlap, R. E. (2011). The politicization of climate change and polarization in the American public's views of global warming, 2001–2010. *The Sociological Quarterly*, *52*(2), 155–194. https://doi.org/10.1111/j.1533-8525.2011.01198.x

McCright, A. M., Marquart-Pyatt, S. T., Shwom, R. L., Brechin, S. R., & Allen, S. (2016). Ideology, capitalism, and climate: Explaining public views about climate change in the United States. *Energy Research & Social Science*, *21*, 180–189. https://doi.org/10.1016/j.erss.2016.08.003

Milfont, T. L., Richter, I., Sibley, C. G., Wilson, M. S., & Fischer, R. (2013). Environmental consequences of the desire to dominate and be superior. *Personality and Social Psychology Bulletin*, *39*(9), 1127–1138. https://doi.org/10.1177/0146167213490805

Peterson, B. E., Doty, R. M., & Winter, D. G. (1993). Authoritarianism and attitudes toward contemporary social Issues. *Personality and Social Psychology Bulletin*, *19*(2), 174–184. https://doi.org/10.1177/0146167293192006

Pratto, F., Sidanius, J., Stallworth, L. M., & Malle, B. F. (1994). Social dominance orientation: A personality variable predicting social and political attitudes. *Journal of Personality and Social Psychology*, *67*(4), 741–763. https://doi.org/10.1037/0022-3514.67.4.741

Santos, J., & Feygina, I. (2017). Responding to climate change skepticism and the ideological divide. *Michigan Journal of Sustainability*, *5*(1), 5-23. https://doi.org/10.3998/mjs.12333712.0005.102

Schultz, P. W., & Stone, W. F. (1994). Authoritarianism and attitudes toward the environment. *Environment and Behavior*, *26*(1), 25–37. https://doi.org/10.1177/0013916594261002

Schwab, N., Harton, H. C., & Cullum, J. G. (2014). The effects of emergent norms and attitudes on recycling behavior. *Environment and Behavior*, *46*(4), 403–422. https://doi.org/10.1177/0013916512466093

Shepherd, S., & Kay, A. C. (2012). On the perpetuation of ignorance: System dependence, system justification, and the motivated avoidance of sociopolitical information. *Journal of Personality and Social Psychology*, *102*(2), 264–280. https://doi.org/10.1037/a0026272

Stanley, S. K., Wilson, M. S., Sibley, C. G., & Milfont, T. L. (2017). Dimensions of social dominance and their associations with environmentalism. *Personality and Individual Differences*, *107*, 228–236. https://doi.org/10.1016/j.paid.2016.11.051

Supran, G., & Oreskes, N. (2017). Assessing ExxonMobil's climate change communications (1977–2014). *Environmental Research Letters*, *12*(8), 084019. https://doi.org/10.1088/1748-9326/aa815f

Swim, J. K., Stern, P. C., Doherty, T. J., Clayton, S., Reser, J. P., Weber, E. U., Gifford, R., & Howard, G. S. (2011). Psychology's contributions to understanding and addressing global climate change. *American Psychologist*, *66*(4), 241–250. https://doi.org/10.1037/a0023220

Taber, C. S., & Lodge, M. (2006). Motivated skepticism in the evaluation of political beliefs. *American Journal of Political Science*, *50*(3), 755–769. https://doi.org/10.1111/j.1540-5907.2006.00214.x

Unsworth, K. L., & Fielding, K. S. (2014). It's political: How the salience of one's political identity changes climate change beliefs and policy support. *Global Environmental Change*, *27*, 131–137. https://doi.org/10.1016/j.gloenvcha.2014.05.002

van Kleef, G. A., Wanders, F., Stamkou, E., & Homan, A. C. (2015). The social dynamics of breaking the rules: Antecedents and consequences of norm-violating behavior. *Current Opinion in Psychology*, *6*, 25–31. https://doi.org/10.1016/j.copsyc.2015.03.013

Weber, E. U. (2016). What shapes perceptions of climate change? New research since 2010: What shapes perceptions of climate change? *Wiley Interdisciplinary Reviews: Climate Change*, *7*(1), 125–134. https://doi.org/10.1002/wcc.377

Wolsko, C., Ariceaga, H., & Seiden, J. (2016). Red, white, and blue enough to be green: Effects of moral framing on climate change attitudes and conservation behaviors. *Journal of Experimental Social Psychology*, *65*, 7–19. https://doi.org/10.1016/j.jesp.2016.02.005

Wong-Parodi, G., & Feygina, I. (2020). Understanding and countering the motivated roots of climate change denial. *Current Opinion in Environmental Sustainability, 42*, 60–64. https://doi.org/10.1016/j.cosust.2019.11.008

4. Media accountability before the climate crisis

Gabriel Valerio-Ureña, Jorge Asprón and Nalleli Salazar

HUMAN, A SOCIAL ANIMAL

Beyond the veracity of information, media has always had the potential to influence people's behavior. One of the most famous examples of this phenomenon took place in 1938, when Orson Welles used to narrate literary passages on his radio program. Prior to Halloween night, Welles decided to narrate *The War of the Worlds* (1898) by George Wells, causing hundreds of people to assume that the Earth was truly being attacked by aliens, and in the face of the hysteria and chaos caused, Welles was forced to apologize the next day.

Surely, at least part of the blame for being highly impressionable lies in our own nature. Being social animals, we build our attitudes under the influence of the information we receive from the environment. What we think, know, and believe depends largely on the information we consume. As tribal animals, our beliefs and much of our feelings and behaviors, are socially determined (Aronson, 2011).

Although social group offers protection, it is also a space where false beliefs might propagate. Social connections can play an important role in disseminating both positive and negative behaviors as they affect every aspect of our daily lives. Social media deeply influences our decisions, actions, feelings, thoughts, and even our desires. It is important to remember that our connections (social network) are not restricted to the people we know directly. Friends of our friends also have an influence on us. The chain may start several contacts away, but it will eventually affect us (Christakis & Fowler, 2009). In the same way, our decisions, actions, feelings, and thoughts can influence people we may not know.

Historically, family and community members were the primary sources of information for building beliefs and developing knowledge. However, with the evolution of human societies and technologies, humans have expanded their contacts circle and, with it, their possible sources of information. Among those sources, we can find "social engineers," namely entities with particular interests who want us to believe certain things that are convenient for them. Politicians and publicists are representative examples of this group (Hadnagy, 2011), taking advantage of such technologies. Social media has become a transformative technology, giving access to communication between people like never before and opening new platforms for public debate on the most diverse matters, such as climate change.

THE CLIMATE CHANGE DEBATE IN THE MEDIA

During the 1970s, the environmental movement began, emerging as a social criticism to the techno-industrial system (Anderson, 2012). It is after this "green" or environmental movement beginning in the 1970s that a wave of related concepts emerged: ecocentrism, biocentrism, ecofeminism, deep ecology (Naess, 1973), ecosophy (Naess, 1993), environmental justice, green politics, and ecological economics (Barkin, Fuente, & Zamora 2012), among many others. This decade is considered the beginning of the so-called Third Industrial Revolution (Rifkin, 2011) and characterized by the implementation of new technologies as well as the change of energy regimes. It was the Second Revolution, when the society of mass consumption arose and a shift toward modern society took place, which marked its impact on the environment and the eventual depletion of non-renewable resources.

Being a central issue in global politics, the discussion on climate change has become dichotomous and polarized: We find strong facts of climate change with baffling evidence growing as we speak (Blunden & Arndt, 2020; FAO, IFAD, UNICEF, WFP, and WHO, 2020; IPCC, 2018). However, we also find that the deniers or skeptics of climate change, with apparent scientific arguments supporting them, are usually "social" organizations or think tanks (Cato Institute, 2019; Heartland Institute, 2020), creating a schism or division in the logic of discourse about climate change and global warming (Hoffman, 2011). Some corporations, such as the American Chamber of Commerce or the American Petroleum Institute, contribute to this narrative, opposing policies regarding carbon emissions (Pooley, 2010). In this regard, we can see that diverse interests, such as the eventual diminishing in the use of fossil fuels, can be behind their apparent affiliation to this stance.

Climate change deniers evolved over time and spread throughout a large segment of the US society as well as other nations. These beliefs have cemented, and individuals without a background in science feel free to write books denying anthropogenic climate change. These publications, which number more than 100 books in the US alone, often lack peer review; this allows their authors to deny climate change and make inaccurate assertions that grossly misrepresent the current state of science on the matter. Even more so, they are often sponsored by companies and institutions that rely on the status quo, resisting the inevitable change that climate science urges us to adopt (Dunlap & Jacques, 2013).

This debate has been observed for a long time through different mass media such as television (Boykoff, 2008), news broadcasts and newspapers (Boykoff & Rajan, 2007), and, more recently, in digital media, which is spearheading the discourse on climate change. Matters such as "amazongate," "glaciergate," and "climategate" are perfect topics for analyzing the most popular blogs, where we can find metaphors such as "science is religion" (science is cult, dogma, science believers, prophets, gurus, preachers, and doomsday prophecies) in their speech, usually written by people affiliated with the conservative right (Hoffman, 2011).

THE ROLE OF DIGITAL MEDIA IN THE CLIMATE CHANGE CRISIS

In modern urban societies, an important number of digital media are social. The new media services are constructed on top of activities of social networks, particularly on the contents produced and shared by users, as opposed to the era before social media, when contents were transmitted and generated by a reduced number of users (i.e., creators, publishers) (Ebner, 2007). In contrast, from the appearance of Web 2.0, most of the content shared in social media is created and distributed by single users (Kollányi, Molnár & Székely, 2007). Given this advantage, social media has grown more and more popular. According to the *Global Digital Report* from We Are Social (2020), active social media users have grown by 14.5% per year, accounting for 45% of the total world population.

Nevertheless, such an apparent democratization of digital spaces also brought new opportunities to those who wished to capitalize on human vulnerabilities. Given that digital spaces have become the primary source of information, the possibility that a part of our collective consciousness about climate change could be manipulated through fake news and misinformation should be a matter of major concern (AVAAZ, 2020).

Social media services are among the most important sources of information in the US as well as in the rest of the world (Ciampaglia & Menczer, 2018). A study (Ogunjinmi & Oluwatosin, 2016) reveals that 93.8% of the participants heard about climate change on social media, whereas 48.6% discussed climate change with their friends on those platforms. However, part of the information that the users of these platforms consume and share is questionable content, such as conspiracy theories, pseudo-science, or fake reports. Therefore, social media has become crucial for potentializing the actual phenomenon of post-truth. Particularly, it is being used by different status quo holders for spreading distinct ideas and beliefs regarding climate change, among other things (AVAAZ, 2020).

Social media is not a source of impartial information. In digital environments, participants are different, each with their own interests and biases. According to Ciampaglia and Menczer (2018) there are three types of bias that make social media a vulnerable environment for intentional and accidental misinformation: bias on machines (such as social media platforms), society (corporations and organizations), and the brain (individuals).

1. Bias on Digital Platforms

Social platforms play a crucial role in politics, the environment, and other social slopes. The optimization of pecuniary opportunities has, as a consequence, a polarized, biased, misinformed, and trending scenario (Grimme et al., 2020). These spaces tend to present biased information in relation to climate change because of several factors, among which are algorithm and consumption of information bias, derived from the sponsors of each social media platform (Bloomfield & Tillery, 2019; Bouie, 2020).

Two scenarios are evident in regard to the bias caused by social media platforms themselves. The first one occurs when the algorithm decides which information to use based on the interests of the hired sponsors, that is, content sponsored by third parties. The second is a scenario in which the presented information does not respond to publicity paid for by third parties but to the interests of the platform itself. For example, a platform can decide to present the information that it assumes will favor more engagement.

Digital platforms, such as social networks, work in accordance with their own private agenda. In the end, they present themselves as advertising landscapes and thus show handcrafted content to their users, based on what the sponsors want them to promote (Mosco, 2014). According to Carrington (2020), Facebook's algorithm has helped to broaden the dissemination of fake information related to climate change. An analytical firm (think tank) found that Facebook ads against climate change (51 climate disinformation ads) have been viewed at least eight million times during the first semester of 2020 (Carrington, 2020). In response to this fact, in February 2020, Facebook created the Climate Science Information Center focused on diminishing fake information around climate change. Despite this, the study found two active campaigns from February to October 2020 with the general intention of misinforming its audience (InfluenceMap, 2020).

On the other hand, AVAAZ, a US-based nonprofit organization that promotes global activism on issues such as climate change, human rights, animal rights, corruption, poverty, and conflict, reported how YouTube leads its users to consume misinformation regarding climate change. The text, titled "Why is YouTube broadcasting climate misinformation to millions?" (AVAAZ, 2020) discloses that on YouTube, when users search for "global warming," up to 16% of the suggested videos from the "Up Next" algorithm contain biased information regarding the topic. The same happens for the term "climate manipulation," with up to 21% of biased information presented as fact-checked and science-proven data. As of the date of publication of the report, 22 million reproductions were or contained explicitly biased climate change information. That assumption, according to AVAAZ (2020), is merely a sample of a specific topic.

In digital economies, where some of the key performance indicators are engagement, network growth, and monetization, the condition of post-truth regarding the information is more present than ever (Harsin, 2018). In an investigation focused on the analysis of 126,000 Twitter publications directly linked to newsworthy facts, it was found that fake news items are more profusely diffused, with greater outreach and more openness, than information categorized as true. Furthermore, it was found that the bots in charge of diffusing fake information with greater speed do so because of the lack of discrimination criteria of the platform itself (Vosoughi, Roy & Aral, 2018). Google (including its search engine and different tools), on its quest for keeping users captivated and spending long periods of time on the platform, has developed an ecosystem of misinformation where the final users can no longer distinguish what type of information they consume (Grimme et al., 2020; Schiermeier, 2019).

2. Bias on Corporations and Organizations

A great deal of the information we consume regarding climate change is a product of the deliberate effort of social collectives to gain ground on the collective image. The current interpretation of climate change also has to do with the epistemological stage of the main discourse in traditional as well as digital media (Baucom & Omelsky, 2017; Tomlinson, 2017). Companies and organizations fight for their voices and discourses to predominate in the collective consciousness, including organizations that look to deny the existence of climate change to companies that accepting its existence, search to turn it into a business opportunity.

Scientific evidence in the post-truth era loses ground in front of non-scientific thinking based on beliefs and emotions rather than science, technology, and profound and verifiable research. Climate change, as a field of study, is one of the many areas that have been affected by this transformation, including its political postures (left versus right biases) and accountability to communication media. Political decisions dictate, in several ways, how the ecological agenda is being shaped. Through the years, the economic system of values (including ecological values) has been modified according to present needs and interests, mainly in favor of global capitalism and groups of interest or political stakeholders (Muradian & Pascual, 2020).

In social media, scientific information can be co-opted and manipulated. For example, a study (Bloomfield & Tillery, 2019) found that some Facebook pages adopt an appearance of credibility by sharing scientific information; however, the authors of the study assert that this is merely a strategy, for they have a network of skeptic collaborators that argue against said content. In this way, they achieve to distance the readers from the original sources of scientific information. Visitors of the webpage use a variety of rhetorical strategies to create an echo of the main topics and discredit alternative points of view. According to the authors of the study, the differences between the strategies used show that the community of deniers of climate change is multifaceted, and it uses social media in order to appear legitimate.

On the other hand, on YouTube it was found that, from an assortment of 200 videos selected based on their thematic relation to "climate change" and "environmental engineering," 107 supported non-scientific postures, 16 related to denial, and 91 presented conspiracy theories. In total, the videos based on scientific evidence (93) had slightly more views than the rest (16.94 million of reproductions against 16.93) despite having a smaller percentage of total videos (Allgaier, 2019).

This is not only a game of the declared deniers searching to reject scientific evidence, but organizations and companies also seek to transform concern for climate change into a business opportunity. The human connection to the climate crisis has been portrayed through consumerist images offering solutions based on green consumerism and recycling patterns (Niemelä-Nyrhinen & Seppänen, 2019; Uusitalo, 2020). A study that analyzed the characteristics of the images found searching on Google detected that most images we find on the internet related to energy-saving are associated with a commercial purpose. This does not necessarily mean that most of the information on the internet related to the aforementioned topic has commercial purposes but only that the contents found through search engines (in this case Google) are of this type (Valerio-Ureña & Rogers, 2019).

Uusitalo (2020) studied the representations of everyday practices surrounding climate change on Instagram. He found that the necessary human actions to solve climate issues have been presented on social media as changes in consumption options and participating in manifestations, therefore connecting human participation with consumerism and political activism. The study also shows that the visualization of climate practices is relatively repetitive and limited to certain established representations related to consumerism and political activism such as vegan food, biodegradable goods, ethical fashion, and climate strikes.

3. Bias on the Individuals

> You know, think about it, when you watch a show from Netflix and you get addicted to it, you stay up late at night. We're competing with sleep, on the margin.
> (Reed Hastings, Netflix's CEO)

Beside the efforts of platforms and different social groups (companies, organizations, and so on) to spread the information most convenient to their interests, people have cognitive biases of their own that leads them to misunderstand the information received or to pay more attention to certain information over others. In other words, they are more vulnerable to biased information (Kahneman, 2013; McIntyre, 2018).

According to Kahneman (2013), social scientists of the 1970s generally accepted two ideas about human nature: (1) people are generally rational, and (2) emotions such as fear, affection, and hate explain most situations in which people move away from rationality. However, he challenged both assumptions by systematically documenting errors in the thought process of normal people. He discovered that the origin of these errors lies more within the process of cognition than in the handling of emotions.

In recent years, heuristics and cognitive biases have been used in many fields such as finance, politics, commerce, and the military. Heuristics are mental shortcuts that generally involve focusing on one aspect of a complex problem and ignoring others. Although they are usually efficient, they sometimes lead to cognitive biases or distortions of perception, which in turn lead a person to make inaccurate judgments or misinterpretations. They are self-defense mechanisms that our brain uses to cope with the difference between what we believe in and what the evidence might present to us (McIntyre, 2018).

It has been observed, for example, that availability heuristics could help explain why some issues are present in the public mind, whereas others fall into oblivion. According to this heuristic, we tend to evaluate the relative importance of certain issues depending on how easily they are brought to our memory, which is largely determined by the degree of coverage they receive in the media. These often-mentioned issues remain on our minds; therefore, one way to make people believe falsehoods is through frequent repetition, because familiarity is not easily distinguishable from the truth (Kahneman, 2013).

Another example is confirmation bias, which refers to our tendency to select information that confirms what we already believe in. Since changing one's mind is "uncomfortable," it is easier to validate information that aligns with our beliefs and to discard information that contradicts us (McIntyre, 2018).

The confirmation bias, according to McIntyre, is very evident when it comes to political positions such as climate change. These psychological tendencies can rob us of the ability to think clearly and prevent us from realizing that we are not thinking clearly (Kahneman, 2013; McIntyre, 2018). This situation is particularly evident in the discussions about climate change that take place on social media.

The strategy to spread doubts regarding scientific studies has been largely used when tobacco and cigarette consumption was held in doubt toward cancer-related diseases (McIntyre, 2018). Factual and objective data face cancelation in the face of emotions and personal beliefs, allowing a multiverse of narrative lines, where TV networks contrast and differ, sometimes in a pendular way, one in front of the other, mostly in accordance with their own agenda. On several occasions, the discussion revolves around whether the information is true or false rather than the central topic, forcing the real issue to be lost in between (Ogunjinmi & Oluwatosin, 2016).

CONCLUSION

The environment is fundamental for covering basic needs as well as for improving our general well-being. We believe that what we think about climate change has an impact on our behaviors regarding the care of the planet and can be influenced by the information we consume through digital media. However, digital media is not necessarily free of particular interests. Much of the information we consume in these media is regulated by commercial or political interests. Considering our natural tendency to be influenced by the group, it is of extreme importance to promote critical thinking and the development of basic computer skills to enable us to decide ourselves what to believe about the climate change crisis.

Sartre already said that "we are what we do with what they made of us." Then, although it could be true that we have an evolutionary predisposition to believe in others, a psychological predisposition in favor of information that aligns with our beliefs, and a permanent exposure to being bombarded with false and/or biased information, no predetermination prevents us from applying critical thinking, especially when we consume digital information.

Perhaps a starting point could be to recognize our own weaknesses. We are all, to some extent, victims of our psychological biases and therefore, vulnerable to the post-truth conundrum. To the extent that we are aware of this vulnerability, we will be more prepared to defend ourselves against false information that seeks to persuade us to believe what is best for some.

We cannot forget that, by participating in digital media, we are part of a network and, as such, we cannot forget certain principles, as Christakis and Fowler (2009) point out: (1) we create and recreate our networks; (2) the place we occupy in the network affects us; (3) other members of the network, with whom we connect, directly affect us; (4) the contacts of our contacts affect us; and (5) networks have a life of their own.

REFERENCES

Allgaier, J. (2019). Science and environmental communication on YouTube: Strategically distorted communications in online videos on climate change and climate engineering. *Frontiers in Communication*, 4(36), 1–15. https://www.frontiersin.org/articles/10.3389/fcomm.2019.00036/full

Anderson, M. (2012). New ecological paradigm (NEP) scale. https://www.researchgate.net/publication/264858463_New_Ecological_Paradigm_NEP_Scale

Aronson, E. (2011). *The Social Animal*. 11th ed. Worth Publishers

AVAAZ (2020). Why is YouTube broadcasting climate misinformation to millions? https://secure.avaaz.org/campaign/en/youtube_climate_misinformation/

Barkin, D., Fuente, M., & Zamora, M. (2012). The significance of a radical ecological economy. *REVIBEC*, 19(n.d.), 1–14. http://www.usfx.bo/nueva/vicerrectorado/citas/ECONOMICAS_6/Economia/70%20d%20barkin.pdf

Baucom, I., & Omelsky, M. (2017). Knowledge in the age of climate change. *South Atlantic Quarterly*, 116(1), 1–18. http://doi.org/10.1215/00382876-3749271

Bloomfield, E. F., & Tillery, D. (2019). The circulation of climate change denial online: Rhetorical and networking strategies on Facebook. *Environmental Communication*, 13(1), 23–34. http://doi.org/10.1080/17524032.2018.1527378

Bouie, J. (2020). Facebook has been a disaster for the world. *The New York Times*. https://www.nytimes.com/2020/09/18/opinion/facebook-democracy.html

Blunden, J., & Arndt, D. S. (Eds.) (2020) State of the climate in 2019. *Bull. Amer. Meteor. Soc.*, 101(8), Si–S429. https://doi.org/10.1175/2020BAMSStateoftheClimate.1

Boykoff, M. T. (2008). Lost in translation? United States television news coverage of anthropogenic climate change. 1995–2004. *Climatic Change*, 86(1–2), 1–11. https://doi.org/10.1007/s10584-007-9299-3

Boykoff, M. T., & Rajan, S. R. (2007). Signals and noise: Mass-media coverage of climate change in the USA and the UK. *EMBO Reports*, 8(3), 207–211. https://doi.org/10.1038/sj.embor.7400924

Carrington, D. (2020, October 8). Climate denial ads on Facebook seen by millions, report finds. *The Guardian*. https://www.theguardian.com/environment/2020/oct/08/climate-denial-ads-on-facebook-seen-by-millions-report-finds

Cato Institute (2019). Climate change doomsday trap. https://www.cato.org/sites/cato.org/files/serials/files/regulation/2019/6/reg-v42n2-2.pdf?queryID=01e51183feb4d586dcfd47cdacbde947

Christakis, N., & Fowler, J. (2009). *Connected: The Surprising Power of Our Social Networks and How They Shape Our Lives*. Little, Brown & Company

Ciampaglia, G., & Menczer, F. (2018). Misinformation and biases infect social media, both intentionally and accidentally. https://theconversation.com/misinformation-and-biases-infect-social-media-both-intentionally-and-accidentally-97148

Dunlap, R. E., & Jacques, P. J. (2013). Climate change denial books and conservative think tanks: Exploring the connection. *American Behavioral Scientist*, 57(6), 699–731. https://doi.org/10.1177/0002764213477096

Ebner, M. (2007). E–learning 2.0 = e–learning 1.0 + Web 2.0? The Second International Conference on Availability, Reliability and Security, 1235-1239 (ARES'07). http://www.informatik.uni-trier.de/~ley/db/indices/a-tree/e/Ebner:Martin.html

FAO, IFAD, UNICEF, WFP, & WHO (2020). The state of food security and nutrition in the world 2020. Transforming food systems for affordable healthy diets. Rome, FAO. https://doi.org/10.4060/ca9692en

Grimme, C., Preuss, M., Takes, F. W., & Waldherr, A. (2020). *Disinformation in Open Online Media*. Springer International Publishing

Hadnagy, C. (2011). *Social Engineering: The Art of Human Hacking*. Wiley

Harsin, J. (2018). Post-truth and critical communication studies. *Oxford Research Encyclopedia of Communication*. https://doi.org/10.1093/acrefore/9780190228613.013.757

Heartland Institute (2020). n.d. Accessed September 13, 2020. https://www.heartland.org/Center-Climate-Environment/About/index.html.

Hoffman, A. J. (2011). Talking past each other? Cultural framing of skeptical and convinced logics in the climate change debate. *Organization & Environment*, 24(1), 3–33. https://doi.org/10.1177/1086026611404336

InfluenceMap (2020). Climate change and digital advertising. https://influencemap.org/report/Climate-Change-and-Digital-Advertising-86222daed29c6f49ab2da76b0df15f76

Intergovernmental Panel on Climate Change (IPCC) (2018). Global warming of 1.5°C: Summary for policymakers. https://report.ipcc.ch/sr15/pdf/sr15_spm_final.pdf

Kahneman, D. (2013). *Thinking, Fast and Slow*. Farrar, Straus & Giroux

Kollányi, B., Molnár, S., & Székely, L. (2007). Social networks and the network society. http://www.ittk.hu/netis/doc/ISCB_eng/04_MKSZ_final.pdf

McIntyre, L. (2018). *Post-Truth*. MIT Press

Mosco, V. (2014). *To the Cloud. Big Data in a Turbulent World*. Paradigm Publications

Muradian, R., & Pascual, U. (2020). Ecological economics in the age of fear. *Ecological Economics*, 169, 1–8. https://doi.org/10.1016/j.ecolecon.2019.106498

Naess, A. (1973). The shallow and the deep, long-range ecology movement: A summary. *Inquiry*, 16(1–4), 95–100. https://doi.org/10.1080/00201747308601682

Naess, A. (1993). *Ecology, Community, and Lifestyle: Outline of an Ecosophy*. Cambridge University Press

Niemelä-Nyrhinen, J., & Seppänen, J. (2019). Kuvajournalismi ja eettisen kuluttamisen haaste. *Media & Viestintä*, 42(3). https://doi.org/10.23983/mv.85780

Ogunjinmi, A., & Oluwatosin, A. (2016). Influence of social media on climate change. Knowledge and concerns. *Nigerian Journal of Agriculture, Food & Environment*, 12(4), 23–30. https://www.researchgate.net/publication/312158825_INFLUENCE_OF_SOCIAL_MEDIA_ON_CLIMATE_CHANGE_KNOWLEDGE_AND_CONCERNS

Pooley, E. (2010). *The Climate War*. Hyperion

Schiermeier, Q. (2019). False statements about climate change trip people up. *Nature*. https://www.nature.com/articles/d41586-019-02637-x?utm_source=%E2%80%A6912&utm_medium=email&utm_term=0_c9dfd39373-6fb58d2862-42158739

Rifkin, J. (2011). *The third industrial revolution: How lateral power is transforming energy, the economy, and the world*. Palgrave Macmillan.

Tomlinson, G. (2017). Two deep-historical models of climate crisis. *South Atlantic Quarterly*, 116(1), 19–31. http://doi.org/10.1215/00382876-3749282

Uusitalo, N. (2020). Unveiling unseen climate practices on Instagram. *Novos Olhares*, 9(1), 120–129. https://www.revistas.usp.br/novosolhares/article/download/171996/161939/

Valerio-Ureña, G., & Rogers, R. (2019). Characteristics of the digital content about energy-saving in different countries around the world. *Sustainability*, 11(17), 4704. MDPI. Agriculturists. http://doi.org/10.3390/su11174704

Vosoughi, S., Roy, D., & Aral, S. (2018). The spread of true & false news online. *Science*, 359(6380), 1146–1151. https://science.sciencemag.org/content/359/6380/1146

We are Social (2020). Digital in 2020. https://wearesocial.com/digital-2020

PART II

Anthropocene literacy

5. A terminology for the Anthropocene
Ernesto Contreras

INTRODUCTION

The original use of the term Anthropocene is to describe the proposed geological epoch following the Holocene. It is characterized by an overwhelming mark of human activity on Earth so as to leave a stratigraphical record. Given the enormous significance of these facts and the observed and potential consequences for the Biosphere, the term has also been adopted to comprise its wider social, economic and cultural implications. This glossary aims at combining the most elementary terms from the Earth, Environmental, Atmospheric and Life Sciences with the increasing attention drawn from the Social and Behavioral Sciences and the Humanities. The purpose is to provide the non-specialist with the basic conceptual tools to navigate the multidisciplinary perspectives of this new epoch. The method follows the Wikipedia Manual of Style due to its popularity and efficiency. Additionally, Google Scholar has been used to weight the relative importance of each term. Finally, an expert consultation exercise lead us to obtain the final selection of terms within the available space. When quoted verbatim from a source, each term is followed by the source, indicated by a consecutive number in squared brackets. A list of all sources is included at the end.

GLOSSARY

Abrupt Climate Change

Sudden (on the order of decades), large changes in some major component of the climate system, with rapid, widespread effects [3].

Actant

Something that acts or to which activity is granted by others. It implies no special motivation of human individual actors, nor of humans in general. An actant can literally be anything provided it is granted to be the source of an action [1].

Adaptation

The process of adjustment to actual or expected climate and its effects. In human systems, adaptation seeks to moderate or avoid harm or exploit beneficial opportunities. In some natural systems, human intervention may facilitate adjustment to expected climate and its effects [2].

Adaptive Capacity

The ability of systems, institutions, humans and other organisms to adjust to potential damage, to take advantage of opportunities, or to respond to consequences [2].

Aerosol

A suspension of airborne solid or liquid particles, with a typical size between a few nanometers and 10 μm that reside in the atmosphere for at least several hours [2].

Afforestation

Planting of new forests on lands that historically have not contained forests [3].

Albedo

The amount of solar radiation reflected from an object or surface, often expressed as a percentage [3].

Anthropocene

Defines Earth's most recent geologic time period as being human-influenced, or anthropogenic, based on overwhelming global evidence that atmospheric, geologic, hydrologic, biospheric and other Earth system processes are now altered by humans. The word combines the root "anthropo," meaning "human" with the root "-cene," the standard suffix for "epoch" in geologic time. The Anthropocene is distinguished as a new period either after or within the Holocene, the current epoch, which began approximately 10,000 years ago (about 8000 BC) with the end of the last glacial period [42].

Anthropogenic Climate Change

Man-made climate change [5].

Anthropogenic Emissions

Emissions of greenhouse gases (GHGs), precursors of GHGs and aerosols caused by human activities. These activities include the burning of fossil fuels, deforestation, land use and land use changes (LULUC), livestock production, fertilization, waste management, and industrial processes [2].

Atmosphere

The gaseous envelope surrounding the Earth, divided into five layers – the troposphere which contains half of the Earth's atmosphere, the stratosphere, the mesosphere, the thermosphere, and the exosphere, which is the outer limit of the atmosphere [2].

Atmospheric Aerosols

Microscopic particles suspended in the lower atmosphere that reflect sunlight back to space [5].

Baseline for Cuts

The year against which countries measure their target decrease of emissions [5].

Biocapacity

The ecosystems' capacity to produce biological materials used by people and to absorb waste material generated by humans, under current management schemes and extraction technologies [4].

Biofuel

A fuel derived from renewable, biological sources, including crops such as maize and sugar cane, and some forms of waste [5].

Biogeochemical Cycle

Movements through the Earth system of key chemical constituents essential to life, such as carbon, nitrogen, oxygen, and phosphorus [3].

Biosphere

The part of the Earth system comprising all ecosystems and living organisms, in the atmosphere, on land (terrestrial biosphere) or in the oceans (marine biosphere), including derived dead organic matter, such as litter, soil organic matter and oceanic detritus [3].

Cap and Trade

An emission trading scheme whereby businesses or countries can buy or sell allowances to emit greenhouse gases via an exchange [5].

Capitalocene

A way of understanding capitalism as a connective geographical and patterned historical system. In this view, the Capitalocene is a geopoetics for making sense of capitalism as a world-ecology of power and re/production in the web of life [6].

Carbon Capture and Storage

The collection and transport of concentrated carbon dioxide gas from large emission sources, such as power plants [5].

Carbon Cycle

The term used to describe the flow of carbon (in various forms, e.g., as carbon dioxide (CO_2)) through the atmosphere, ocean, terrestrial and marine biosphere and lithosphere [2].

Carbon Dioxide

A gas in the Earth's atmosphere. It occurs naturally and is also a by-product of human activities such as burning fossil fuels. It is the principal greenhouse gas produced by human activity [5].

Carbon Footprint

The amount of carbon emitted by an individual or organization in a given period of time, or the amount of carbon emitted during the manufacture of a product [5].

Carbon Neutral

A process where there is no net release of CO_2 [5].

Carbon Offsetting

A way of compensating for emissions of CO_2 by participating in, or funding, efforts to take CO_2 out of the atmosphere [5].

Carbon Sequestration

The process of storing carbon dioxide. This can happen naturally, as growing trees and plants turn CO_2 into biomass (wood, leaves, and so on). It can also refer to the capture and storage of CO_2 produced by industry. See Carbon capture and storage [5].

Carbon Sink

Any process, activity or mechanism that removes carbon from the atmosphere [5].

Carrying Capacity

The average population density or population size of a species below which its numbers tend to increase and above which its numbers tend to decrease because of shortages of resources. The carrying capacity is different for each species in a habitat because of that species' particular food, shelter, and social requirements [7].

CFCs

The short name for chlorofluorocarbons – a family of gases that have contributed to stratospheric ozone depletion, but which are also potent greenhouse gases [5].

Cities of Refuge

Cities where humans and nonhuman (climates refugees) will migrate to seek refuge due to global warming and other environmental changes [8].

Climate

The statistical description in terms of the mean and variability of relevant quantities over a period of time ranging from months to thousands of years. The classical period is three decades, as defined by the World Meteorological Organization (WMO). These quantities are most often surface variables such as temperature, precipitation, and wind. Climate in a wider sense is the state, including a statistical description, of the climate system [3].

Climate Change

A pattern of change affecting global or regional climate, as measured by yardsticks such as average temperature and rainfall, or an alteration in frequency of extreme weather conditions [5].

Climate Crisis

Anodyne references to "climate change" and "global warming" are being scorned by those who think it's time for more drastic talk, and action, on the environment [44].

Climate Feedback

A process that acts to amplify or reduce direct warming or cooling effects [3].

Climate Model

A quantitative way of representing the interactions of the atmosphere, oceans, land surface, and ice [3].

Climate Refugee

Those people who have been forced to leave their traditional habitat, temporarily or permanently, because of marked environmental disruption (natural and/or triggered by people) that jeopardized their existence and/or seriously affected the quality of their life [9].

Climate Sensitivity

The amount of global surface warming projected in response to a doubling of atmospheric CO_2 concentrations compared to the pre-industrial levels of 280 ppm [47].

Climate System (or Earth System)

The five physical components (atmosphere, hydrosphere, cryosphere, lithosphere, and biosphere) that are responsible for the climate and its variations [3].

Commons

Is a constant emergence of common social, economic and environmental relations and practices. They are spaces of experimentation, for both theorizing and practising, providing a lens for critique and affirmation, a method for resistance and creation. They might be purely affective, tethered to a specific location, or an unpredictable entangling of both [10].

Contaminant

Any physical, chemical, biological, or radiological substance found in air, water, soil, or biological matter that can have a harmful effect on people, animals, or plants [3].

Coral Bleaching

The process in which a coral colony, under environmental stress expels the microscopic algae (zooxanthellae) that live in symbiosis with their host organisms (polyps) [3].

Critical Posthumanism

Is a theoretical approach which maps and engages with the "ongoing deconstruction of humanism." It differentiates between the figure of the "posthuman" and "posthumanism" as the social discourse (in the Foucauldian sense) which negotiates the pressing question of what it means to be human under the conditions of globalization, technoscience, late capitalism and climate change [11].

Decolonial Critique

A form of the epistemic struggle aims to move towards intercultural dialogue and decolonial understanding [12].

Deforestation

The permanent removal of standing forests that can lead to significant levels of carbon dioxide emissions [5].

Desertification

Land degradation in arid, semi-arid, and dry sub-humid areas resulting from various factors, including climatic variations and human activities [3].

Earth

Is the third planet from the Sun [13].

Ecocide

Is the extensive damage to, destruction of or loss of ecosystem(s) of a given territory, whether by human agency or by other causes, to such an extent that peaceful enjoyment by the inhabitants of that territory has been or will be severely diminished [14].

Ecological Footprint

A measure of how much area of biologically productive land and water an individual, population or activity requires to produce all the resources it consumes and to absorb the waste it generates, using prevailing technology and resource management practices [4].

Ecosystem

The interacting system of a particular biological community and its non-living environmental surroundings, or a class of such systems (e.g., forests or wetlands) [3].

Environmentalism

Can be described as a social movement or as an ideology focused on the welfare of the environment. Environmentalism seeks to protect and conserve the elements of Earth's ecosystem, including water, air, land, animals, and plants, along with entire habitats such as rainforests, deserts and oceans [15].

Existential Risks

Are events that could lead to human extinction or civilizational collapse. These events include runaway climate change, pandemics, nuclear warfare and artificial intelligence related disasters to name a few [46].

Exosphere

This is the upper limit of our atmosphere. It extends from the top of the thermosphere up to 10,000 km (6,200 mi) [43].

Extinction

In biology, the dying out or extermination of a species. Extinction occurs when species are diminished because of environmental forces (habitat fragmentation, global change, natural disaster, overexploitation of species for human use) or because of evolutionary changes in their members (genetic inbreeding, poor reproduction, decline in population numbers). [16].

Feedback Loop

In a feedback loop, rising temperatures on the Earth change the environment in ways that affect the rate of warming. Feedback loops can be positive (adding to the rate of warming), or negative (reducing it) [5].

Feedback Mechanisms

Factors which increase or amplify (positive feedback) or decrease (negative feedback) the rate of a process [3].

Fluorinated Gases

Powerful synthetic greenhouse gases such as hydrofluorocarbons, perfluorocarbons, and sulfur hexafluoride that are emitted from a variety of industrial processes [3].

Forcing Mechanism

A process that alters the energy balance of the climate system, that is, changes the relative balance between incoming solar radiation and outgoing infrared radiation from Earth [3].

Fossil Economy

An economy characterized by self-sustaining growth predicated on growing consumption of fossil fuels, and therefore generating a sustained growth in emissions of carbon dioxide. Thus defined, the concept refers to an expansion in the scale of material production realized through expansion in the combustion of coal, oil and/or natural gas [17].

General Circulation Model (GCM)

A global, three-dimensional computer model of the climate system which can be used to simulate human induced climate change [3].

General Ecology

A new ecological paradigm, which incorporates different ecological studies and their relations with the techne, the society, the power and others [18].

Geological Epoch

Unit of geological time during which a rock series is deposited. It is a subdivision of a geological period, and the word is capitalized when employed in a formal sense (e.g., Pleistocene Epoch). Additional distinctions can be made by appending relative time terms, such as early, middle, and late. The use of epoch is usually restricted to divisions of the Paleogene, Neogene, and Quaternary periods [19].

Geological Sequestration

The injection of carbon dioxide into underground geological formations [5].

Geological Time Scale

Is the "calendar" for events in Earth history. It subdivides all time into named units of abstract time called – in descending order of duration – eons, eras, periods, epochs, and ages. The enumeration of those geologic time units is based on stratigraphy, which is the correlation and classification of rock strata. The fossil forms that occur in the rocks, however, provide the chief means of establishing a geologic time scale, with the timing of the emergence and disappearance of widespread species from the fossil record being used to delineate the beginnings and endings of ages, epochs, periods, and other intervals [20].

Geopolitics

Analysis of the geographic influences on power relationships in international relations. The word geopolitics was originally coined by the Swedish political scientist Rudolf Kjellén about the turn of the twentieth century, and its use spread throughout Europe in the period between World Wars I and II (1918–39) and came into worldwide use during the latter. In contemporary discourse, geopolitics has been widely employed as a loose synonym for international politics [21].

Global Energy Budget

The balance between the Earth's incoming and outgoing energy [5].

Global Heating

Another term for global warming, emphasizing the degree of the phenomenon and the urgency of the problem [45].

Global Warming

The steady rise in global average temperature in recent decades, which experts believe is largely caused by man-made greenhouse gas emissions. The long-term trend continues upwards, they suggest, even though the warmest year on record, according to the UK's Met Office, is 1998 [5].

Great Acceleration

Refers to the most recent period of the proposed Anthropocene epoch during which the rate of impact of human activity upon the Earth's geology and ecosystems is increasing significantly. While the proposed start dates for the Anthropocene span the Industrial Revolution and earlier, the Great Acceleration begins after World War II. This concept has been further extended to refer to the rate of change in technology and society as a whole [22].

Great Divergence

Coined by American historian Kenneth Pomeranz, designates the industrial boom that has separated Europe from China since the nineteenth century. According to Pomeranz, it was the unequal geographical distribution of coal resources and the conquest of the New World that gave the decisive impetus to the European economy [4].

Greenhouse Effect

The insulating effect of certain gases in the atmosphere, which allow solar radiation to warm the earth and then prevent some of the heat from escaping [5].

Greenhouse Gases (GHGs)

Natural and industrial gases that trap heat from the Earth and warm the surface. The Kyoto Protocol restricts emissions of six greenhouse gases: natural (carbon dioxide, nitrous oxide, and methane) and industrial (perfluorocarbons, hydrofluorocarbons, and sulfur hexafluoride) [5].

Heat Island

An urban area characterized by temperatures higher than those of the surrounding non-urban area [3].

Heat Waves

A prolonged period of excessive heat often combined with excessive humidity [3].

Holocene

Formerly Recent Epoch, younger of the two formally recognized epochs that constitute the Quaternary Period and the latest interval of geologic time, covering approximately the last 11,700 years of Earth's history. The sediments of the Holocene, both continental and marine, cover the largest area of the globe of any epoch in the geologic record, but the Holocene is unique because it is coincident with the late and post-Stone Age history of humankind [23].

Hydrocarbons

Substances containing only hydrogen and carbon. Fossil fuels are made up of hydrocarbons [3].

Hydrologic Cycle

The process of evaporation, vertical and horizontal transport of vapor, condensation, precipitation, and the flow of water from continents to oceans [3].

Indigenous Knowledge

The understandings, skills and philosophies developed by societies with long histories of interaction with their natural surroundings [24].

Ionosphere

Is an abundant layer of electrons and ionized atoms and molecules that stretches from about 48 kilometers (30 miles) above the surface to the edge of space at about 965 km (600 mi), overlapping into the mesosphere and thermosphere. This dynamic region grows and shrinks based on solar conditions and divides further into the sub-regions: D, E and F; based on what wavelength of solar radiation is absorbed. The ionosphere is a critical link in the chain of Sun–Earth interactions. This region is what makes radio communications possible [43].

IPCC

The Intergovernmental Panel on Climate Change is a scientific body established by the United Nations Environment Programme and the World Meteorological Organization [5].

Kin

A wild category that all sorts of people do their best to domesticate. Making kin as oddkin rather than, or at least in addition to, godkin and genealogical and biogenetic family troubles important matters, like to whom one is actually responsible [25].

Kyoto Protocol

A protocol attached to the UN Framework Convention on Climate Change, which sets legally binding commitments on greenhouse gas emissions. Industrialized countries agreed to reduce their combined emissions to 52 percent below 1990 levels during the five-year period 2008–12 [5].

LULUCF

Refers to Land Use, Land-Use Change, and Forestry [5].

Megacities

Cities with populations over 10 million [3].

Mesosphere

The mesosphere starts just above the stratosphere and extends to 85 kilometers (53 miles) high. Meteors burn up in this layer [43].

Methane

Is the second most important man-made greenhouse gas. Sources include both the natural world (wetlands, termites, wildfires) and human activity (agriculture, waste dumps, leaks from coal mining) [5].

Natural Source

A term used to describe any air emission's source of natural origin. Examples include volcanoes, wildfires, wind-blown dust, and releases due to biological processes [3].

Nitrogen Cycle

The natural circulation of nitrogen among the atmosphere, plants, animals, and microorganisms that live in soil and water [3].

Nonhuman Agency

Sometimes means "the agency of that which subtends the human," and sometimes "the agency of entities other than human beings" [26].

Object-Oriented Ontology (OOO)

Heidegger-influenced school of thought that rejects the privileging of human existence over the existence of nonhuman objects [27].

Overshoot

Global overshoot occurs when humanity's demand on nature exceeds the biosphere's supply, or regenerative capacity. Such overshoot leads to a depletion of Earth's life-supporting natural capital and a build-up of waste. At the global level, ecological deficit and overshoot are the same, since there is no net-import of resources to the planet. Local overshoot occurs when a local ecosystem is exploited more rapidly than it can renew itself [28].

Ozone

The triatomic form of oxygen (O_3) is a gaseous atmospheric constituent. Depletion of stratospheric ozone, due to chemical reactions that may be enhanced by climate change, results in an increased ground-level flux of ultraviolet (UV-) B radiation [3].

Particulate Matter (PM)

Very small pieces of solid or liquid matter such as particles of soot, dust, fumes, mists or aerosols [3].

Parts Per Million by Volume (ppmv)

Number of parts of a chemical found in one million parts of a particular gas, liquid, or solid [3].

Permaculture

An agricultural system or method that seeks to integrate human activity with natural surroundings so as to create highly efficient self-sustaining ecosystems [29].

Permafrost

Ground (soil or rock and included ice and organic material) that remains at or below 0°C for at least two consecutive years [2].

Planetary Boundaries

Are human-determined values of the control variable set at a "safe" distance from a dangerous level (for processes without known thresholds at the continental to global scales) or from its global threshold. These are: Climate change, ocean acidification, stratospheric ozone depletion, interference with the global phosphorus and nitrogen cycles, rate of biodiversity loss, global freshwater use, land system change, aerosol loading and chemical pollution [30].

Planet Equivalent

Or the number of planets it would take to support humanity's needs at any given time [4].

Posthumanism

The idea that humanity can be transformed, transcended, or eliminated either by technological advances or the evolutionary process; artistic, scientific, or philosophical practice which reflects this belief [31].

Ppm (350/450)

An abbreviation for parts per million, usually used as short for ppmv (parts per million by volume). The Intergovernmental Panel on Climate Change (IPCC) suggested in 2007 that the world should aim to stabilize greenhouse gas levels at 450 ppm CO_2 equivalent in order to avert dangerous climate change [5].

Pre-industrial Levels of Carbon Dioxide

The levels of carbon dioxide in the atmosphere prior to the start of the Industrial Revolution. These levels are estimated to be about 280 parts per million (by volume) [5].

Promethean Worldview

Considers humans as exceptional beings without limits, without any ethics of environmental restraint [32].

Radiative Forcing

The strength of drivers is quantified as Radiative Forcing (RF) in watts per square meter (W/m2) as in previous IPCC assessments. RF is the change in energy flux caused by a driver and is calculated at the tropopause or at the top of the atmosphere [2].

Reflectivity

The ability of a surface material to reflect sunlight including the visible, infrared, and ultraviolet wavelengths [3].

Relative Sea Level Rise

The increase in ocean water levels at a specific location, taking into account both global sea level rise and local factors, such as local subsidence and uplift [3].

Renewable Energy

Is energy created from sources that can be replenished in a short period of time. The five renewable sources used most often are: biomass (such as wood and biogas), the movement of water, geothermal (heat from within the earth), wind, and solar [5].

Resilience

The capacity of a social-ecological system to continually change and adapt yet remain within critical thresholds [33].

Rewilding

The planned reintroduction of a plant or animal species and especially a keystone species or apex predator (such as the gray wolf or lynx) into a habitat from which it has disappeared (as from hunting or habitat destruction) in an effort to increase biodiversity and restore the health of an ecosystem [34].

Sink

Any process, activity or mechanism that removes a greenhouse gas (GHG), an aerosol or a precursor of a GHG or aerosol from the atmosphere [2].

Sixth Extinction

An ongoing current event in which a large number of living species are threatened with extinction or are going extinct because of environmentally destructive human activities [35].

Solar Radiation

Radiation emitted by the Sun. It is also referred to as short-wave radiation [3].

Spheres

For the Russian mineralogist and geologist Vladimir Vernadsky, who devised the concept of biosphere in 1926, Planet Earth is made up of the intermeshing of five distinct spheres – the lithosphere, the rigid, rock outer layer; the biosphere, comprising all living beings; the atmosphere, the envelope of gases known as air; the technosphere resulting from human activity; and the noosphere, the part of the biosphere occupied by thinking humanity, including all thoughts and ideas. Other authors have since added to this list the notions of hydrosphere (all the water present on the planet) and cryosphere (ice) [4].

Stratosphere

The stratosphere starts just above the troposphere and extends to 50 kilometers (31 miles) high. The ozone layer, which absorbs and scatters the solar ultraviolet radiation, is in this layer [43].

Subsiding/Subsidence

The downward settling of the Earth's crust relative to its surroundings [3].

Survival

The act or fact of living or continuing longer than another person or thing [36].

Sustainable Development

Is development that meets the needs of the present without compromising the ability of future generations to meet their own needs [37].

Technodiversity

The diversity of technological artifacts in technoecosystems [38].

Technofossils

Are the remains of technological objects [4].

Technosoil

Soils with technical origin, produced entirely by humans [39].

Technosphere

The physical part of the environment that is modified by human activity [4].

Terrestrial

Of or relating to the Earth or its inhabitants [40].

Thermosphere

The thermosphere starts just above the mesosphere and extends to 600 kilometers (372 miles) high. Aurora and satellites occur in this layer [43].

Tipping Point

A threshold for change, which, when reached, results in a process that is difficult to reverse [5].

Troposphere

The troposphere starts at the Earth's surface and extends 8 to 14.5 kilometers high (5 to 9 miles). This part of the atmosphere is the densest. Almost all weather is in this region [43].

Ultraviolet Radiation (UV)

The energy range just beyond the violet end of the visible spectrum. Although ultraviolet radiation constitutes only about 5 percent of the total energy emitted from the sun, it is the major energy source for the stratosphere and mesosphere, playing a dominant role in both energy balance and chemical composition [3].

Vibrant Matter

The capacity of things – edibles, commodities, storms, metals – not only to impede or block the will and designs of humans, but also to act as quasi agents or forces with trajectories, propensities, or tendencies of their own [41].

SOURCES

[1] Latour, B. (1990). On actor-network theory. A few clarifications plus more than a few complications. *Soziale Welt*, *47*(4), 369–381.
[2] Pachauri, R. K., & Meyer, L. A. (2015). *Climate change 2014: synthesis report: contribution of working groups I, II and III to the Fifth Assessment Report of the Intergovernmental Panel on Climate Change* (First, Vol. 151). IPCC.
[3] EPA. (2016, September 29). *Glossary of Climate Change Terms*. https://19january2017snapshot.epa.gov/climatechange/glossary-climate-change-terms.html.
[4] UNESCO. (2018, April 27). *A lexicon for the Anthropocene*. UNESCO. https://en.unesco.org/courier/2018-2/lexicon-anthropocene.
[5] BBC. (2014, April 13). *Climate change glossary*. BBC News. https://www.bbc.com/news/science-environment-11833685.
[6] Moore, J. *Who is responsible for the climate crisis?* https://www.maize.io/en/content/what-is-capitalocene.
[7] The Editors of Encyclopædia Britannica. (2019, September 25). *Carrying capacity*. Encyclopædia Britannica. https://www.britannica.com/science/carrying-capacity.
[8] *cities of refuge*. The Anthropocene Atlas of Geneva. https://head.hesge.ch/taag/en/glossaire/cities-of-refuge/.
[9] Apap, J. (2019). The concept of "climate refugee": Towards a possible definition. *European Parliamentary Research Service*.
[10] Weber, L. G. (2019). Commons, the. In Braidotti, R., & Hlavajova, M. (eds.), *Posthuman glossary* (pp. 83–86). Bloomsbury Academic.
[11] Herbrechter, S. (2013). *Posthumanism: a critical analysis*. Bloomsbury.
[12] Vázquez, R. (2012). Towards a decolonial critique of modernity. Buen Vivir, relationality and the task of listening. *Capital, Poverty, Development, Denktraditionen im Dialog: Studien zur Befreiung und interkulturalität*, *33*, 241–252.
[13] NASA. (2019, March 22). *Overview*. NASA. https://solarsystem.nasa.gov/planets/earth/overview/.
[14] Higgins, P., Short, D., & South, N. (2013). Protecting the planet: a proposal for a law of ecocide. *Crime, Law and Social Change*, *59*(3), 251–266 https://doi.org/101007/s10611-013-9413-6.
[15] Lovelady, D. M. *Environmentalism*. Environmentalism | Learning to Give. https://www.learningtogive.org/resources/environmentalism.
[16] Gittleman, J. L. (2020, August 27). *Extinction*. Encyclopædia Britannica. https://www.britannica.com/science/extinction-biology.
[17] Malm, A. (2013). The origins of fossil capital: From water to steam in the British cotton industry. *Historical Materialism*, *21*(1), 15–68 https://doi.org/101163/1569206x-12341279.
[18] Hörl, E. (2019). General ecology. In Braidotti, R., & Hlavajova, M. (eds.), *Posthuman glossary* (pp. 172–175). Bloomsbury Academic.

[19] The Editors of Encyclopædia Britannica. (2009, September 11). *Epoch*. Encyclopædia Britannica. https://www.britannica.com/science/epoch-geologic-time.
[20] The Editors of Encyclopædia Britannica. (2020, June 2). *Geologic time*. Encyclopædia Britannica. https://www.britannica.com/science/geologic-time.
[21] Deudney, D. H. (2013, June 12). *Geopolitics*. Encyclopædia Britannica. https://www.britannica.com/topic/geopolitics.
[22] Stockholm Resilience Centre. *Great Acceleration*. Globaïa. https://globaia.org/great-acceleration.
[23] Agenbroad, L. D., & Fairbridge, R. W. (2018, December 7). *Holocene Epoch*. https://www.britannica.com/science/Holocene-Epoch.
[24] UNESCO. (2018). *What is Local and Indigenous Knowledge: United Nations Educational, Scientific and Cultural Organization*. What is Local and Indigenous Knowledge | United Nations Educational, Scientific and Cultural Organization. http://www.unesco.org/new/en/natural-sciences/priority-areas/links/related-information/what-is-local-and-indigenous-knowledge.
[25] Haraway, D. J. (2016). *Staying with the trouble: making kin in the chthulucene*. Duke University Press.
[26] Bowden, S. (2015). Human and Nonhuman Agency in Deleuze. In Stark, H., & Roffe, J. (eds.), *Deleuze and the non/human* (pp. 60–80). Palgrave Macmillan.
[27] Graham, H. (2003). Tool-being: Heidegger and the metaphysics of objects. *Choice Reviews Online*, *40*(7). https://doi.org/105860/choice.40-3935.
[28] Global Footprint Network. TEST CONTENT – Global Footprint Network. https://www.footprintnetwork.org/about-us/support-our-work/test-content/.
[29] Merriam-Webster. *Permaculture*. Merriam-Webster. https://www.merriam-webster.com/dictionary/permaculture.
[30] Rockström, J., Steffen, W., Noone, K., Persson, Å., Chapin III, F. S., Lambin, E., … & Nykvist, B. (2009). Planetary boundaries: exploring the safe operating space for humanity. *Ecology and Society*, *14*(2), 472–475.
[31] Lexico Dictionaries. *Posthumanism: Definition of Posthumanism by Oxford Dictionary on Lexico.com also meaning of Posthumanism*. Lexico Dictionaries | English. https://www.lexico.com/definition/posthumanism.
[32] Ischwark. (2013, December 10). *Glossary*. The Anthropocene. https://blogs.nelson.wisc.edu/es113-310-1/glossary/.
[33] *resilience*. The Anthropocene Atlas of Geneva. https://head.hesge.ch/taag/en/glossaire/resilience/.
[34] Diccionario Cambridge inglés. Significado de REWILDING en el Diccionario Cambridge inglés. https://dictionary.cambridge.org/es/diccionario/ingles/rewilding.
[35] Wagler, R. (1970, January 1). *Anthropocene extinction*. Access Science. https://www.accessscience.com/content/anthropocene-extinction/039350.
[36] Merriam-Webster. *Survival*. Merriam-Webster. https://www.merriamwebster.com/dictionary/survival.
[37] United Nations. (1987). *Our common future*. Oxford University Press.
[38] *technodiversity*. Technodiversity dictionary definition | technodiversity defined. https://www.yourdictionary.com/technodiversity.
[39] Boivin, P. *Technosol and Anthrosol*. The Anthropocene Atlas of Geneva. https://head.hesge.ch/taag/en/glossaire/technosol-and-anthrosol/.
[40] Merriam-Webster. *Terrestrial*. Merriam-Webster. https://www.merriam-webster.com/dictionary/terrestrial.

[41] Bennett, J. (2010). *Vibrant matter: A political ecology of things*. Duke University Press.
[42] Ellis, E. (2013, September 3). *Anthropocene*. Anthropocene – The Encyclopedia of Earth. https://editors.eol.org/eoearth/wiki/Anthropocene.
[43] Zell, H. (2015, March 2). *Earth's Atmospheric Layers*. NASA. https://www.nasa.gov/mission_pages/sunearth/science/atmosphere-layers2.html.
[44] Vickers, E. (2019, September 17). *Is Climate Change a Crisis Yet? Why Terms Matter*. Bloomberg.com. https://www.bloomberg.com/news/articles/2019-09-17/when-is-change-a-crisis-why-climate-terms-matter-quicktake.
[45] *Definition of global heating: New word suggestion: Collins Dictionary*. Definition of global heating | New Word Suggestion | Collins Dictionary. https://www.collinsdictionary.com/submission/21177/global heating.
[46] Tann, H., & Evatt, A. *Alliance: Humanities & Existential Risk*. TORCH. https://www.torch.ox.ac.uk/alliance-humanities.
[47] Hausfather, Zeke (2018, June 19). Explainer: How scientists estimate "climate sensitivity." https://www.carbonbrief.org/explainer-how-scientists-estimate-climate-sensitivity.

6. A directory of digital resources about the Anthropocene
Paulo David Soasti-Bareta

INTRODUCTION

This chapter seeks to offer a selection of digital resources that might help the non-specialist navigate the multi-dimensional information layers the Internet carries about the Anthropocene. It includes a list of relevant international agencies, research centres, environmental activism networks as well as educational resources. These are listed by resource name, followed by a quick interpretation of their actions facing the Anthropocene and a link to the respective website. Such collection of digital resources should facilitate the creation of relevant relationships between the actants of the arts, technoscience, designs, late capitalism, climate crisis and post-humanism, as drivers of this new approach to knowledge for the Anthropocene. The method used was weighted Google prominence followed by a Delphi consultation of a set of experts. Each item is followed by the source, indicated by a consecutive number in squared brackets. A list of all sources is included at the end. No ranking intention is underlying this exercise and we are aware that important omissions are likely to have occurred. We regard this as an initial list to build upon for more specific purposes.

RESOURCES

10 in 10

10 in 10 indicates that its "mission is to tackle ten global challenges in ten years". Their ten global challenges include: Climate Crisis, Water, Agriculture & Food, Gender Inequality, Youth Unemployment, Social Housing, Mental Health, Early Years (Children Under 5), Structural Racism, and Artificial Intelligence [1].

10000 changes

This project promotes commitment to rethinking plastics as part of everyday life. It raises the so-called "New 3R's model: Refuse, Replace and Reimagine". 10000 changes invites citizens to rethink their relations with plastic, through the principles of the circular economy and the sharing economy [2].

350

An international movement of citizens working to end the age of fossil fuels and build a world of community-led renewable energy for all. It is named after 350 parts per million—the safe concentration of carbon dioxide in the atmosphere [3].

Alliance of World Scientists (AWS)

AWS is a new international assembly of scientists. They state that "to avoid widespread misery caused by catastrophic damage to the biosphere, humanity must practice more environmentally sustainable alternative to business-as-usual." They react to the "scientists' unique responsibility as stewards of human knowledge and champions of evidence-based decision-making" [4].

Anthropocene and Digital Technologies

This resource offers an experimental way of navigating through the website by combining traditional keyword searches with the results of network analysis on subjects related to the Anthropocene phenomenon [5].

Anthropocene Arts—beyond Higgs

In constructing an "Anthropocene Thinking," this initiative aims at developing a conscious, mature attitude to develop human relationship with Earth. Such process involves abandoning the "largely naive behaviour of dealing with the Earth in a subordinate way to our own basic needs alone" [6].

Anthropocene Curriculum

This initiative explores "frameworks for critical knowledge and education in our transition into a new geological epoch". It draws together heterogeneous knowledge practices, from academics, artists, and activists to co-develop curricular experiments that collectively respond to this geo-global crisis [7].

Anthropocene Knowledge

Part of the Anthropocene umbrella project sponsored by MPIWG. "This research area addresses questions of relevance for a historical epistemology of Anthropocene-related practice fields, such as (pre-)industrial chemistry, Earth system and climate sciences, risk assessment, historiography, and the current transformation of scholarship" [8].

Broken Nature: Design Takes on Human Survival

Broken Nature exhibits material explorations visualizing the links between objects, the planet, and people. Restorative design is introduced as a means to make sense of global posthumanist problems in the Anthropocene [9].

Cape Farewell

Cape Farewell is a "project to instigate a cultural response to the climate challenge". They claim to bring together artists, scientists, and communicators to "stimulate a cultural narrative that will engage and inspire a sustainable and vibrant future society". This engagement is intended "to evolve and amplify a creative language to communicate the urgency of the global climate challenge" [10].

Center for Climate and Energy Solutions (C2ES)

Formerly Pew Research, C2ES has a mission: "to advance strong policy and action for reducing greenhouse gas emissions, promote clean energy, and strengthen resilience to climate impacts". To achieve it they claim to conduct their operation on three main areas: politics, cutting carbon emissions, and education on climate basics [11].

Centre for Humans & Nature

This organization allegedly questions the traditional economic concept of nature as a raw material for human use, and the human control over nature. They believe this disintegrative perspective ignores current scientific knowledge and distorts our sense of self, nature, community, economy, and democracy. They propose new approaches to coexistence between humans and nature [12].

Centre for International Environmental Law (CIEL)

This Centre focuses on the environmental aspects of law applying political and legal power to defend the environment, as well as the promotion of human rights and access to a sustainable society. They claim to achieve their goals through legal research, advocacy, educational programs, and legal training, centring their operations on engaging with local non-profit partners across the world [13].

Climate Advocacy Lab

Designed for climate practitioners—advocates, social scientists, data experts and funders—the Lab promotes effective tools and tactics for engaging US citizens on climate emergency. The Interactive Tools provide easy access to helpful climate crisis information, while the Resource Library, research and articles relate to public engagement efforts [14].

Climate and Clean Air Coalition

This organization helps their partners "to create policies and practices that will deliver substantial reductions in short-lived climate pollutant emissions". They support actions on the ground through their eleven initiatives: Heavy-Duty Vehicles, Oil and Gas, Waste, Bricks, Hydrofluorocarbons, Household Energy, Agriculture, Efficient Cooling, Supporting National Action and Planning, Finance, Assessments, and Health. Resources and training are provided through their Action Program to Address the 1.5°C Challenge [15].

Climate Change Library

Offers legal information regarding Anthropocene's issues, addresses topics of sustainable development, climate crisis, among others from an international legal perspective. It offers a list of regulations, books, journals, statistical data, and organizations. It is carried out by Prof. Whitney Curtis [16].

Climate Science Special Report (CSSR)

Published by the U.S. Global Change Research Program, deals with the climate variables that evidence global climate heating. It proposes detection methods that characterize temperature, precipitation, ocean level, drought, floods, fires, ocean acidification, and large-scale circulation variables, in order to develop climate models, scenarios, and future projections of climate crisis [17].

Climate Clock

A mixture of design and technology that seeks to raise awareness about carbon consumption we have nowadays. It shows two numbers, the first in red, it's a timer that indicates how much time we have until we consume our "carbon budget," this is our deadline to keep the heating threshold below 1.5°C; the second number, in green, indicates the world percentage of energy from renewable sources. This is our lifeline. "Simply put, our challenge is to get our lifeline to 100% before our deadline reaches 0" [18].

Climate Emergency EU

An open letter to global leaders with demands to act on the Climate Emergency [98].

Common Sense Education (CSE)

CSE offers educational resources to cope with the future consequences of the Anthropocene. The platform breaks things down into a set of causes, effects and potential solutions on climate change [19].

Connect4Climate

Seeks to promote leadership, collective actions and disruptive solutions that change the course of actions that have brought us to the Anthropocene. It's promoted by a partnership between the World Bank, the Italian Ministry of Environment, and the German Federal Ministry of Economic Cooperation and Development [20].

Copernicus Europe's Eyes on Earth

Copernicus is the European Union's Earth Observation Program. Copernicus offers geo information services on atmosphere, land, and marine monitoring, climate change assessment, emergency management, and security, all based on satellite images [21].

Covering Climate Now (CCNow)

A "global journalism initiative committed to more and better coverage of climate breakdown". They aim at producing quality journalistic material based on scientific facts related to climate crisis, using a ten-criteria guidance to assess accomplishment of such purpose [22].

Dezzeen

A popular architecture and design online magazine. They offer a thematic section focused on the Anthropocene, including a selection of articles based on projects performed and presented by practitioners from around the world [23].

Earth Law

Earth Law's mission is to transform the law to recognize "nature's rights to exist, evolve and thrive". A legal figure that they seek to promote is a characterization of Ecocide as a crime in international law [24].

Ecosia

Ecosia sponsors a global tree plantation program by using the profits they make from offering search engine capabilities through a free browser extension [25].

Ecowarriors.app

Ecowarriors propose a gamification experience related to teaching the public on waste sorting and recycling. Users learn history and geography while meeting endemic animals capable of helping them to clean their environment and fight pests. The goal is to induce kids to act upon the climate emergency [26].

Environmental Protection Agency (EPA)

EPA mission "is to protect human health and the environment". To accomplish this goal, the agency exerts political and economic power in coordination with other US federal agencies using the best scientific information available [27].

Extinction Rebellion

Extinction Rebellion is an international movement that uses non-violent civil disobedience in an attempt to induce government actions towards halting mass extinction and minimizing the risk of social collapse [28].

Fridays for Futures

A movement started by Greta Thunberg, a teenager who sat in front of the "Swedish Parliament to protest against lack of action on the climate crisis". This was a three-week civil action but set the precedent for a protest model around the world every Friday which went viral on social networks using the hashtag #FridaysForFuture [29].

Friends of the Earth

Seek to challenge the current economic models and globalization mindset, while promoting social justice and environmental solutions for the problems of humanity. Their work addresses topics such as: "climate justice and energy, economic justice and resisting neo-liberalism, food sovereignty, forest and biodiversity, human rights defenders, gender justice and dismantling patriarchy" [30].

Future Earth

Future Earth harnesses the experience and reach of thousands of scientists and innovators from across the globe collaborating to achieve a more sustainable world [31].

Global Diversity Foundation (GDF)

Claims to protect the environment and increase people's well-being by providing support for livelihoods of communities to maintain their cultural landscapes, their biodiversity while developing their capacities and institutions [32].

Global Footprint Network

Global Footprint Network "envisions a future where all can thrive within the means of our one planet". They claim their vision can be achieved through their mission: "to help end ecological overshoot by making ecological limits central to decision-making" [33].

Great Transition Initiative

This initiative provides "an online forum of ideas and an international network for the critical exploration of concepts, strategies, and visions for a transition to a future of enriched lives, human solidarity, and a resilient biosphere" [97].

Green Dreamer

Green Dreamer is a blog focused on "our exploration paths to holistic healing, ecological regeneration, and true abundance and wellness for all". This digital resource presents a list of 35 environmental organizations for a suitable future, and ways citizens can get involved. They present this list on six categories as follows: "Social & environmental justice; conservation & reforestation; wildlife protection; ocean conservation; climate justice; ecological agriculture" [34].

Half Earth Project

The Half-Earth Project is "a scientific initiative promoting the conservations of half the land and sea to safeguard the bulk of biodiversity, including ourselves" [35].

Humanities, Arts, Science, and Technology Alliance and Collaboratory (HASTAC)

This organization is an actor network focused on education for the Anthropocene, their interdisciplinary approach to collaboration creates a "community of humanists, artists, social scientists, scientists, and technologists changing the way we teach and learn" [36].

Haus der Kulturen der Welt (HKW)

The Anthropocene has become a fixed element of contemporary debates where HKW re-explores artistic positions, scientific concepts, and spheres of political activity, asking: How do we grasp the present and its accelerated technological upheavals? What will tomorrow's diversified societies look like? And what responsibilities will the arts and sciences assume in this process? [37].

Inhabiting the Anthropocene

A resource aiming to contribute to conversations about the Anthropocene. They focus on habitability as the process that changed our planet. They also imagine how habitability can contribute to improve the environment according to the climate change new reality [38].

Institute for Public Policy Research (IPPR)

IPPR is a UK think tank. They approach the phenomenon of the Anthropocene from the point of view of new economy, the social and political sciences and studies of science and technology to face problems such as "poverty, unemployment, or those in need for the reason of youth, age, ill-health, disability, financial hardship, or other disadvantage". They state their commitment towards the "advance environmental protection and sustainable development," as well as to "advance the arts, culture, heritage or science" [39].

International Environmental Organizations

Wikipedia offers an alphabetically ordered list of 186 international environmental organizations. It has 10 subcategories presenting actions and organizations performing their operations on intergovernmental environmental, climate change, forestry sustainability, nature conservation, and funds for nature [40].

IPCC (Intergovernmental Panel on Climate Change)

IPCC is the UN agency to evaluate the scientific aspects of climate emergency. They are widely regarded as the consensus view on the climate crisis. Their goal is "to provide governments, with scientific information in order to develop climate policies" [41].

List of Environmental Organizations

Wikipedia offers a list of environmental organizations related "to conservation or environmental movements that seek to protect, analyse or monitor the environment against misuse or degradation from human forces" during the Anthropocene [42].

Lund University Centre for Sustainability Studies (LUCSUS)

Lund University Centre for Sustainability Studies (LUCSUS) is a sustainability research, and teaching center working to, "understand, explain, and catalyse social change and transformations in relation to material limits in the biosphere" [43].

Material Practices: Earth in the Making

Another part of the Anthropocene umbrella project sponsored by MPIWG. "This part investigates the development of particular knowledge economies that concern resource extraction and resource flows, the transformations of energy systems, climate change, the changing interfaces between the environment, architecture and human bodies as well as the particular human-environment interaction within agricultural practices and traditions" [44].

Max Planck Institute for the History of Science (MPIWG)

The Max Planck Institute for the History of Science (MPIWG) "is dedicated to the study of history of science and aims to understand scientific thinking and practice as historical phenomena". They sponsor initiatives such as "The Anthropocene Project," "Material Practices: Earth in the Making," "Anthropocene Knowledge," "Anthropocene and Digital Technologies," "Technosphere," and "Anthropocene Curriculum" [45].

Mercator Research Institute on Global Commons and Climate Change (MCC)

This institution's objective "is to provide solution-oriented policy pathways for governing the global commons to enhance sustainable development and human well-being". They collaborate with climateclock.world [46].

NatGeo

The National Geographic Society claims to use the power of science, exploration, education, and storytelling to illuminate and protect the wonder of our world. NatGeo has published a broad selection of articles about the Anthropocene and its effects [47].

National Academies Press (NAP)

NAP publishes reports from fields such as: Engineering, Sciences, and Medicine. They offer reports on the characterization of measurable variables on climate emergency that might lead to a better understanding of key issues such as the effects of climate crisis on the Arctic, the urban impact of global heating, and prospective scenarios [48].

National Aeronautics and Space Administration (NASA)

Using technologies of remote tracking carried by a fleet of satellites and airborne and ground-based facilities, NASA is able "to monitor a broad spectrum of variables on land, air, and space on long-term data records, used by scientists worldwide to tackle some of the questions raised by the impact of human activity on the planet and our future" [49].

Natural Resources Defence Council (NRDC)

This organization states on its website: "NRDC works to safeguard the earth—its people, its plants and animals, and the natural systems on which all life depends". Their programs consider a dozen areas, partnering with businesses, elected leaders, and communities on issues society faces nowadays, such as: Climate & Clean Energy, Healthy People & Thriving Communities, International, Litigation, Nature, and Science [50].

Nature

A prestigious weekly international journal of science. It offers a wide Anthropocene coverage from humanities, sciences, engineering and technology, design and arts [51].

OLCreate: Climate Change Resources

Open Learn Create offers free, open courses and resources seeking to educate the public on necessary changes to survive the Anthropocene. Topics such as: Energy, ecopsychology, business and economics related to climate emergency and sustainability are included [52].

Open Educational Resources OER Commons

OER Commons offers 131 resources including courses, lectures, readings, interactive multimedia, working guides for teachers and students, on topics related to the Anthropocene. They provide everyday activity suggestions in order to induce environmental actions, from children to adults [53].

Organisation for Economic Co-operation and Development (OECD)

The OECD has focused on the following topics related to the Anthropocene: International Climate Crisis Framework, Finance and Investment, Policy Coherence and Economic Analysis, Climate Crisis Mitigation, Adaptation to Climate Breakdown, Environment and Environmental Policy, Green Growth, Taxation and other Market-Based Instruments, Agriculture & Fisheries, Energy, Transport, and Tourism. The OECD also offer an Application Programming Interface (API) that provides access to their data bases [54].

Organizations Relating to Climate

The Union of International Associations (UIA) offers a list of 50 organizations relating to climate crisis, climate justice, climate adaptation, political organizations, energy and natural resources, youth action, all focused on the Anthropocene [55].

Post Carbon Institute

An Institute with a mission "to lead the transition to a more resilient, equitable, and sustainable world by providing individuals and communities with the resources needed to understand and respond to the interrelated ecological, economic, energy, and equity crises of the 21st century" [99].

Proceedings of the National Academy of Sciences (PNAS)

An academic organization based in the USA. PNAS publishes science news, papers, podcasts, and profiles. On the Anthropocene arena, they cover three main areas: Physical Sciences, Social Sciences and Biological Sciences [56].

Rapid Transition Alliance

This organization argues about the climate crisis speed and our inability to keep up with it. They "offer examples of empirical evidence hope for change whose speed and scale will steer Humanity to remain within global boundaries". According to them, "these are clear, quantifiable changes in our values, behaviours, attitudes, and use of resources, energy, technology, finance and infrastructure that can happen and guide what we do over the next five to ten years" [57].

Resilience Alliance (RA)

RA is based on an international multidisciplinary approach "to advance the understanding and practical application of resilience, adaptive capacity, and transformation of societies and ecosystems to cope with change and support human well-being" [58].

Restoration for Sustainable Coastal Ecosystems (RESCuE)

RESCuE is a research initiative focused on the global efforts on conservation, and rehabilitation of coastal zones, as well as sustainable management of natural resources on coastal areas around the world [59].

ScienceDirect

This academic publisher produces two journals dedicated to the Anthropocene: "The Lancet Planetary Health" and "Anthropocene" [60].

Scientists Warning

Scientists Warning Foundation is a non-profit organization made up of an activist network of researchers, scientists, and global citizen scientists. They gather evidence to report and educate on climate crisis issues [61].

Society of Environmental Journalists (SEJ)

Offers resources and vital support to journalists of all media, who face the coverage of complex environmental issues. They have published a list of organizations dedicated to environmental issues, as well as media coverage of environmental events, related to the Anthropocene [62].

Stockholm Resilience Centre (SRC)

The SRC is a "research institution on resilience and sustainability science". They claim to "believe in the importance of reconnecting to the biosphere" since people and nature are truly intertwined in what they refer to as social-ecological systems. They stand for a development model with an increased understanding of nature's role for mankind's survival and well-being [63].

Sunrise Movement

A political movement which seeks to stop climate change by adhering to the Green New Deal and "create millions of good jobs in the process". They argue three theoretical axes of political action to achieve this goal: (a) Establish dialogues between citizens and communities to create moral protests; (b) Capitalize on the support of supporters who will fight for their shared vision of a more just future; (c) Build networks between movements and groups that share the vision of a government that fights for dignity and justice for all [64].

Systemiq Earth

They claim to be a "company with the heart of an NGO," their vision states: "a thriving planet where sustainable economic systems drive prosperity for all." By December 2020 they release a report about the countries' contributions to achieve the Paris agreements, which "shows how corporates, finance, government and the public are creating the potential for a low-carbon economy to scale rapidly in the 2020s" [65].

Technosphere

A project looking at the "socio-epistemic and technological frameworks that have co-evolved in the interplay of material practices and specific knowledge economies throughout centuries of intensified human-environment interaction". They offer a critical view of our relationship with the natural environments; our human impacts and the role technologies play in this dynamic [66].

TerraBrasilis

A platform developed by the Brazilian Space Research National Institute (INPE) to provide access, query, analysis, and dissemination of spatial data generated by Brazilian government environment monitoring programs. The portal is open to the international public [67].

The Anthropocene Project

Offers digital experiences on Anthropocene's impact on the environment, combining information technologies, such as virtual and augmented reality, along with visual arts—film and photography and scientific research. Its objective is to show the influence of human activities on the state, dynamics, and future of the planet. The project is endorsed by the RCGS [68].

The Breakthrough Institute

Focuses on three areas of impact of the Anthropocene: Food, Energy and Environmental Conservation. They claim to be a "research centre that identifies and promotes technological solutions to environmental and human development challenges" [69].

The Climate Mobilization (TCM)

TCM pursue a rapid restoration to safe climate levels and seek to create a just and democratic society. They visualize a democratic society embedded in a safe climate environment where "the living world flourish for generations to come" [70].

The Dark Mountain Project

An editorial project that walks away from the stories that prevent western societies "seeing clearly the extent of the ecological, social and cultural unravelling that is now underway". They claim to create art which does not take human centrality as their main driver, but as one more actor in the complex play of environmental systems [71].

The Earth Institute

The Earth Institute seeks for a better understanding of earth environmental systems from environmental and social sciences. They claim, "to use this knowledge in order to develop policy and practical solutions to Anthropocene challenges." They have prioritized their operations on the following areas: water, climate, energy, urbanization, hazard and risk reduction, global health, agriculture, ecosystems, and peace and justice [72].

The Ecologist

An environmental affairs platform which offers free content such as news, comments, stories, features, "articles in academic ecology and discursive articles touching on theoretical approaches that help the public to understand nature and the impact of societies on the environment" [73].

The Green New Deal

This movement and political platform look forward into a new, sustainable economy that is environmentally sustainable, economically viable and socially responsible. It seeks to "solve the climate crisis by combining quick action to get to net-zero greenhouse gas emissions and 100% renewable energy by 2030" [74].

The Guardian

A journalistic organization with a stance that highlights aspects of the Anthropocene such as climate crisis, overpopulation, ocean acidification, carbon footprint and greenhouse effect, storms, floods, severe droughts and fires, polar melt, among others; It also presents articles showing actions that have been carried out to alleviate the effects of the Anthropocene [75].

The Hartwell Paper

The Hartwell paper stresses the need for human dignity as a guiding principle for climate policy. It sets out three objectives around the concept of human dignity: (i) Ensuring energy access to everyone; (ii) To ensure a way of developing without affecting the essential Planet's systems; (iii) To ensure that our societies have the means and capacities to withstand and overcome any climate risks that may arise [76].

The Nation

The Nation empowers readers to fight for justice and equality for all. One of its editorial lines addresses the impacts of human activity on the planet and the influence on issues such as climate emergency, global heating, environmental policies, and the arts and their interaction in the Anthropocene [77].

The New Weather Institute (NWI)

The New Weather Institute seeks to find solutions to climate problems from the perspective of a fair economy and within planetary boundaries. To do this they use the figure of the rapid transition, which in practice, is about identifying quantifiable successful cases of change in "values, behaviour, attitudes and use of resources, energy, technology, finance and infrastructure," which can occur and guide our way of life in a ten-year time frame [78].

The Oxygen Project

The purpose of this project is to raise awareness and defend the oxygen-producing ecosystems. After a conversation with Jacques Cousteau and decades of acquired knowledge, Rutherford Seydel, founder of the organization, "realized phytoplankton was being overlooked by the scientific community" as the most efficient carbon sink and decided to start his own activist organization [79].

The Royal Canadian Geographic Society (RCGS)

RCGS focuses on the cultural link of citizens with Canadian geography, through the dissemination and support of various cultural and educational initiatives. They offer online educational material, finance projects related to the environment and culture, such as: 10000changes.ca and theanthropocene.org [80].

The UK Labour Party

Offers information about the LP's actions towards the Anthropocene, on topics such as: Economy and Energy, Innovation, Education towards STEM, Transport, Environment, Animal welfare, Green industrial revolution [81].

The UN Climate Change Learning Partnership (UN CC: Learn)

This UN resource includes four relevant sections: (i) Courses, (ii) Library, (iii) Good Learning Practices Map and (iv) Climate Watch. Course topics include: Adaptation, Cities, Climate Crisis, Energy, Finance, Gender, Green Economy, Health, Science and REDD+, the UN Programme on Reducing Emissions from Deforestation and Forest Degradation. The second section offers a comprehensive library on analytical-technical documents, Policy and Activity reports, audio-visual resources, and guidance documents. The third section offers study cases outlining good learning practices from projects done by the UN. The last section offers "a go-to resource for media, policymakers, and researchers to visualize and download data on global and national emissions trends, track climate commitments, and more" [82].

The Washington Post

This USA newspaper has continuous coverage on the Anthropocene on two sections: "Climate & Environment" and "Climate Solutions" focused on news, technology, politics, economy, food production, carbon print, waste reduction, human impact on the environment [83].

The World Bank (WBG)

WBG claims "to work to reduce poverty and build prosperity in developing countries". On the Anthropocene topic, it offers a series of reports and data related to global heating, climate emergency solutions, sustainable futures, and economic development [84].

The World Economic Forum (WEF)

"The WEF is an international organization that engages prominent political, business, cultural and other figures to discuss global, regional and industry agendas". They offer a network of platforms focused on a prospective view of technologies, governance, energy, new economy, consumption, and society [85].

The World Future Council (WFC)

"WFC works to pass on a sustainable planet with just and peaceful societies to our children and grandchildren". To achieve this, they declare "to focus on identifying, developing, highlighting, and spreading effective, future-just policies for current challenges humanity is facing" [86].

The World Weather Attribution (WWA)

WWA analyses and discloses the potential impacts of events as a consequence of climate breakdown, "such as storms, extreme rainfall, heatwaves, cold spells, and droughts" [87].

Think Tank Networks

This initiative studies the political influence of think tanks and how they increase their power through networks. "Think Tank publish policy reports, participate in expert panels, organize lobby events, and are often welcomed guests in the media. By establishing networks, Think Tank often function as strategic connection hubs in the political mobilization process" [88].

Tyndall Center for Climate Change Research

The Tyndall Center's mission is "to provide evidence to inform society's transition to a sustainable, low-carbon and climate-resilient future". This center is a conglomerate of English universities that covers social and natural sciences as well as engineering, seeking to develop sustainable value propositions to face climate change [89].

UK Student Climate Network

UKSCN is a group of young people exercising their right to protest, operating in English and Wales. Their activism includes "campaigning, working with other organisations and providing support to over 100 local groups". Their aim is to educate new activists about climate emergency and the adjustments that have become mandatory in the Anthropocene [90].

Union of Concerned Scientists (UCS)

"UCS is a non-profit organization founded by scientists and students at the Massachusetts Institute of Technology". Their work is based on values such as: Science, Democracy, Justice, Integrity, Action. Their activity areas cover the fields of climate, energy, transport, food, nuclear weapons, science, and democracy [91].

United Nations Framework Convention Climate Change (UNFCCC)

UNFCCC is the United Nations entity tasked with supporting the global response to the threat of climate emergency. "The ultimate objective of all agreements under the UNFCCC is to stabilize greenhouse gas concentrations in the atmosphere at a level that will prevent dangerous human interference with the climate system, in a time frame which allows ecosystems to adapt naturally and enables sustainable development" [92].

United Planet

This organization is focused on the creation of a global community committed to fighting climate breakdown, through a multicultural approach [93].

VOX

An online magazine with coverage of topics related to the Anthropocene such as: climate emergency, renewable energy, conservation, ancient endogenous cultural knowledge, environmental activism and environmental policies [94].

World Capital Institute (WCI)

"An independent international Think Tank whose purpose is to further the understanding and application of knowledge as the most powerful leverage of development." The WCI has recently shifted its focus from Knowledge Based Development and Knowledge Cities to Knowledge for the Anthropocene and City Preparedness for the Climate Crisis [100].

World Meteorological Organization (WMO)

WMO is dedicated to international cooperation and coordination "on the state and behaviour of the Earth's atmosphere, its interaction with the land and oceans, the weather and climate it produces, and the resulting distribution of water resources" [95].

World Wildlife Fund (WWF)

WWF has a presence in over 100 countries, "At every level, we collaborate with people around the world to develop and deliver innovative solutions that protect communities, wildlife, and the places in which they live. WWF works to help local communities conserve the natural resources they depend upon; transform markets and policies toward sustainability; and protect and restore species and their habitats. Our efforts ensure that the value of nature is reflected in decision-making from a local to a global scale" [96].

SOURCES

[1] https://www.xinx.co/
[2] https://10000changes.ca
[3] https://350.org/
[4] https://scientistswarning.forestry.oregonstate.edu/
[5] https://www.mpiwg-berlin.mpg.de/project/anthropocene-digital
[6] https://eusg.org
[7] https://www.anthropocene-curriculum.org/
[8] https://www.mpiwg-berlin.mpg.de/project/knowledge-anthropocene
[9] http://www.brokennature.org
[10] https://capefarewell.com/
[11] https://www.c2es.org/
[12] https://www.humansandnature.org/
[13] https://www.ciel.org/
[14] https://climateadvocacylab.org/
[15] https://ccacoalition.org
[16] https://eguides.barry.edu/c.php?g=519696&p=3553904
[17] https://science2017.globalchange.gov/chapter/executive-summary/
[18] https://climateclock.world/
[19] https://www.commonsense.org/education/top-picks/climate-change-resources-for-students-and-teachers
[20] https://www.connect4climate.org/
[21] https://www.copernicus.eu/en
[22] https://www.coveringclimatenow.org/
[23] https://www.dezeen.com/tag/anthropocene/
[24] https://www.earthlaw.org/
[25] https://www.ecosia.org/
[26] https://www.ecowarriors.app/
[27] https://www.epa.gov/
[28] https://extinctionrebellion.uk/
[29] https://fridaysforfuture.org
[30] https://www.foei.org/
[31] https://futureearth.org/
[32] https://www.global-diversity.org/
[33] https://www.footprintnetwork.org/
[34] https://greendreamer.com/
[35] https://www.half-earthproject.org/
[36] https://www.hastac.org/
[37] https://www.hkw.de/de/index.php
[38] https://inhabitingtheanthropocene.com/
[39] https://www.ippr.org/

[40] https://en.wikipedia.org/wiki/Category:International_environmental_organizations
[41] https://www.ipcc.ch/
[42] https://en.wikipedia.org/wiki/List_of_environmental_organizations
[43] https://www.lucsus.lu.se/about-lucsus
[44] https://www.mpiwg-berlin.mpg.de/project/earth-making
[45] https://www.mpiwg-berlin.mpg.de/
[46] https://www.mcc-berlin.net/
[47] https://www.nationalgeographic.org/encyclopedia/anthropocene/
[48] https://www.nap.edu/
[49] https://www.nasa.gov/topics/earth/index.html
[50] https://www.nrdc.org/
[51] https://www.nature.com/search?q=Anthropocene
[52] https://www.open.edu/openlearncreate/course/index.php?categoryid=350
[53] https://www.oercommons.org/browse?f.keyword=global-warming
[54] http://www.oecd.org/
[55] https://uia.org/s/sub/en/1900000197/
[56] https://www.pnas.org/
[57] https://www.rapidtransition.org
[58] https://www.resalliance.org/
[59] https://rescue-pro.net/
[60] https://www.sciencedirect.com/
[61] https://www.scientistswarning.org/
[62] https://www.sej.org/initiatives/international-agencies/international-agencies
[63] https://www.stockholmresilience.org/
[64] https://www.sunrisemovement.org/
[65] https://www.systemiq.earth/
[66] https://www.hkw.de/en/programm/projekte/2015/technosphere/technosphere_start.php
[67] http://terrabrasilis.dpi.inpe.br/en/home-page/
[68] https://theanthropocene.org
[69] https://thebreakthrough.org/
[70] https://www.theclimatemobilization.org/
[71] https://dark-mountain.net
[72] https://www.earth.columbia.edu/
[73] https://theecologist.org/
[74] https://www.gp.org/green_new_deal
[75] https://www.theguardian.com
[76] http://eprints.lse.ac.uk/27939/1/HartwellPaper_English_version.pdf
[77] https://www.thenation.com
[78] http://www.newweather.org
[79] https://www.theoxygenproject.com
[80] http://www.rcgs.org/
[81] https://labour.org.uk/manifesto-2019/
[82] https://www.uncclearn.org/resources/
[83] https://www.washingtonpost.com/
[84] https://www.worldbank.org
[85] https://www.weforum.org/
[86] https://www.worldfuturecouncil.org
[87] https://www.worldweatherattribution.org/

[88] http://thinktanknetworkresearch.net/
[89] https://www.tyndall.ac.uk/
[90] https://ukscn.org/
[91] https://www.ucsusa.org/
[92] https://unfccc.int/
[93] https://www.unitedplanet.org
[94] https://www.vox.com/energy-and-environment
[95] https://public.wmo.int/
[96] https://www.worldwildlife.org/about/
[97] https://greattransition.org
[98] https://climateemergencyeu.org
[99] https://www.postcarbon.org
[100] https://www.worldcapitalinstitute.org

7. Educating for the Anthropocene

Audrey Groleau, Chantal Pouliot, Isabelle Arseneau

The state of planet Earth, its environments and human communities is worrying. Human activities are so destructive that a vast majority of scientists fear we have reached a point of no return. As inequalities grow, oceans are acidifying, extreme climatic phenomena are more and more frequent, ecosystems are being tested and species are disappearing.

Most of us agree that education is a significant part of the solution to the Earth's natural limits exceeding. In *Anthropocene: The Human Epoch* (Baichwal, de Pencier, & Burtynsky, 2018), rapper Hassan Shakur evokes his life in the biggest dump of Nairobi:

> I was born in Pumwani
> I grew up in Dandora City
> Ah, nobody by my side
> Life is to hustle
> Education is the key to success.

Since we watched it, his words inhabit us: *education is the key to success.*

AN EDUCATION THAT AIMS TO BE CRITICAL AND TRANSFORMATIVE, AND THAT PROMOTES SOCIOPOLITICAL ACTIONS

Students are already taught about the environmental crisis (climate change, loss of biodiversity, water pollution, etc.) in schools, sometimes about its economic and social aspects, sometimes about its scientific aspects, but most of the time about small gestures they should perform in their everyday life. Current educational approaches are varied, but many propose to teach more or less contextualized notions in an encyclopedic perspective rather than aspiring to train critical citizens capable of taking sociopolitical actions. However, in recent years, many education researchers have highlighted the importance to teach public speaking and, even, dissension. They have also showed that exploring socioscientific issues (as the environmental crisis) allows students to develop their value system, to examine the arguments formulated, to learn contextualized knowledge, to make critical use of expertise, and so on (Albe, 2009). The state of the world makes it necessary to offer an education that trains citizens not only to be able to understand the issues related to the production and usage of technoscience, but also to be critical, to get engaged in debates and political decision-making processes and to act for the well-being of other species (El Halwany et al., 2020; Sjöström & Eilks, 2018). In other words, to educate for the Anthropocene with sensitivity and relevance, in-class activities must "aim for the development of an emancipated relationship with political, industrial and scientific power, for the courage to express one's disagreement with decisions that cause or allow environmental or health injustice, and for a real capacity of action and responsible entrepreneurship" (Pouliot & Barroca-Paccard, 2020, p. 148; free translation).

Furthermore, educating for the Anthropocene involves accompanying students throughout their learning of emotionally charged and destabilizing knowledge, that is, *difficult knowledge* (Garrett, 2017). If this leads them to better comprehend what is at stake in the situations they examine, it can also affect their trust relationship with political leaders, health authorities, and industrial projects promoters. Briefly, taking charge of that difficult knowledge in the classroom is political, in the sense of the distribution and exercise of power, and is necessary, not optional.

If educating for the Anthropocene is undoubtedly an important mission entrusted to teachers, it is also challenging and sometimes perceived risky to address socially acute questions related to environmental and health issues in the classroom. Indeed, teaching those questions comes with a range of possible drifts, from a relativistic drift (considering that all points of view are equal) to an authoritative drift (considering that reflections surrounding current issues should be delegated to experts), or the risk of being blind to the differences between citizen knowledge, expert knowledge and social knowledge (Groleau, 2019; Legardez, 2006).

However, there are some strong principles on which we can rely to facilitate this teaching and make it relatively serene for both students and teachers. Here are seven of them.

1. **Familiarize students with the effects of human activities on Earth.** It gives them the opportunity to identify structural causes of the environmental crisis and the effects of human activities on individuals, environments and societies; on living and non-living actors (Bencze, 2017; Pierce, 2013).
2. **Accompany students in interdisciplinary problematization of situations that truly interest them.** Those situations, because they are usually complex and contextualized in local realities, systematically concern

diverse disciplines, for example sociology, engineering, law, sciences, mathematics, philosophy, health, etc.[1] This makes possible the use of a variety of knowledge in the investigation surrounding authentic situations (Bencze & Carter, 2011; Maingain & Dufour, 2002).
3. **Bring students to carry out actions and concrete socioenvironmental projects anchored in their territorial realities.** Young people can elaborate these actions or participate in existing actions in their community. Thus, they should be exposed to (or take part in) relevant solutions that involve current expert knowledge, and that are carried by actors committed in this knowledge production. This is about moving away from small individual gestures, which finally often lead to despair or cynicism, and rather think of diverse-scale collective actions that can have a significant impact on the situations and make sense to students (Bader et al., 2014).
4. **Highlight asymmetries and injustices.** Places where power and resource asymmetries are played out are often those where citizens and communities can act and significantly transform the situations. By making the situations more symmetrical, we give actors opportunities to develop their capacities and to take part in relevant and effective actions (Bencze & Carter, 2011).
5. **Encourage students in the development of an emancipated relationship to knowledge**, namely:
    - A relationship to oneself (identity dimension) as able to learn and understand complex (economic, ethical, scientific, political) issues;
    - A relationship to others (social dimension) that is both open to diverse discourses and critical vis-à-vis dishonest arguments;
    - A relationship to the world (epistemic dimension) in which the action horizon aims towards the common good and the protection of species (including humankind) and of environments (Charlot, 1997; Pouliot, Bader, & Therriault, 2010).
6. **Recognize and push further students' critical position-making capacities.** Quite often, educational discourse depicts citizens as suffering from knowledge, interest and comprehension deficits towards technoscientific development and environmental impacts, although research in sociology of science and science education clearly shows that citizens are able to appropriate knowledge, to build an opinion and to generate useful knowledge (Bencze & Pouliot, 2016).
7. **Present concrete cases of successful citizen mobilizations.** Class explorations of citizen initiatives help to understand how citizens succeeded in constituting strong networks of social actors who maintained the object of their concerns on the political agenda and led industrial entities to modify their practices. Cases are numerous and are often well documented

through general media, children's literature, specialized books, documentaries or fiction movies that are inspired by them (Pouliot, 2019).

Even if curricula differ from jurisdiction to jurisdiction, the principles presented in the previous section appear in many countries' orientations regarding citizenship, science or environmental education. In the following section, we discuss different teaching strategies to materialize those principles.

EDUCATING FOR THE ANTHROPOCENE: SOME CONCRETE STRATEGIES FOR TEACHERS

Various teaching strategies allow learning of disciplinary knowledge, reflecting on the nature of disciplines and producing relevant citizen and disciplinary knowledge. There are several ways to accompany students through the development of their action potential, as illustrated by those few examples.

Graphic Novels

Many graphic novels are about health or environmental issues. One of us (Chantal) documented the journey of citizens committed in the socioscientific issue surrounding nickel dust in the air of residential areas near Québec City's port (in Canada). With colleagues, she produced a graphic novel (Zouda et al., 2019) entitled *Ban the Dust: A Graphic Novel About Citizen Actions to Eliminate Urban Dust Pollution*, telling this story. The novel is specifically addressed to teachers and their students.[2] It is relevant for teaching and learning because it highlights many ways citizens can engage in current issues that interest them (Bencze & Pouliot, 2016). For example, in this graphic novel, the protagonists hired scientists to analyze the dust samples they collected around their house and in other neighborhoods; they launched a website to collect information and make public the actions carried out (and the upcoming ones); they made recommendations concerning the management of metal transhipment; they took part in two collective actions, and so on. Such a graphic novel shows roles, capacities, and responsibilities that can be associated with different social groups. The students could then produce a representation (cartography, mind map, etc.) of those roles, responsibilities and capacities. They could also be invited to lead an enquiry on a current local issue that is interesting to them and present the fruit of their labour in an original way, as in a graphic novel.

Opinion-building and Action-research Processes Leading to Social Actions

Opinion-building processes allow students to enquire about an issue, discuss it, build their opinion and perform a social action in line with the point of view they have developed. Social actions can take different forms. For example, Justine Dion-Routhier's students, who attend a Québec City elementary school, carried out a process that led them to write a letter to an elected official. They also got their letter published in local newspapers (Dion-Routhier, 2016a; 2016b; Turcotte, 2016). With this activity, Dion-Routhier made her students work on the 4-billion-litre sewage discharge in the St. Lawrence River (in 2015, this issue was locally known as *the flushgate*) near Montréal[3] due to the building of an expressway. Students decided to dig into the impacts of the discharge on the fish living in the river: their breathing, their food supply and their reproduction. At the end of the process, they wrote recommendations in the event that another sewage discharge in the St. Lawrence River was envisaged.[4] Finally, they sent a letter summarizing the process they followed and their recommendations to the then mayor of Montréal, Denis Coderre. The Mayor replied to the students, congratulating them for their interesting investigation and encouraging them to pursue a career in biology or in water management.

In the same vein, the STEPWISE [Science and technology education for the well-being of individuals, societies and environments] theoretical and pedagogical framework (Bencze & Carter, 2011) leads students to choose an issue that interests them and to carry out an action-research about it. They are invited to invest a part of their social capital for the common good through social action. For example, Toronto (in Canada) high-school students decided to investigate the issues surrounding gene patenting in the context of breast cancer tests and research (Sklarzyk et al., 2014) and to prepare two video clips "to spread awareness among [their] peers and the wider community" (p. 1): one explaining Angelina Jolie's choice to undergo a double mastectomy; the other one presenting different aspects of gene patenting. This action-research process can also be led outside of school. For example, Calabrese Barton and Tan (2014) documented how three girls participated in the transformation of the roof of the youth center they attended into a green roof.

Documentaries

Documentaries (e.g. *Anthropocene: The Human Epoch* (Baichwal et al., 2018); *An Inconvenient Truth* (Guggenheim, 2006); *An Inconvenient Sequel: Truth to Power* (Cohen & Shenk, 2017)). Those documentaries (and others) carry images and reflections that are both beautiful and difficult. They are particularly interesting to familiarize students with the effects of human activities. They also illustrate particularly well the *difficult knowledge* notion, namely overwhelming but necessary knowledge. They can be used as the start of a course sequence in which students choose a way to express themselves about the content of one of the documentaries (by producing a sculpture, a comic strip, a song, etc.). Supported by their teachers, students could also enhance their comprehension of one of the situations presented in the documentaries, conceive social action scenarios and perform one of them.

Court Hearings

Many cases of sanitary or environmental crisis are judicialized. Because those cases are often linked to class actions, court hearings are generally public. Going to the courtroom with students allows them to get familiarized with the places and with the ins and outs of class actions, and to hear citizen testimonials. The rendered judgment becomes even more interesting, as it can subsequently be analyzed by students, for example by making an inventory of the stakeholders (and their roles, capacities, and interactions), the problems, the issues and the solutions linked to the class action examined.

Meetings with Engaged Citizens and Environmental Organization Representatives

Those social actors can be invited to meet students in the classroom. Such a visit, preceded by preparatory activities, allows the development of saner relationship to environmental action, as those meetings are inspiring, bring hope and highlight effective aspects of the social mobilization. Those models can bring young people to realize that citizen leaders are accessible and their actions diverse.

CONCLUSION

Education for the Anthropocene creates a space (writ large, as it can occur outside of school, e.g. in community or activist groups) for allowing students to acquire critical perspectives for individual and collective action. It is political, as it is closely linked with the examination of the distribution and exercise of power by different social groups, including citizens and students or pupils themselves. Furthermore, it aims at a citizenship education that trains young people to take part in debates and decision-making processes (Barma & Guilbert, 2006) and to understand complex issues, discuss them and take social actions (Bencze & Carter, 2011).

The principles and teaching strategies proposed in this chapter have in common to counteract the despair and cynicism that come with small individual gestures (Bader et al., 2014; Connell et al., 1999). Furthermore, they avoid depoliticizing the encountered situations and offer tools to overcome the disarray that often accompanies the acquisition of difficult knowledge by highlighting the fact that many mobilizations result in effective and positive changes.

We strongly believe that going against epistemological anesthesia (Darré, 1999), that is the idea that citizens should hush up their own knowledge, feelings and opinions to delegate the reflections, discussions and decision-making processes to experts and governments, is the first step to the development of young people's capacities, and to their positioning as legitimate interlocutors in the context of the environmental and social crisis.

Today's young people are tomorrow's adults. Education must offer them a real possibility to fight the environmental crisis, in the short, medium and long term. In a nutshell, Hassan Shakur is right to sing out *education is the key to success*. We need to act now (Pouliot, Arseneau, & Groleau, 2020).

NOTES

1. For a discussion of the links between complexity and interdisciplinarity, see Thompson Klein (2004).
2. A pedagogical guide has also been conceived to accompany the graphic novel (Zouda et al., n.d.).
3. Montréal is located on an island of the St. Lawrence River. Québec City is on the north shore of the St. Lawrence River, 250 km downstream from Montréal.
4. The discharge was planned; it was not an emergency due to a breakage in the sewage system.

REFERENCES

Albe, V. (2009). *Enseigner des controverses*. Rennes: Presses universitaires de Rennes.

Bader, B., Morin, É., Therriault, G., & Arseneau, I. (2014). Rapports aux savoirs scientifiques et formes d'engagement écocitoyen d'élèves de quatrième secondaire face aux changements climatiques. *Revue francophone du développement durable*, *4*(novembre), 171–190.

Baichwal, J., de Pencier, N., & Burtynsky, E. (2018). *Anthropocene: The Human Epoch*. Canada. Retrieved from https://theanthropocene.org/

Barma, S., & Guilbert, L. (2006). Différentes visions de la culture scientifique et technologique: défis et contraintes pour les enseignants. In A. Hasni, Y. Lenoir, & J. Lebeaume (Eds.), *La formation à l'enseignement des sciences et des technologies au secondaire dans le contexte des réformes par compétences* (pp. 11–39). Québec: Presses de l'Université du Québec.

Bencze, L. (Ed.) (2017). *Science and Technology Education Promoting Wellbeing for Individuals, Societies and Environments*. Cham: Springer International Publishing.

Bencze, L., & Carter, L. (2011). Globalizing students acting for the common good. *Journal of Research in Science Teaching*, *48*(6), 648–669. https://doi.org/10.1002/tea.20419

Bencze, L., & Pouliot, C. (2016). Battle of the bands: toxic dust, active citizenship and science education. *Journal for Activist Science and Technology Education*, *7*(1), 1–21.

Calabrese Barton, A., & Tan, E. (2014). "It changed our lives": activism, science, and greening the club/community. In L. Bencze & S. Alsop (Eds.), *Activist Science and Technology Education* (pp. 491–508). Dordrecht: Springer.

Charlot, B. (1997). *Du rapport au savoir: éléments pour une théorie*. Paris: Anthropos.

Cohen, B., & Shenk, J. (2017). *An Inconvenient Sequel: Truth to Power*. United States. Retrieved from https://aninconvenientsequel.com/

Connell, S., Fien, J., Lee, J., Sykes, H., & Yencken, D. (1999). "If it doesn't directly affect you, you don't think about it": a qualitative study of young people's environmental attitudes in two Australian cities. *Environmental Education Research*, *5*(1), 95–113. https://doi.org/10.1080/1350462990050106

Darré, J.-P. (1999). *La production de connaissance pour l'action: arguments contre le racisme de l'intelligence*. Paris: Éditions de la maison des sciences de l'homme.

Dion-Routhier, J. (2016a). Des écoliers écrivent à Denis Coderre. Retrieved from http://www.lapresse.ca/le-soleil/opinions/points-de-vue/201601/28/01-4944821-des-ecoliers-ecrivent-a-denis-coderre.php

Dion-Routhier, J. (2016b). Déversement: des élèves de 4e-5e année s'adressent à Denis Coderre. Retrieved from http://quebec.huffingtonpost.ca/justine-dion-routhier/deversement-eleves-4e-5e-annee-ecole-les-primeveres_b_9132396.html

El Halwany, S., Zouda, M., Pouliot, C., & Bencze, J. L. (2020). Teacher Candidates' Relationships to Knowledge and to Their Practices for Critical and Activist stse Education. In K. W. Clausen & G. L. Black (Eds.), *The Future of Action Research in Education: A Canadian Perspective* (pp. 171–193). Montréal & Kingston: McGill-Queen's University Press.

Garrett, H. J. (2017). *Learning to be in the World with Others: Difficult Knowledge and Social Studies Education*. New York: Peter Lang.

Groleau, A. (2019). Éviter autant la dérive relativiste que la dérive autoritariste en classe de sciences et de technologie à l'ère des fausses nouvelles. *Spectre*, *48*(3), 18–20.

Guggenheim, D. (2006). *An Inconvenient Truth*. United States.

Legardez, A. (2006). Enseigner des questions socialement vives. Quelques points de repère. In A. Legardez & L. Simonneaux (Eds.), *L'école à l'épreuve de l'actualité. Enseigner les questions vives* (pp. 19–31). Paris: ESF éditeur.

Maingain, A., & Dufour, B. (2002). *Approches didactiques de l'interdisciplinarité*. (G. Fourez, Ed.). Bruxelles: De Boeck & Larcier s.a.

Pierce, C. (2013). *Education in the Age of Biocapitalism: Optimizing Educational Life for a Flat World*. New York: Palgrave Macmillan.

Pouliot, C. (2019). Éducation aux démarches d'enquête citoyennes. In J. Simonneaux (Ed.), *La démarche d'enquête: Une contribution à la didactique des questions socialement vives* (pp. 115–128). Dijon: Educagri.

Pouliot, C., Arseneau, I., & Groleau, A. (2020). Climate crisis, science, and education. *BioScience*, *70*(6), 445–446.

Pouliot, C., Bader, B., & Therriault, G. (2010). The notion of the relationship to knowledge : A theoretical tool for research in science education. *International Journal of Environmental & Science Education*, *5*(3), 239–264.

Pouliot, C., & Barroca-Paccard, M. (2020). Pistes d'activités à réaliser en classe. In P. Provost & T. Lefèvre (Eds.), *Collection Des Universitaires: Tome 1, 2019–2020. Des Universitaires descendent de leur tour d'ivoire* (p. 148). Québec: Regroupement Des Universitaires.

Sjöström, J., & Eilks, I. (2018). Reconsidering different visions of scientific literacy and science education based on the concept of Bildung. In Y. J. Dori, Z. Mevarec, & D. Baker (Eds.), *Cognition, Metacognition, and Culture in STEM Education: Learning, Teaching and Assessment* (pp. 65–88). Cham: Springer International Publishing.

Sklarzyk, J., Jameson, E., Abdullahi, N., & Shah, M. (2014). The gene patenting controversy. *Journal for Activist Science and Technology Education, 5*(1), 1–6.

Thompson Klein, J. (2004). Interdisciplinarity and complexity: an evolving relationship. *Emergence: Complexity and Organization, 6*(1–2), 2–10.

Turcotte, C. (2016). Profil de Justine Dion-Routhier. Commencer sa carrière dans la controverse. *Spectre, 45*(3), 31–33.

Zouda, M., El Halwany, S., Milanovic, M., Padamsi, Z., Qureshi, N., Schaffer, K., & Bencze, L. (n.d.). *Ban the Dust: A Graphic Novel About Citizen Actions to Eliminate Urban Dust Pollution. A Complementary Pedagogical Booklet*. Retrieved from https://drive.google.com/file/d/1Wdqpl8VYlogvgTYi6WYkV3NHadZ7ljUX/view

Zouda, M., Schaffer, K., Pouliot, C., Milanovic, M., El Halwany, S., Padamsi, Z., … Bencze, L. (2019). *Ban the Dust: A Graphic Novel About Citizen Actions to Eliminate Urban Dust Pollution* (p.34). Toronto: Critical science education collective.

8. Localization and globalization of core adaptive knowledge

Alexander K. Lautensach

The Anthropocene is the name for a geologic era that began with the appearance of quantifiable human-made environmental constituents such as increased atmospheric carbon dioxide levels (first unequivocally noticeable in the 1800s), the accumulation of microplastic particulates (from the 1950s), and the presence of specific radioisotopes that could only have been generated in the nuclear explosions beginning in the 1940s (Crutzen & Steffen 2003; Burtynsky et al. 2018). Other changes include exponential increases in populations, consumption, economic output, emissions, climate change parameters and other pollution indicators, summarized as the *Great Acceleration* (Steffen et al. 2015). In addition, a catastrophic mass extinction of species (Kolbert 2014) and worldwide deterioration of ecosystems (IPBES 2019) indicate that the human impact on the biosphere has reached a disastrous magnitude. The conclusion by many academic analysts is that "[t]he Anthropocene is functionally and stratigraphically distinct from the Holocene" (Waters et al. 2016).

In this new geologic age humanity finds itself facing numerous challenges that lack precedents both in kind and extent. The challenges are influenced by two aspects, based on John Searle's (1995) model describing how humans tend to construct and assess their realities. On the one hand we are faced with ontologically objective manifestations of our global situation within the biosphere and the associated regional environmental changes (Monastersky 2015; Lewis & Meslin 2018). They include the physical characteristics of the Anthropocene as mentioned above. On the other hand, our perceptions of those manifestations and changes are influenced by ontologically subjective patterns of interpretation and culturally contingent belief systems, such as our varying disposition to recognize and interpret crisis situations. Those subjectivities have both contributed to the anthropogenic causation of the Anthropocene and are being reinforced by our ways of experiencing and interpreting it. Examples include our widespread desensitization to eco-catastrophe (Rees 2019), our dissociation from non-human nature and our bouts of violent behaviour against it (Hawkins 2020). Given the seriousness of those challenges and the likelihood of crisis events, effective approaches and strategies to address them would qualify as core adaptive knowledge.

This chapter focuses on the tasks of discovering, clarifying, communicating, differentiating and applying core adaptive knowledge at the global and local levels. We begin with the questions, what knowledge is most helpful, in what ways, and at what levels? To determine and to prioritize what knowledge counts as adaptive and actionable (Arnott et al. 2020), we shall employ three criteria proposed by Dewulf and coworkers (2020): it must facilitate desirable achievements (consequentiality), it must be morally and practically acceptable (appropriateness), and it must make sense (meaningfulness).

Understanding the aim of that undertaking represents an important category of adaptive knowledge: We need to be clear on what is at stake and how fundamentally important to all of humanity it is that we weather this challenge. Since our beginnings as a species, the collective aim of humanity has been sustainable human security, to ensure our survival at an acceptable quality (Potter 1988) – even if that aim was not always enunciated in those terms. Human security rests on the four pillars of sociopolitical, economic, health-related and environmental security (Lautensach & Lautensach 2020a). Together those pillars describe the necessary conditions for welfare at local through global levels. A sizeable body of literature has been devoted to identifying what human security means to people at those levels (UN Human Security Unit 2016; Lautensach & Lautensach 2020a). To summarize this first, perceptual category of adaptive knowledge: Humanity as a whole must become aware of the unprecedented seriousness of our collective situation and its existential threats to human security (Rees 2019). Failing that, very little else will matter.

A second, strategic category of adaptive knowledge focuses on what the specific challenges to human security are and how they might be addressed. Learning in this category is to take place at the global level as well as in diverse regional and local contexts. The Great Acceleration (Steffen et al. 2015) is driven by five continuous self-reinforcing trends, variable regionally in their impacts but evident in their global extents: the growth of populations and of their consumption patterns; pollution with all its effects, including on the global climate; arms races and general militarization; and increasing socio-economic disparity worldwide (Daly & Cobb 1994: 21; Gaffney & Steffen 2017). To a large extent these trends reinforce each other. Together they have caused humanity to enter global ecological *overshoot* since the 1980s, a condition marked by the global demand of natural resources and ecosystem services exceeding the capacity of the biosphere and local ecological support structures to sustainably provide them (Wackernagel & Beyers 2019).

The reality of overshoot has been illustrated by many authors as comparisons of ecological footprints with available biocapacities (Wackernagel et al. 2002), as the transgression of environmental boundaries in the context of specific activities (Rockström et al. 2009), as shortfalls in terms of minimum requirements for social welfare (Raworth 2017), and as the pedagogically powerful model of Earth Overshoot Day (GFN 2020). Those methods of quantifying overshoot can be applied globally to all of humanity, regionally to local populations, or at the levels of communities, families and individuals. They thus represent examples of how adaptive knowledge, which in this case describes the size of the challenge, can be presented at both the global and regional levels.

One insidious aspect of the Great Acceleration is that under the dominant ideology of the *Conventional Development Paradigm* (CDP) (Hall 2004; Lautensach & Lautensach 2020b), its upward-trending curves are viewed primarily as signs of progress, while the underlying menace is neglected. Strategic adaptive knowledge in this context includes the precautionary interpretation of those graphs, to recognize overshoot for its ecological challenges, and to translate global challenges at the regional level into concrete agenda for proactively coping.

In light of the ecological manifestations of overshoot, urgent questions about sustainable human security can be addressed at the regional level, and how communities might mitigate, to the extent that is still possible, a general collapse imposed by nature. The prospect of imminent collapse is supported by abundant ecological precedents and the models that they informed (Dobkowski & Wallimann 2002; Rees 2019). Those models predict that every animal population that has overshot the biocapacity of supporting ecosystems faces regulatory mechanisms in the forms of epidemics, resource shortages, aggression, infertility and the like – mechanisms through which the population size is brought back below its sustainable maximum. There is no conceivable reason to assume that *Homo sapiens* is exempt from this natural law; in fact, the demise of many regional populations and cultures through history confirmed that we have never been exempt (Diamond 2005). The catastrophic manifestations of ecological overshoot are now evident at regional and global levels, indicating the power of its self-reinforcing driving forces and the momentum of the underlying dynamics. They will most likely result in composite scenarios marked by varying extents of collapse, resilience, cooperation and anarchy, as outlined by Raskin (2016) in a spectrum of scenarios (see Chapter 26 by Sampson).

This strategic category of actionable, adaptive knowledge consists of organized policy changes to head off nature's implacable way of dealing with our overshoot. One recommendation, based on the much discussed $I=PAT$ relationship, states that a reduction of our impact I depends on reductions in population size P, in per capita consumption A, and in the technological impact per person T (Gaffney & Steffen 2017). Yet, population as a topic is still culturally taboo in most circles. The treasured ideal of consumption enjoys similar immunity, despite widespread lip service to 'turning towards a simpler life'. The desperate efforts by governments, industries and consumers alike in the wake of COVID to restore the economic status quo ante under the guise of some misguided view of 'normal' could not have been more revealing. Under the CDP, the driving factors of overshoot appear immune to efforts at downscaling and reducing our impact on the biosphere. In light of that failure up to now, it is safe to conclude that such a deliberate downscaling or de-growth regime is not going to happen in time. Instead, the most likely scenario is a series of 'transition events' (the first example being the COVID pandemic), eventually resulting in limited collapse on a global scale, accompanied by more severe collapse events in some particularly fragile regions (Motesharrei et al. 2014; Cumming & Peterson 2017; O'Neill et al. 2018). This conclusion is the basis for Deep Adaptation as described by Monios and Wilmsmeier in Chapter 11 – describing programs for coping with what can be reasonably expected under a perspective of precaution and a minimum of wishful thinking.

According to Jem Bendell (2018), Deep Adaptation is based on the three principles of resilience, relinquishment and restoration, following the maxim of utility. Resilience prioritizes whatever knowledge, beliefs, values, and dispositions serve the purpose of adaptation, mitigation and the capacity to adapt to changing circumstances in order to survive while retaining valued norms and behaviours. Relinquishment refers to norms and behaviours that may have become counterproductive; it follows the question, "what do we need to let go of in order not to make matters worse?" Targeted for relinquishment are traditions that exert undue pressures on the biosphere, contributing unduly to our collective footprint – along with ludicrous aspirations towards planet-wide domination over 'nature' and its 'management' by unlimited numbers of people. Restoration involves the rediscovery of attitudes and approaches to life that have become eroded by modernity – such as seasonal diets, community focus, and regional self-sufficiency. The guiding question here is "what do we need to bring back to help us with the coming difficulties and tragedies?" Those principles qualify as categories of actionable, adaptive knowledge according to Dewulf et al.'s (2020) three criteria: they promise attractive outcomes; their demands seem appropriate; and they make sense from an experience of crisis. Learning forms the basis of all three principles, beginning with acknowledging their global importance and extending into local applications.

Not all plans towards 'sustainability' pass this test. An example of a well-organized political action program that nevertheless cannot qualify as useful actionable knowledge is presented by the much-publicized Sustainable Development Goals (SDGs) (UN 2015). They are widely accepted, because they promise to achieve just about all conceivable benefits for all living people. However, they suffer from internal contradictions and a negligence of basic ecology that render them unlikely to succeed. Analyses using the

Donut Economics model (O'Neill et al. 2018) and footprint accounting (Ewing et al. 2010: 21–22; WWF 2012: 61; Shropshire 2019) indicated that they do not allow for acceptable (by their own standards) compromise solutions between human development (SDGs#1,2,3,8,9,11,12) and environmental limits (SDGs#13,14,15). Recent reports confirm that the SDGs are already falling short of their targets in some categories (e.g. FAO et al. 2020). Thus, the SDGs as a policy program fail Dewulf's criteria of consequentiality and meaningfulness despite their cosmetic appeal. The associated prospects for learning are largely confined to the status quo and the modern paradigm; they do not address the likelihood of fundamental changes in environmental conditions, human survival, modes of governance, and social organization. Most importantly, the official SDGs fail to address population growth, nor do they seem to recognize its essential role in our Anthropocene predicament.

That crucial failing is addressed in a set of 'real SDGs' that reconcile with Deep Adaptation (Lautensach & Lautensach 2020b). While the official SDGs and the underlying CDP advocated as their guiding principle 'the greatest welfare for the greatest number', this revised set of SDGs is grounded on a new principle: *minimum sufficient welfare for the greatest sustainable number* (Lautensach & Lautensach 2020a). In accordance with similar models in the literature (Crutzen & Steffen 2003; Ehrlich 2009; Wackernagel et al. 2017; Club of Rome 2020), it addresses two facts that excessively large populations tend to collide with: loss in their quality of life, even under extremely equitable conditions, and self-reinforcing environmental deterioration. With the help of the Deep Adaptation agenda, along with an ecocentric environmental ethic, each of the seventeen SDGs was revised to present a realistic chance of success. Their demands match those of the climate protest movements spearheaded by Greta Thunberg and other positive deviants (Thunberg et al. 2020), who have been pointing to the chronic failure of governments worldwide to counteract the dominant denial and complacency (Lautensach 2020). Some of the requisite learning is to build consciousness and an awareness of emergency and duty at the global level – but mainly it is directed at developing regional and local strategies towards counterhegemony, sufficiency and just survival. This may not be achievable without a transition to an ecocentric environmental ethic.

In spite of their shortcomings, the official SDGs (like their authors, the UN) still represent the only official comprehensive attempt at the global level – humanity's first concerted effort to work towards an organized transition to sustainability instead of waiting for natural adjustments to happen to us. It has served to galvanize global consciousness and solidarity in many parts of the world that used to be dominated by concerns for development at national or regional levels, not in the global dimension. As an underlying moral principle, the UN has long advocated and worked towards realizing a vision of human security under the aims of 'freedom from want, freedom from fear, and freedom to live in dignity' (Annan 2005; UN Human Security Unit 2016). As such, even in their present form the official SDGs convey some useful adaptive knowledge in the forms of global, cross-cultural solidarity and a sense of purpose. The 'real SDGs' merely redirect that initiative towards achievable local and global solutions, to build a succession of scenarios by which a smaller, wiser humanity can safely make its way through the Anthropocene, guided by respect and kindness towards all life forms.

Our discussion of strategic adaptive knowledge so far has revolved either around humanity as an amorphous whole in the first-person plural, or as populations of individuals with unspecified cultural affiliations. Although such a simplified approach is by no means uncommon in the social and natural sciences, a more differentiated analysis seems called for. The learning of new knowledge is only ever effective if it is considered culturally acceptable by those whom it is meant to benefit. To remediate that oversimplification, to take into account the manifold heterogeneities in the human species, and to ensure effective learning in diverse localities, the variable of culture needs to be introduced.

Clifford Geertz (1972: 261) defined culture as "the shared patterns that set the tone, character, and quality of people's lives". This can refer to ways in which people collectively interpret information about the world, how they interpret and make sense of their observations, how they structure their discourse about such topics, how they arrive at decisions and plans, how they conceptualize new phenomena within the frameworks of their values and beliefs, and many other aspects that describe the interaction of people with the new realities of the Anthropocene. As the applications and implementations of adaptive knowledge must include diverse place-based solutions and strategies, as well as universal guidelines, cultural differences in the forms of biogeographical adaptions, value priorities, belief systems, and worldviews will make a crucial difference for a successful transition to a secure and sustainable future.

The diverse regional reactions to the COVID pandemic, as the first globally experienced transition event, illustrate that point (Nikiforuk 2020). Even when leaving aside differences of social-economic systems and infrastructures, not all countries appeared equally adept at coping – at cooperatively implementing reasonable containment regimes, and at learning how to modify their life styles quickly for everybody's benefits (Horton 2020; Speckhard et al. 2020). Two conclusions can be drawn: First, those regional differences in resilience, flexibility, and collective learning reflect at least in part a diversity of cultural capabilities to cope with critical transition events, partly grounded in differential proclivities for collective learning. Second, not all cultures seem equally deserving of external support, especially in cases where they denied or belittled the pandemic in the face of massive evidence. This points to a deeper distinction that disqualifies certain moralities from egalitarian considerations in cases wherever they neglect or belittle universal values and norms such as human rights, justice and dignity. Gbadegesin (2009: 32) recognized this in his refutation of cultural relativism in a multicultural society: "[W]e need fully to understand and appreciate the viewpoint of a particular standard before we judge it as inadequate."

Returning to our earlier question how regional diversity affects the discovery and application of actionable, adaptive knowledge, we can identify cultural differences as a crucial factor that determines the efficacy with which that knowledge is acquired and acted upon. As different individuals in a classroom learn in different ways and at different rates, so do cultures in the global community differ in their collective learning. Under the global imperatives imposed by the Anthropocene, that process of cultural learning may determine the survival of a society. In the past, those societies that collapsed in the face of daunting environmental challenges did so because they were unable to learn and adapt fast enough, to develop sufficient resilience (Diamond 2005). Those mechanisms of collective learning and adapting are culturally shaped to a crucial extent. Cultural learning also includes the ability of cultures to learn from each other and to develop cooperatively the necessary strategies. It requires flexibility, openness and a willingness to abandon traditions that have become counterproductive or harmful (Orr 2004), paving the way for Deep Adaptation at regional levels (Lautensach 2020).

This proclivity for cultural learning represents a third, pedagogical category of adaptive knowledge, a set of skills, attitudes and dispositions that mirror at the collective level the concept of Learning II in individuals – Bateson's (1973/2000) model of *learning to learn*. Current differences in cultural learning will partly determine the cultural composition of that smaller, wiser humanity that will, hopefully, survive the Anthropocene under acceptable circumstances. The origins and causes of those differences in collective abilities of cultures to learn remain an open question. They are partly rooted in native homelands, partly carried along by migrating populations (Speckhard et al. 2020) – as is much of culture on the whole. Yet, the transition to sustainability involves less competition than cooperation: The manifold benefits of communities learning from each other's successes and failures necessitate global connectivity and communication. The much-advocated reliance on local solutions must not be allowed to mutate to a flight into isolation, lest we forego the potential of synergy (Selby & Kagawa 2015). Nevertheless, the learning required in this pedagogical context is primarily local.

Cultural learning is occasionally amplified to astounding extents through the influence of positive deviants, as in the example of Greta Thunberg. Sara Parkin (2010: 1) defined a positive deviant as "a person who does the right thing for sustainability, despite being surrounded by the wrong institutional structures, the wrong processes and stubbornly uncooperative people". Thus, a phenomenon that distinguishes individual learners can bridge into the collective domain and boost the learning of entire populations. In the same way, positive deviants can help with the acquisition and application of adaptive knowledge at both levels, local and global. The onus is on education systems and on teachers to nurture, protect and support the emergence of positive deviants in spite of cultural obstacles and the decidedly mixed track record of public education (Lautensach 2020).

What needs to be learned, individually and collectively, is subsumed under the three categories of adaptive knowledge we identified – perceptual, strategic and pedagogical. They include factual knowledge, skills, beliefs, values, attitudes, interests and ideals that all inform our decision-making. Among the values are included ecocentric norms that prioritize the welfare of the biosphere (sometimes referred to as Gaia), the principle of keeping within sustainable boundaries to ensure human security, and respecting universal rights and justice for humans and non-humans. Those learning outcomes have been compiled and organized into curricula for sustainability (e.g. Lautensach 2020; see also Chapter 7).

How those learning outcomes are to be regionally apportioned and adapted to local needs is determined by an assessment of the regional situation regarding overshoot (Wackernagel et al. 2002); particular practices and traditions contributing to environmental impact or footprint of the region (GFN 2020); population size, affluence, technology according to $I=PAT$; particular environmental boundaries that are being transgressed in the region (Rockström et al. 2009); particular shortfalls in social needs that need to be amended in the region (Raworth 2017); and what needs to be done to stabilize, replenish and restore ecological integrity and biodiversity in that region. These efforts are likely to collide with population ceilings, carrying capacities that are negotiable only to a degree. Even if people are prepared to reduce their individual footprints, dropping treasured cultural traditions perhaps, and making do with much less material affluence, the ultimate transition to a sustainable future may not be possible without a reduction of the population in a region. This certainly applies to the global level; between regions, at least there remains the possibility of mutualistic complementation, of trading strengths and needs, which enriches the dualistic approach of 'glocalism' (Hempel 1996; Rees 2014). Ultimately, and especially at a time of failing governance at the national level (Kupchan's [2012] 'crisis of governability'), the success of our transition efforts will depend to a crucial extent on local and regional efforts.

The interrelationships between global and local dimensions are reflected in the three categories of adaptive knowledge that describe different requirements in the overall learning that a successful transition will require. The learning of perceptual knowledge focuses on universal limits and global imperatives and contributes to the development of global solidarity and cooperation. In contrast, efforts to improve our performance as learners take place primarily at the local, culturally contingent level; the learning of such pedagogical knowledge relies on individual growth as well as on communities and place-based cultures developing and learning from each other. In between we placed the learning of strategic knowledge, which takes place at all levels, depending on the scope of particular innovations, transformative practices and reforms.

The learning of locally relevant adaptive knowledge, along with globally essential goals and strategies, will be the only kind of growth that can pass the test of sustainability. Much of this learning will rely on organized, formal systems of education whose efficacy depends on the quality of transformative teaching involved, its cultural contexts, and the extent of political support it receives. But beyond formal education, individuals, communities and entire cultures must learn, and learn to learn faster, how best to uphold human security while our environmental support structures are increasingly eroded. This transformative goal represents the greatest educational hurdle humanity has ever faced.

REFERENCES

Annan, K. 2005. In Larger Freedom: Towards Development, Security, and Human Rights for All. United Nations. https://www.un.org/en/events/pastevents/in_larger_freedom.shtml (22 August 2020).

Arnott, J.C., Mach, K.J. & Wong-Parodi, G. (eds). 2020. Advancing the Science of Actionable Knowledge for Sustainability. Special Issue of *Current Opinion in Environmental Sustainability* 42: A1–A6, 1–82.

Bateson, G. 1973. *Steps to an Ecology of Mind.* Paladin. Second revised edition: GB (edited by Mary Catherine Bateson). 2000. *Steps to an Ecology of Mind: Collected Essays in Anthropology, Psychiatry, Evolution and Epistemology.* University of Chicago Press.

Bendell, J. 2018. Deep Adaptation: A Map for Navigating Climate Tragedy. IFLAS Occasional Paper 2 www.iflas.info (27 July). Institute of Leadership & Sustainability (IFLAS), University of Cumbria, UK. http://www.lifeworth.com/deepadaptation.pdf (29 July 2018).

Burtynsky, E., Baichwal, J. & De Pencier, N. 2018. *Anthropocene.* Art Gallery of Ontario, Goose Lane in collaboration with the National Gallery of Canada.

Club of Rome. 2020. *Planetary Emergency 2.0: Securing a New Deal for People, Nature and Climate.* Winterthur, CH: The Club of Rome & Potsdam Institute for Climate Impact research. https://clubofrome.org/wp-content/uploads/2020/08/Planetary_Emergency_Plan_2.0-.pdf (20 August 2020).

Crutzen, P.J. & Steffen, W. 2003. How Long Have We Been in the Anthropocene Era? *Climatic Change* 61(3): 251–257.

Cumming, G.S. & Peterson, G.D. 2017. Unifying Research on Social-Ecological Resilience and Collapse. *Trends in Ecology & Evolution* 32(9): 695–713. https://doi.org/10.1016/j.tree.2017.06.014 (22 August 2020).

Daly, H.E. & Cobb Jr., J.B. 1994. *For the Common Good: Redirecting the Economy toward Community, the Environment, and a Sustainable Future* (2nd edn). Beacon Press.

Dewulf, A., Klenk, N., Wyborn C. & Lemos, M.C. 2020. Usable Environmental Knowledge from the Perspective of Decision-Making: The Logics of Consequentiality, Appropriateness, and Meaningfulness. *Current Opinion in Environmental Sustainability* 42: 1–6.

Diamond, J. 2005. *Collapse: How Societies Choose to Fail or Succeed*. Penguin Books.

Dobkowski, M.N. and Wallimann, I. 2002. *On the Edge of Scarcity: Environment, Resources, Population, Sustainability and Conflict*. Syracuse University Press.

Ehrlich, P.R. 2009. Ecoethics: Now Central to All Ethics. *Bioethical Inquiry* 6: 417–436.

Encyclopedia of Earth. 2012. *Welcome to the Anthropocene*. http://www.anthropocene.info (29 May 2019).

Ewing, B., Moore, D., Goldfinger, S. et al. 2010. *The Ecological Footprint Atlas 2010*. Global Footprint Network. https://www.footprintnetwork.org/content/images/uploads/Ecological_Footprint_Atlas_2010.pdf (22 August 2020).

FAO (Food and Agriculture Organisation of the UN). 2020. *The State of Food Security and Nutrition in the World 2020. Transforming food systems for affordable healthy diets*. FAO, IFAD, UNICEF, WFP & WHO. https://doi.org/10.4060/ca9692en (22 August 2020).

Gaffney, O. & Steffen, W. 2017. The Anthropocene Equation. *The Anthropocene Review* 1–9.

Gbadegesin, S. 2009. 'Culture and Bioethics'. In *A Companion to Bioethics* (2nd ed.), H. Kuhse & P. Singer (eds). Wiley-Blackwell, Ch. 3, 24–35.

Geertz, C. 1972. *The Interpretation of Cultures*. Basic Books.

GFN (Global Footprint Network). 2020. About Earth Overshoot Day. https://www.overshootday.org/about-earth-overshoot-day/ (6 May 2020).

Hall, C.A.S. 2004. Sanctioning Resource Depletion: Economic Development and Neoclassical Economics. In *Ecojustice – The Unfinished Journey*. Gibson, William, E. (ed.). SUNY Press. 201–212.

Hawkins, R. 2020. Our War Against Nature. In *Human Security in World Affairs: Problems and Opportunities*, A.K. Lautensach & S.W. Lautensach (eds) (2nd ed.). UNBC & BCcampus (n.p.)

Hempel, L.C. 1996. *Environmental Governance: The Global Challenge*. Island Press.

Horton, R. 2020. *The COVID-19 Catastrophe: What's Gone Wrong and How to Stop It Happening Again*. Polity.

IPBES (Intergovernmental Science-Policy Platform on Biodiversity and Ecosystem Services). 2019. *2019 Global Assessment Report on Biodiversity and Ecosystem Services*. https://ipbes.net/news/Media-Release-Global-Assessment (22 August 2020).

Kolbert, E. 2014. *The Sixth Extinction: An Unnatural History*. Henry Holt & Co.

Kupchan, C.A. 2012. The Democratic Malaise. *Foreign Affairs* 91 (1) (Jan/Feb): 62–67.

Lautensach. A.K. 2020. *Survival How? Education, Crisis, Diachronicity and the Transition to a Sustainable Future*. Ferdinand Schoeningh / Brill.

Lautensach, A.K. & Lautensach, S.W. (eds) 2020a. *Human Security in World Affairs: Problems and Opportunities* (2nd ed.). UNBC & BCcampus. Ch. 21. https://opentextbc.ca/humansecurity/.

Lautensach, S.W. & Lautensach, A.K. 2020b. What Future for International Development in the Anthropocene? *Indian Journal of Politics and International Relations* (special 2019 issue).

Lewis, S. & Meslin, M. 2018. *The Human Planet: How We Created the Anthropocene.* Pelican-Penguin.

Monastersky, R. 2015. Anthropocene: The Human Age. *Nature* 519 (7542): 144–147. http://www.nature.com/news/anthropocene-the-human-age-1.17085 (14 November 2017).

Motesharrei, S., Rivas, J. & Kalnay, E. 2014. Human and Nature Dynamics (HANDY): Modeling Inequality and Use of Resources in the Collapse or Sustainability of Societies. *Ecological Economics* 101: 90–102.

Nikiforuk, A. 2020. Brazil's Descent into COVID-19 Hell: Plus Check-Ins with Five More Nations Fighting the Pandemic in Different Ways. *The Tyee* (10 June) https://thetyee.ca/News/2020/06/10/Brazil-Descent-COVID-Hell/?utm_source=national&utm_medium=email&utm_campaign=110620 (22 August 2020).

O'Neill, D.W., Fanning, A.L., Lamb, W.F. & Steinberger, J.K. 2018. A Good Life for all Within Planetary Boundaries. *Nature Sustainability* 1: 88–95.

Orr, D.R. 2004. *Earth in Mind: On Education, Environment and the Human Prospect.* Island Press.

Parkin, S. 2010. *The Positive Deviant: Sustainability Leadership in a Perverse World.* Earthscan.

Potter, V.R 1988. *Global Bioethics: Building on the Leopold Legacy.* Michigan State University Press.

Raworth, K. 2017. *Doughnut Economics: Seven Ways to Think Like a 21st Century Economist.* Chelsea Green Publishing, Cornerstone, Penguin.

Raskin, P. 2016. *Journey to Earthland: The Great Transition to Planetary Civilization.* Paul Raskin / Telus Institute.

Rees, W.E. 2014. Avoiding Collapse: An Agenda for Sustainable Degrowth and Relocalizing the Economy. Canadian Centre for Policy Alternatives.

Rees, W.E. 2019. End Game: The Economy as Eco-Catastrophe and What Needs to Change. *Real-World Economics Review* 87: 33–50.

Rockström, J., Steffen, W., Noone, K. et al. 2009. A Safe Operating Space for Humanity. *Nature, 461* (September 24): 472–475.

Searle, J.R. 1995. *The Construction of Social Reality.* The Free Press.

Selby, D. & Kagawa, F. (eds) 2015. *Sustainability Frontiers: Critical and Transformative Voices from the Borderlands of Sustainability Education.* Barbara Budrich Publishers.

Shropshire, A. 2019. Human Development & National Eco-Footprints: A Visual Orientation. https://towardsdatascience.com/human-development-national-eco-footprints-a-visual-orientation-9adf86618d4f (22 August 2020).

Speckhard, A., Mahamud, O. & Ellenberg, M. 3 April 2020. When Religion and Culture Kill: COVID-19 in the Somali Diaspora Communities in Sweden. *Homeland Security Today* https://www.hstoday.us/subject-matter-areas/counterterrorism/when-religion-and-culture-kill-covid-19-in-the-somali-diaspora-communities-in-sweden/ (22 August 2020).

Steffen, W., Broadgate, W., Deutsch, L. et al. 2015. The Trajectory of the Anthropocene: The Great Acceleration. *The Anthropocene Review*, 2(1): 81–98.

Thunberg, G., Neubauer, L., De Wever, A. & Charlier, A. 2020. After Two Years of School Strikes, the World is Still in a State of Climate Crisis Denial. *The Guardian*

(19 August) https://www.theguardian.com/commentisfree/2020/aug/19/climate-crisis-leaders-greta-thunberg (19 August 2020).

UN (United Nations). 2015. Sustainable Development Goals: 17 Goals to Transform Our World. New York: UN Department of Public Information. Based on UNGA resolution A/RES/70/1 (25 September). https://www.un.org/sustainabledevelopment/sustainable-development-goals/ (22 August 2020).

UN Human Security Unit. 2016. *The Human Security Handbook*. United Nations Trust Fund for Human Security. https://procurement-notices.undp.org/view_file.cfm?doc_id=212254 (1 June 2020).

Wackernagel, M. & Beyers, B. 2019. *Ecological Footprint: Managing Our Biocapacity Budget*. New Society Publishers.

Wackernagel, M., Hanscom, L. & Lin, D. 2017. Making the Sustainable Development Goals Consistent with Sustainability. *Frontiers in Energy Research* 18 (July) (18) doi: 10.3389/fenrg.2017.00018 http://journal.frontiersin.org/article/10.3389/fenrg.2017.00018/full (22 August 2020).

Wackernagel, M., Schulz, N.B., Deumling, D. et al. (2002). Tracking the Ecological Overshoot of the Human Economy. *Proceedings of the National Academy of Sciences* (USA) 99 (14): 9266–9271. http://www.pnas.org/cgi/reprint/99/14/9266.pdf (22 August 2020).

Waters, C.N., Zalasiewicz, J., Summerhayes, C. et al. 2016. The Anthropocene is Functionally and Stratigraphically Distinct from the Holocene. *Science* (8 Jan) 351 (6269): aad2622. doi: 10.1126/science.aad2622.

WWF (World Wildlife Fund). 2012. *Living Planet Report 2012*. Worldwide Fund for Nature & WWF International. https://d2ouvy59p0dg6k.cloudfront.net/downloads/lpr_living_planet_report_2012.pdf (22 August 2020).

PART III

Anthropocene economics

9. The end of Holocene economics[1]
Richard Heinberg

In this chapter, I briefly survey the evolution of human economies during the past 11,000 years. This evolution occurred in a few clearly defined increments, each closely correlated with a substantial shift in ways of harvesting energy. We will see how the scale of economic processes depends on available energy, and how societies' increasing energy usage has often resulted in greater economic inequality. We will also see how expansion of trade, based largely on the increased usage of money and debt, has transformed societies and human relations. We will review ways in which economic theory, which developed primarily during the past 200 years, has been shaped by rapid increases in fossil-fuel energy capture and the development of new technologies for using these energy sources. Finally, we will explore how and why economic expansion has led humanity to an ecological bottleneck.

HUNTER-GATHERER ECONOMIES

Throughout most of our species' period of existence, we lived by harvesting energy from our environment in the forms of wild foods and combustible biomass. The latter was a significant boon: although it is uncertain exactly when we humans harnessed fire, it is clear that this innovation provided significant benefits, giving us the ability to stay warm, ward off predators, cook food, transform materials, and alter landscapes to increase their productivity for human purposes. Meanwhile, simple stone tools leveraged muscle power to enable us to kill prey (and other humans) at a distance, as well as to make still more tools, such as clothing, boats, and shelters, which allowed us to spread into new environments.

Hunter-gatherers maintained what anthropologists call *gift economies* (Mauss & Halls, 2000; Graeber, 2001). People simply shared whatever they had. Within-group cooperation was essential to survival, and so innate competitive drives (especially among males) were moderated through ritual and custom, while conditions of mutual indebtedness helped maintain a generally cooperative attitude on everyone's part. Trade is an inherently competitive activity, since each trader tries to get the best deal possible, even at the expense of other traders. So, while trade certainly existed among hunter-gatherers, it was confined to relations between members of different communities.

The community was essentially like a family. Freeloading was occasionally a problem, and when it became a drag on the rest of the community it was punished by subtle or not-so-subtle social signals—ultimately, ostracism. For bullies, even capital punishment was an option. But otherwise, no one kept score of who owed what to whom. Gendered division of labor was only partial, with women often supplying the bulk of the calories consumed; accordingly, women enjoyed social status roughly equal to that of men in most societies.

Economic inequality hardly existed among hunter-gatherers: no one starved unless everyone was starving. However, competition between groups often led to bloodshed, and archaeological and ethnographic evidence collected in the past couple of decades suggests that interpersonal violence among foragers was widespread (Wrangham & Peterson, 1996).

Horticultural Economies

Humans began domesticating plants and animals at the start of the Holocene, and perhaps earlier (certainly so in the case of the dog). As some hunter-gatherer groups in particularly abundant regions began living in semi-permanent villages, population pressure probably forced them to intensify food production, which meant planting and hoeing.

Gardening and village life enabled the production of a seasonal surplus, which could be stored in preparation for winter or as a hedge in case of a poor harvest next year. But production of a significant surplus required that people work harder than was otherwise necessary. Within the group, one individual (nearly always male, according to available evidence) set an example, working hard and encouraging his relatives and friends to do so as well. The payoff for all this toil would come in the form of parties thrown occasionally by the Big Man and his crew.

The Big Man wielded influence, gained prestige, and could represent his group at inter-group ceremonies, where he competed for status with other regional Big Men. But being a Big Man required that he give away nearly all of his wealth every year; moreover, other males in the group were always aspiring to be Big Men, too, so there was never any assurance that he could maintain his status. Even though one individual had acquired status, at least temporarily, society was still highly egalitarian.

Horticultural societies that survived into the 20th century, including ones in New Guinea, tended to persist in a near-constant state of warfare. Enslaved war captives sometimes provided forced labor, but equal redistribution among other group members tended to preserve economic equality.

Complex horticultural societies with permanent villages and hereditary chiefs, who enjoyed unequal levels of wealth and prestige, appeared somewhat later, perhaps 8,000 years ago, and persisted in many places until European colonial conquests (Harris & Johnson, 2007: 67–69, 104).

The Economies of Agrarian States

The primary energy innovation of the agrarian era consisted of production of grain crops. Grains (wheat, barley, millet, rice, and maize) have a unique capacity to act as a concentrated store of food energy that is easily collected, measured, and taxed. Grains' taxability is especially consequential: it enabled the emergence, beginning roughly 6,000 years ago, of a new political entity—the state; and with it came systemic economic inequality.

The formation of the first states appears to have occurred in places of relative natural abundance such as river deltas, where human population had become particularly dense. Population pressure and constant warfare led to the consolidation of societies around strong war chiefs, who eventually took on the role of divine kings. Agrarian states also formed around cities—permanent communities with streets, grand public buildings usually (Scott, 2017).

The surplus energy of grains enabled full-time division of labor. Work tended to be sharply divided by gender, and the social and economic status of women declined dramatically. The great majority (often 90 percent or more) of the populace consisted of peasants working the land, or captive slaves working in mines, quarrying stone, felling timber, dredging, rowing ships, and engaging in similar sorts of forced drudgery (slavery was universally practiced in early agrarian societies, and was a primary component of the social mechanism by which solar energy was channeled to do work) (Nikiforuk, 2012). The rest of the population at first consisted mostly of the royal family, their attendants, priests, and full-time soldiers. Writing was invented primarily to keep economic records, but soon served other purposes. Gradually, the professional classes expanded to include scribes, accountants, lawyers, merchants, craftspeople of various descriptions, and more.

Technologically, the agrarian state depended upon one innovation above all others—the plow, which facilitated the planting of field crops. Working hard, usually with the help of traction animals such as oxen, a farming family could produce a surplus of grain that the king claimed on behalf of the state as a tax. In return, the state provided protection from raids and also held grain in storage in case of poor harvests. States also competed with each other to develop more effective weapons. Competition for better weapons in turn drove innovations in metallurgy—leading to the adoption first of copper, then of bronze, and finally iron for crafting blades and other tools.

Economically, the agrarian state functioned as a wealth pump, with surplus production from the peasantry and enslaved persons continually being funneled via taxes upward to the king, the king's family, the priesthood, and the aristocracy—which at first consisted primarily of elite soldiers and their families. In addition, the idea and practice of land ownership—unknown previously—became an essential means of organizing relations between families, and between people and the state. Ownership demanded an entirely new way of thinking about the world, one in which sympathies with nature would recede and numerical calculation would play an increasing role.

Two key economic innovations of the agrarian state were money and debt. Economists often describe money as a neutral medium of exchange; but, in reality, money is better thought of as quantifiable, storable, transferrable, and portable social power (Heinberg, in press). Trade was increasingly occurring within the broadened borders of society, and, in order to keep the process of exchange from eroding community solidarity, the state created and regulated markets. The palace and the temple made laws specifying where and when markets could operate and set the terms on which sellers should extend credit to buyers. The state also created and enforced standard units of measurement (money also co-evolved with number systems, arithmetic, and geometry).

The origin of the practice of charging of interest on loans is difficult to identify; while it probably existed in ancient Mesopotamia, clear written documentation only goes back to about 2,500 years ago (Graeber, 2011: 400). Interest on loans continually and subtly siphoned wealth from borrowers to those in position to loan money, thereby strengthening the agrarian "wealth pump." People whose main occupation was investment or trade gradually became more powerful, while those whose main occupation was growing food, making things, or taking care of others gradually became relatively poorer.

All of these innovations were mutually reinforcing. Full-time division of labor tended to stoke technological change, as professionals had the resources with which to experiment and invent, refining weapons and farm tools. New weapons enabled the growth of military power. And new technologies also contributed to economic change by generating new professions and new sources of wealth.

From the origin of the first state until roughly the 19th century of the current era, kingdoms rose and fell, sometimes expanding to become empires. Technologies became more sophisticated, and wars of conquest ultimately enabled a few European states to gain political and economic control of much of the rest of the world. Still, the basic mechanisms of production and distribution, while undergoing constant refinement, remained essentially the same until the widespread adoption of fossil fuels starting in the 19th century.

Fossil Fuel Economies

The central energy innovation of the fossil fuel era was, of course, the adoption of machines running on coal, oil, and natural gas, thereby largely replacing muscle power in agriculture, manufacturing, transportation, and other economic activities. Fossil fuels represent millions of years of ancient sunlight transformed by natural processes into forms that are storable and transportable. The energy required to access fossil fuels was minimal: in the early years of the petroleum industry, for example, the ratio of energy returned on the energy invested in prospecting and drilling for oil was greater than 50:1 (far greater than the averaged 3:1 energy profit of subsistence agriculture). Further, the sheer quantity of energy that could be unleashed was unprecedented: since 1820, global per capita usage of energy has increased eight-fold, even while population increased nearly eight-fold as well.

Fossil-fueled industrialism nearly took off in China around the year 1000 (McNeill, 1982: 24–62). The essential ingredients were present, including systematic technological innovation, pervasive and increasing usage of coal, and an economic system favoring privatization and investment. However, the Chinese government viewed industrial expansion as a threat to its own power and prevented further development. So, instead, the industrial revolution got its start in Britain several centuries later.

The increasing adoption of powered machinery, starting with a simple steam engine for pumping water out of flooded British coalmines, quickly replaced animal and human muscle power in many onerous occupations, thereby eventually contributing to the abolition of slavery. Further, powered farm machines gradually displaced much of the rural workforce, resulting in a continuing trend toward urbanization. Previously, the great majority of people were needed on the land; now, manufacturing required more hands to guide the new machines. Although wage labor had existed since ancient times, the availability of urban jobs now became central to the rise of a burgeoning middle class. Since new jobs to operate and manage machinery could be filled by people of either gender, women entered the workforce in large numbers and began to seek equal pay and equal political rights.

Science and technology partnered to produce new materials and processes that were dependent on fossil fuels. Among the results were better sanitation and medicines, leading to a lowering of human mortality. At the same time, chemical fertilizers and pesticides and new crop varieties dramatically increased farm yields. Population grew from about one billion in 1820 to nearly 8 billion a mere two centuries later.

Powered mining and manufacturing operations were capable of churning out products at vastly higher rates than was previously possible. The result was a crisis of overproduction, which contributed to the Great Depression of the 1930s. Industrial, financial, and political elites collaborated to develop a solution—the economic system that came to be known as *consumerism* (Ewen, 2001). Pervasive and sophisticated advertising, which took advantage of the new science of psychology, motivated the masses to desire and acquire more products. Meanwhile, a significant expansion of consumer credit enabled households to purchase more than could be afforded with cash. The result was a persistent increase in consumption, which drove job creation, higher tax revenues, and growing profits for industrialists. Government agencies began referring to citizens as "consumers," signifying a key new role in the economy.

The growth of production and consumption was now seen as essential to the continued health of the economy—which was for the first time constantly monitored and measured, using new indices such as the unemployment rate and gross domestic product (GDP). Meanwhile, trade expanded rapidly, based on motorized shipping and rail distribution. Rapidly growing commerce required an expanding money supply and new instruments of debt—including various kinds of financial derivatives. In the 20th century, most nations shifted from using currency based on precious metals to currencies created through bank loans.

The wealth pump of taxation, profit, and interest continued to generate economic inequality in fossil-fueled societies, and rising inequality occasionally led to political turmoil. The desire to avoid political unrest motivated government efforts to manage inequality through progressive taxation and various redistributive programs. Meanwhile, unions sought higher wages and better working conditions from employers. Still, as of 2017, a mere eight individuals controlled as much wealth as the poorer half of humanity—a level of inequality never before imaginable (Oxfam, 2017).

Environmental historians, notably J. R. McNeill and Peter Engelke (2014), have termed the fossil fuel period the Great Acceleration. Not only were population and production expanding rapidly, but also the environmental harms caused by resource extraction, production, and waste dumping. Habitat for other species disappeared, contributing to substantial declines in most wild animal and plant species. Nonrenewable resources, including minerals and fossil fuels themselves, began depleting at accelerating rates. And pollution posed increasing risks for humans and other species—most ominously, carbon dioxide pollution from the burning of fossil fuels, which increasingly destabilized the global climate.

THE EVOLUTION OF ECONOMIC THINKING

As economies changed, so did the ways in which people thought about economic activity. The first economists were ancient Greek and Indian philosophers (Plato discussed government-created money in his *Laws*), but little of real substance was added to the discussion during the next two thousand years, as the economies of complex societies maintained their agrarian basis. It was with the start of the Great Acceleration that economic thinking really heated up.

"Classical" economic philosophers Adam Smith (1723–1790), Thomas Robert Malthus (1766–1834), and David Ricardo (1772–1823), who lived at the very start of the fossil fuel era, introduced concepts such as supply and demand, division of labor, and the balance of international trade. These pioneers set out to discover natural laws in the day-to-day workings of economies so as to create a science of economics on a par with the emerging disciplines of physics and astronomy. Economic philosophers could point to price as arbiter of supply and demand, acting everywhere to allocate resources far more effectively than could any human manager or bureaucrat—a principle surely as universal and impersonal as the force of gravitation. Isaac Newton had shown there was more to the motions of the stars and planets than could be found in the book of Genesis; similarly, Adam Smith was revealing more potential in the principles and practice of trade than had ever been realized through the ancient, formal relations between princes and peasants.

The classical theorists' followers adopted the mathematics and some of the terminology of science. Unfortunately, however, they failed to incorporate into economics the basic self-correcting methodology that is science's defining characteristic. Economic theory required no falsifiable hypotheses and demanded no repeatable controlled experiments (which would be hard to organize in any case). Economists began to think of themselves as scientists, while in fact their discipline was, and remains, a branch of moral philosophy.

In theory, the market was a beneficent quasi-deity tirelessly working for everyone's good by distributing the bounty of nature and the products of human labor as efficiently and fairly as possible. But, in fact, everybody wasn't benefiting equally or (in many people's minds) fairly from economic expansion. The market worked especially to the advantage of those for whom making money was a primary interest in life (bankers, traders, and investors), and who happened to be clever and lucky. It also worked nicely for those who were born rich and who managed not to squander their birthright. Others, who were more interested in growing crops, teaching children, or taking care of the elderly, or who were forced by circumstance to give up farming or cottage industries in favor of factory work, seemed to be getting less and less—certainly as a share of the entire economy, and often in absolute terms. Was this fair? That was a moral and philosophical question. In defense of the market, many economists said that it *was* fair: merchants and factory owners were making more because they were increasing the general level of economic activity; as a result, everyone else would also benefit eventually. But other economists outside the mainstream (Thomas Piketty is a prominent recent example) have produced evidence that the current setup of capitalist markets contains rules and institutions that inevitably produce an ever-wider gap between haves and have-nots.

Importantly, early economic philosophers had some inkling of natural limits and anticipated an eventual end to economic growth. The essential ingredients of the economy were understood to consist of *labor*, *land*, and *capital*. There was on Earth only so much land (which, in these theorists' minds, stood for all natural resources), so of course at some point the expansion of the economy would cease. Both Malthus and Smith explicitly held this view.

But, over time, land came to be disregarded as one of the economy's basic theoretical components, leaving only labor and capital, under the justification that (in the words of economists William Nordhaus and James Tobin, 1973: 522), "… reproducible capital is a near perfect substitute for land and other exhaustible resources." This seemed sensible because capital was becoming so productive that there was less apparent economic need for land. However, capital was becoming more productive primarily because society had stumbled upon extraordinary new energy sources in the form of fossil fuels, which made every productive process (including agriculture) much cheaper. The proper response of economists should have been to begin including energy as a factor of production, but economists didn't do that; instead, they omitted land from their equations.

This line of thinking eventually led to the absurd assertion, by celebrated economist Robert Solow (1974: 11), that, "The world can, in effect, get along without natural resources." More recently, prize-winning economist William Nordhaus (1993: 11–25) has said that climate change isn't a terribly serious problem because it just affects land, and only capital and labor really count (more specifically, he argues that even if a destabilized climate makes agriculture impossible, that's not so bad because agriculture accounts for only a small fraction of global GDP). Nordhaus fails to make the obvious connection: without agriculture there would be far fewer humans, no civilization, and therefore no capital or labor in any meaningful sense (Fix, 2020).

Starting with Adam Smith, the idea that continuous "improvement" in the human condition was possible had come to be generally accepted. At first, the meaning of "improvement" (or "progress") was kept vague, perhaps purposefully so. Gradually, however, "improvement" and "progress" came to mean "growth" in the current economic sense of the term—abstractly, an increase in GDP, but, in practical terms, an increase in production and consumption. Even a moment's thought reveals the absurdity of assuming constant exponential increase of any physical quantity within a bounded system such as Earth. Nevertheless, most economists continue to see growth as good, and lack of growth as a problem to be solved—despite the role that economic expansion plays in worsening levels of greenhouse gas emissions, resource depletion, and habitat destruction.

Two further economic developments made it easier to loot nature and harder to understand what was actually happening. One of these is summed up in the word *externality*. In economics, an externality is the impact of an economic transaction that is not priced into that transaction. No one sets out to produce negative externalities, in the sense that no one pollutes just for the sake of polluting. Pollution is a byproduct of doing business, and industry typically assumes that society as a whole will either learn to live with the mess or pay to clean it up. Only rarely does industry foot the cleanup bill (that's what might be called internalizing the externality). Most of the time, industry profits while nature bleeds and society pays. In most economic theory, negative externalities are largely disregarded; if they were all to be quantified and internalized, much of the profit from industrial processes would disappear.

The other development making it easier to loot nature had to do with private ownership of resources. Ownership of land began, as we have seen, with the first state societies. During the Middle Ages, enclosure of common lands had added to the wealth of the aristocracy while impoverishing peasants. Most modern economic theorists have simply assumed a justification for exclusionary property rights—thereby once again playing the role of moral philosophers while pretending to be scientists. If a corporation buys land that happens to contain a major coal deposit, the corporation *owns* that coal and can mine and sell it. (In some cases, corporations can even buy rights to resources below land owned by others.) But no business made the coal, or the soil above it. Industrialists simply claim ownership by paying a fraction of real value, and then profit from the extraction of whatever valuable minerals may exist. American economist Henry George proposed back in the 1870s (and Native Americans and other indigenous peoples have always agreed) that land should be the common property of all, and that other species should have the right to habitat and survival. Workers should own the products of their labor, but no one should unilaterally own our common inheritance of nature's bounty. If economists had adopted this alternative moral philosophy, profits would again have been harder to achieve, though innumerable environmental harms would have been avoided.

Late Holocene economic theory could be considered a religion, with an origin myth (the discredited notion that, prior to the invention of money, everyone lived by barter), a priesthood (credentialed economists), and a deity (the "invisible hand" of the market). However, it can also be thought of simply as an ideology in service of plunder.

THE BOTTLENECK[2]

Here is economic history compressed into one sentence: As societies have grown more complex, larger, more far-flung, and diverse, the tribe-based gift economy has shrunk in importance, while the trade economy has grown to dominate most aspects of people's lives, and has expanded in scope to encompass the entire planet.

The casualties from expanding production and trade have been piling up as never before. A recent report (WWF, 2020) claims that two-thirds of the world's wildlife has disappeared in the last half-century. Antarctic and Greenland ice sheets are melting at a frightening pace. Worsening wildfires have been devastating the western U.S., Siberia, and Australia. And the oceans appear to be dying as a result of acidification, oxygen loss, overfishing, and plastic pollution.

Can we continue to grow the economy, but do so harmlessly? The pathway to this goal, according to many mainstream environmental organizations, is to deploy alternative energy sources at sufficient scale so as to replace fossil fuels. There are two reasons to doubt that doing so would either enable continued growth or render our economy harmless.

First, can alternative energy sources maintain economic growth? Costs for solar panels and wind turbines have continually fallen. Unfortunately, however, the difficulties of a complete transition from fossil fuels to renewables cannot be boiled down to a question of cost per unit of electricity produced by solar versus coal. Renewables and fossil fuels are very different sources of energy, requiring different systems to manage and use them. Therefore, the transition will require a great deal of investment in infrastructure beyond panels and turbines themselves.

The intermittency of sunshine and wind imposes the need for energy storage technologies, for much greater redundancy of energy sources, for more robust transmission grids, and for infrastructure to turn electricity into fuels for technologies that will be hard to electrify (such as long-distance airplanes, big farm machinery, and high-heat industrial processes like cement making). All of these will cost energy as well as money, and in the early stages of the transition the energy will be coming mostly from fossil fuels. If other uses of energy are not curtailed somewhat, the energy transition will generate a burst of new carbon emissions. At least two independent analyses of the costs and opportunities of the transition from fossil fuels to alternative energy sources conclude that overall energy usage and commercial activity will have to decline, perhaps significantly, in order for the transition to succeed.[3]

Would alternative energy sources render the economy relatively harmless? Again, only if the economy were to shrink in size; in addition, some of its fundamental operational principles would need to be transformed. Climate change is not the only environmental harm being caused by the economy at its current scale. Resource depletion and pollution are inherent in extractive and productive processes that undergird industrial-scale production.

Creating a sustainable economy will require humanity to use renewable resources at less than their natural rate of replenishment, to extract and use nonrenewable resources only to the degree that they can be recycled, to clean up any mess as thoroughly as possible, and to avoid significant inequality so as to maintain social cooperation. This is a recipe for a modest society of gardeners, not a global industrial growth machine.

Thus, whatever economies emerge from the bottleneck are likely to be smaller, more locally organized, and wary of growth. Ideally, humanity could achieve this outcome without first passing through a period of collapse due to worsening climate change, resource depletion, agricultural failure, famine, conflict over dwindling habitable area, and population die-off. However, a relatively peaceful passage through the bottleneck would require policy makers to question basic assumptions about economic growth, the role and impacts of money and debt in society, and private ownership of natural resources.

NOTES

1. Portions of this chapter are drawn from the author's previously published work: *The End of Growth* (New Society, 2011), and "What If Preventing Collapse Isn't Profitable?" as well as *Power: On the Origins of Climate Change and Social Inequality* (in press).
2. For this concept I am indebted to the late William R. Catton, Jr. See his book *Bottleneck: Humanity's Impending Impasse.* Xlibris, 2009.
3. See Richard Heinberg and David Fridley, *Our Renewable Future: Laying the Path for 100 Percent Clean Energy.* Island Press, 2016. Full text available at www.ourrenewablefuture.org. Accessed September 2, 2020. See also Carey King, "An Integrated Biophysical and Economic Modeling Framework for Long-Term Sustainability Analysis: The HARMONY Model." *Ecological Economics*, Vol. 169, March 2020. https://doi.org/10.1016/j.ecolecon.2019.106464. Accessed September 2, 2020.

REFERENCES

Ewen, S. (2001). *Captains of Consciousness: Advertising and the Social Roots of Consumer Culture* (25th ed.). Basic Books.

Fix, B. (2020, July 11). Can the World Get Along Without Natural Resources? *Economics from the Top Down.* https://economicsfromthetopdown.com/2020/06/18/can-the-world-get-along-without-natural-resources/. Accessed July 5, 2021.

Graeber, D. (2001). *Toward an Anthropology of Value: The False Coin of Our Own Dreams.* Palgrave Macmillan.

Graeber, D. (2011). *Debt: The First 5000 Years.* Melvillehouse.

Harris, M., & Johnson, O. (2007). *Cultural Anthropology* (7th ed.). Pearson.
Heinberg, R. (in press). *Power: On the Origins of Climate Change and Social Inequality*.
Mauss, M., & Halls, W.D. (2000). *The Gift: The Form and Reason for Exchange in Archaic Societies*. W. W. Norton.
McNeill, J. R., & Engelke, P. (2014). *The Great Acceleration: An Environmental History of the Anthropocene Since 1945*. Harvard University Press.
McNeill, W. (1982). *The Pursuit of Power*. University of Chicago Press.
Nikiforuk, A. (2012). *The Energy of Slaves: Oil and the New Servitude*. Greystone.
Nordhaus, W. (1993). Reflections on the Economics of Climate Change. *Journal of Economic Perspectives*, 7(4), 11–25. https://doi.org/10.1257/jep.7.4.11
Nordhaus, W. D., & Tobin, J. (1973). Is Economic Growth Obsolete?. In W. Moss (Ed.), *The Measurement of Economic and Social Performance* (pp. 509–564). National Bureau of Economic Research.
Oxfam. (2017). *An Economy for the 99%: It's Time to Build a Human Economy that Benefits Everyone, Not Just the Privileged Few* [Briefing paper]. https://policy-practice.oxfam.org.uk/publications/an-economy-for-the-99-its-time-to-build-a-human-economy-that-benefits-everyone-620170. Accessed July 5, 2021.
Scott, J. C. (2017). *Against the Grain: A Deep History of the Earliest States*. Yale University Press.
Solow, R. (1974). The Economics of Resources or the Resources of Economics. *The American Economic Review*, 64(2), 1–14. http://www.jstor.org/stable/1816009
Wrangham, R., & Peterson, D. (1996). *Demonic Males: Apes and the Origins of Human Violence*. Houghton Mifflin.
WWF. (2020). *Living Planet Report 2020*. https://livingplanet.panda.org/en-us/

10. Precursors of an economics for the Anthropocene
Daniel Dahm and Günter Koch

THE BIG CHANGE DURING THE LAST THREE DECADES

In the last 30 years, the imagination of people in their world has changed fundamentally to one of sustainable development for a future worth living in. Since the Rio Conference in 1992, known as the "World Summit", which marked the beginning of a series of international follow-up conferences (Conference of Parties = COP) (UNCED, 1992), civil society and politics have initiated a lot of programmes. Never before has there been such a broad consensus that the common foundations of humanity's life require appreciation and protection. Also, the understanding that in a limited planetary sphere (ancient Greek: σφαῖρα sphaira = shell, ball) everyone depends on each other, that is, we cannot look any more at ourselves in isolation from the many others as has never been so pronounced in modern times. In fact, a true global consciousness has found its way into people's everyday thinking, and the media reflections of such cross-border ecological interdependence also reflects this. Even though economic conflicts, nationalisms and ethnic, religious and social cleavages dominate the news. Even though consumer society is globalising in the turmoil and doubts about this progress have rarely been so evident. In an increasingly complex (Koch, 2013) networked world, efforts are being made everywhere to create and promote a more peaceful and life-enhancing society and economy.

At the same time, we are learning that the speed with which the transformation to a sustainable society and economy is taking place is insufficient. Even faster than we are changing our minds as human beings, we are confronted with climatic and ecological, but also technological changes, most of which we can only understand in their basic features. In parallel we are experiencing a dramatic divide between poor and rich and international migrations on a historically unprecedented scale. In the first quarter of the 21st century, more and more people around the globe are witnessing an imminent turn of an era. The standards of social and economic success that have been taught and learnt and which many blindly followed, have so far promised a good life through diligence, work and assertiveness, but also through the accumulation of money and goods and the desire to continuously have more and more of it (Fromm, 1976).

As is pointed out in other chapters of this book, such traditional and learned beliefs can only be maintained today if disregarding the reality of life. The limits of consumerist lifestyles, the resilience of democratic societies and natural habitats are much evident. The side effects of material- and energy-intensive models of prosperity and their economic realisation are too obvious and drastic. Whether it is the destruction of the earth's ecological and climatic integrity, or the social, cultural and political consequences of this disastrous modernity – the triggering processes and drivers of the consumption of resources and ecological substance and the pollution and destruction of nature are well known and undeniable. Humanity's ideas and interpretations fail in the face of reality.

This socio-economic and political failure in terms of the possibilities and limits of use of the foundations of life follows a negative response to the previously preconceived claims of human development. The underlying social and political claims and fantasies now seem strangely backward-looking and even reactionary in a Biedermeier-style (Barea, 1992). At the very least, they are unrealistic and ideological, far removed from rationality and objectivity. The neoliberalistic economic theory and practice of the last decades and its foundations (Hayek,1944; Baird, 2008) perverted the enlightened idea of economy and markets and turned it upside down. What was supposed to create prosperity, freedom, justice and peace resulted in a sweeping attack on the living space of the living. Sustainability in the sense of future viability became the most important paradigm of human development before the measurable changes in the planetary ecosystem.

THE VENUE OF THE ANTHROPOCENE

The massive ecological effects of our activities have triggered a global change which, as enormous civilising challenges, demands a life-sustaining and life-enhancing cultivation of our common future. From a geological point of view, human activities and their effects are also changing the earth's climate, material, landscape and geochemical balance in such a dominant way and are shaping the morphogenesis of the landscape so drastically, that they can be traced in the sedimentation of the last centuries. It is appropriate to dedicate a separate geochronological section to this, the *Anthropocene*. This was expressed by the atmospheric chemist Paul J. Crutzen in his famous essay "Geology of mankind" published 2002 in the scientific magazine *Nature* (Crutzen, 2002). Even though Crutzen himself is not a geoscientist, his finding that human activities have left their mark deep in the stratigraphy (= geological layer sequences due to deposits) of the Earth's history is difficult for any geoscientist to dismiss.

The geological section of the Anthropocene is distinct from the Holocene, which began around 11,700 years ago with the retreat of the glaciers and has been detectable around the globe since around 1950. The main features of anthropogenically induced sedimentation are soot particles resulting from the burning of fossil fuels, whose emissions have been rising steeply since the 1960s.

The Venue of the *Anthroposphere*

With the emergence of man, a new sphere has risen in the last 1.4 million years, at first slowly and gently. Shaped by the diversity and simplicity of cultural meanings, social conceptions, political and economic strategies, cultural communities, institutions and practices, and much more, it represents a spatial abstract of the spiritual-cultural sphere of human beings, their exchanges and actions. It is the self-constituting space of culture in all its forms. "Culture or civilisation – in the broadest ethnographic sense of the word – is that complex whole which comprises knowledge, faith, art, morality, law, custom, practice and all other abilities and habits…" (Tylor, 1871, p. 1). In the cultural network of relationships between the immaterial-spiritual reality of planet earth and its material-bio-ecological foundation, the change in global living spaces is structured and made possible, which has been profoundly reshaping the earth's landscapes recently since the middle of the 20th century. Filled with cultural symbolisms and figures, interspersed with ideas, meanings and relationships, jointly invented and handed down traditions of the past, Homo Sapiens created a new, immaterial reality, which is constantly being re-negotiated and updated, artistically and scientifically documented and described, and socially, politically and economically realised. It is anthropogenic and deeply rooted within the bio-geo-sphere.

It can justifiably be called the anthroposphere, the genuine human sphere.

The anthroposphere is Homo Sapiens' collectively designed space for interpretation and creation, an immaterial sphere of the world that contains the spiritual cosmos of humanity. A new, genuinely creative spatial dimension, open in principle through information and permanent re-arrangement in itself, in which all living things are constantly re-designed and permeate each other. It is unlimited in its interrelationships and possibilities of interaction, quasi-infinite, but physically and spatially it is naturally limited to the planet. We experience the limitations of the anthroposphere both through the human physicality, which sets limits, and through the materially, biochemically and thermally connected, but materially, energetically and geographically limited habitats, which make up our home planet and shape it ecologically. Generated by intelligent individuals and communities, the anthroposphere is the result, expression and unintended reflection of the correlated experiences, sensations, ideas and perceptions of currently living and past generations in a diversely vital world. Its many aspects of design, which are constantly changing and re-interpreted, include human–nature relationships, hierarchies, gender roles, cultures of conflict and communication, languages and knowledge, beliefs, values and consumer habits, and much more.

Despite and by means of its immateriality, the anthroposphere causes the manifold interactions with the material-energetic processes and manifestations of the earth. It is cultural orientations and social instruments and institutions that create immaterial and material infrastructures and directly transforms the bio-geo-sphere through economic activities and value-added processes. Economic thinking and striving thereby depict processes of social negotiation, lifestyle ideas, teachings, knowledge, traditions, values and much more. It is the everyday welfare, our ideas of prosperity and consumption models, collaborations and hierarchical striving, the search for care, communication and individual and collective security and participation, our passions, dreams and visions and the perpetual striving for a good, fulfilled and happy life by which we individually and collectively shape our earth.

The anthroposphere confronts us as an im-material spatial entity, a spherically appearing sphere, which is composed and structured of diversely connected, individual, cultural, social, political, economic and ecological manifestations (a more correct denotation would actually be "im-manifestations"). It is polycentric in itself: each individual creates and shapes the anthroposphere in mutual exchange with other individuals and the bio-geo-spheric, ecological habitats that embed it. In its wholeness it is a reflection of all human individuals, communities, their identifications, ideas and cultures in all their plurality.

A REORIENTATION OF THE ECONOMY IN THE ANTHROPOSPHERE TOWARDS ECOLOGICAL STANDARDS

A reorientation of economic activity towards ecological principles such as diversity and difference, and freedom and solidarity is the necessary response to the climate-ecological and cultural-humanitarian situation of humanity. A peaceful, fair and liveable future can succeed, but this requires a clear turning away from the economic-political ideologies and feasibility fantasies of old 20th-century design. Economies and their agents must re-structure themselves to enable the global process towards sustainable development as required by UN's Sustainable Development Goals (UN SDGs). At the heart of such understanding of sustainability are the preservation and development of commons (= common goods).

Commons often do not receive the esteem they deserve but are perceived as unprotected and can be exploited with the goal to privatise their benefits for particular interests, while burdens are left in them and socialised. Such shifting (= externalisation) of damages and burdens, for example to other geographical areas as well as to future generations as, for example, will be to the commons, and the private accumulation of profits and benefits result in conflictual re-distribution dynamics that are counterproductive towards any sustainable development (Goydke & Koch, 2020).[1]

Si'ahl, Chief of the American-Indian tribe of Duwamish, in order to explain his lack of understanding of the idea of owning the world indigenous to us, in a speech addressed to the 14th President of the USA, Franklin Pierce, in 1855, formulated this as follows:

> How can one buy or sell heaven – or the warmth of the earth? This idea is strange to us. If we do not possess the freshness of the air and the glitter of the water – how can you buy it from us? […] Everything is connected. What attacks the earth, attacks the sons of the earth. Man did not create the fabric of life; he is only a fibre in it. Whatever you do to the fabric, you do to yourselves. (Si'ahl, 1855)

Sustainability requires intact and strong commons (Ostrom, 1990). They are the prerequisite for sustainable social developments, represent the basis of life and provide the production basis for economic activity, agricultural and forestry productivity and the long-term availability of resources, and can enable a peaceful, just and liveable future through sensitive and good management.

A BASIS FOR DEFINING ECOLOGICAL ECONOMICS

On the basis of an analysis of the man-made causes of the current climate crisis, we propose a new benchmark as an indicator for sustainable development: the Sustainability Zeroline. Our insights leading to this proposal are based on an empirical and then theorised foundation developed by us. We derive the need for and a proposal of a connecting sustainability benchmark from an analysis of a series of existing evaluation methods and instruments, in this chapter not further elaborated in depth, but explained elsewhere (Dahm, 2019). Our compound conclusion is envisaged to resume in a call for a re-adjustment of the political-regulatory framework and economic actions for a life-sustaining development of the material and immaterial foundations of life.

The analysis in the empirical dimension follows the measurable results of economic practice and entrepreneurial strategies. Its content is embellished by the hostile orientation of contemporary economics and the phenomenology of a serious crisis that is not only looming but has become acute due to the coronavirus pandemic at the beginning of this decade. With the proposal of a sustainability benchmark defined by us in this way, practical orientations for political and economic regulations, methods and argumentation are to be offered, as well as supporting a discursive objectification. Furthermore, strategic decisions for sustainable management are to be accompanied by arguments as well as practical impulses for re-structuring entrepreneurial and political practice are to be given. In addition to politics, the real economic, infrastructural and ecological foundation of companies, of investments and of industrial production are becoming key functions in the big transformation process in our days. The persistence of the status quo must be consistently overcome. For this purpose, a clear and sharp benchmark for sustainability as we propose is overdue.

The Sustainability Zeroline can be summarised and condensed into a pseudo-mathematical formula as follows:

$$(internalisation + compensation + good\ impact) - (externalisation\ of\ negative\ effects) \leq 0$$

As said, the methodological development and definition of the Sustainability Zeroline combines empirical analysis with the theoretical foundations of a series of studies and offers one of several alternative development paths for sustainability that we believe to be promising.

THERE IS AN ALTERNATIVE

The fact that the terms economy and ecology originate from the same etymological root *oikos* (ancient Greek: οἶκος = household, economic community) and refer to the same meaning, no longer plays a role in our times. The word eventually leads to a *"there is no alternative"* mentality – once proclaimed by then British Prime Minister Margaret Thatcher (Thatcher, 1980) and adopted in the early days of our decade by the then German Chancellor Angela Merkel and even more by her follow-up candidates. This mindset is hardly questioned anymore by a large number of citizens because it is understood as being normal. The economic model serving as the basis for the vast majority of today's economic theories remains unquestioned. The experience of such present becomes more and more dominant, fostered through informative media flooding. The experience of nature or historical memory is further pushed into the background, completely forgetting that already some 200 years ago a polymath, Alexander von Humboldt, argued on the risk of destroying nature by interfering into it thereby destructing its ecological balance (Jackson, 2019). The perception of the here and now determines the reflection of one's own and collectively experienced existence. In this way, cultural development can temporarily be frozen in fixed interpretations of reality which no longer allow for any alternative – in case of doubt, the earth is interpreted still to be flat. The actual interdependence of reality, which is in a state of constant change and transformation, moves out of sight until finally we fail due to the inevitable entropic gradient of a dynamically changing world, simply because we have forgotten our ability to transform and adapt. This becomes paradigmatic in the interplay of economic strategies and economic practice to the natural, ecological living environment, which is and remains both the basis of life and the basis of any production. Truth is that nature continues to create the basis for production in the earth's household, which people use, cultivate and consume. Economy and ecology are not at all contradictory rather than interdependent; they have never been separable; they are and remain intrinsically linked. According to Frithjof Capra "The interdependence of all life processes is the essence of all ecological relationships" (Capra, 1999, p. 344).

A BASIC REQUIREMENT: THE ELIMINATION OF NEGATIVE EXTERNALITIES

With the introduction of the concept of the Sustainability Zeroline, a normative demand is claimed.

Business organisations mediate directly between the interests of particular groups, individuals and communities, and at the same time influence the material and energetic (pre-)conditions of the bio-geo-sphere. Enterprises thus have a decisive moderating and shaping role. In cooperation with other actors, companies produce and mediate goods and flows of goods starting with agriculture and forestry, by resource extractions, by means of infrastructures, digital networks and services, and thus directly transform habitats, cultural forms and institutions. In the anthropogenic and bio-ecological complex of life, the economic functional areas and qualities become practically effective, are interrelated and result from each other.

The Sustainability Zeroline is followed by the need for the complete integration of all negative externalisations into the entrepreneurial processes, so that there is zero outsourcing of negative effects from an economic activity. It results in the need for radical internalisation of ecological costs and, for efficiency improvements, to reduce emissions. This consequence becomes much more effective in order to prevent negative effects on social systems, cultural areas and, especially, on ecosystems. It requires a consistent micro- and macroeconomic restructuring of infrastructures, institutions, investment strategies, entrepreneurial and investor performance measurement and the way profit expectations are to be met.

However, if one aims beyond this at sustainability, the mere exclusion of the externalisation of losses is not sufficient. In addition, it will become necessary to systematically aim at the externalisation of benefits as well as to align the corporate purpose and economic value-added process with such goals and to strategically restructure economic activity accordingly.

In practice, future-oriented restructuring of companies will require the comprehensive technological conversion, rehabilitation and renewal of the infrastructures for production, logistics and distribution. Particular emphasis is placed on maximising operational energy and material efficiency and resource and energy productivity. This includes closing the value-added cycles with recycling stages with as little loss as possible and, ideally, real eco-efficiency (Braungart & McDonough, 2002), in which the waste from the first recycling stage becomes food for the next. This complete internalisation of all production costs, including all non-monetarily measurable sources of added values, does not stop at the company's own production site, its production line and workshops, but affects all service providers, suppliers and underlying logistics chains, energy and resource sources, extraction stages and site connections.

The transformation process towards a technological and infrastructural restructuring takes time and is cost-intensive. However, it may be possible to offset these costs by fully or partially offsetting all ecological and socio-cultural costs and slowing down the process of progressive destruction. In this way, small-scale transformation steps can be taken without having to remain inactive in the shattering process of destroying our livelihoods and to continue to drive it forward. (Re-)Investment in the social and cultural commons, for example, in social cohesion, health, education, participation processes, art and (everyday) culture, is also an indispensable requirement, and they serve as appropriate compensatory investments. After all, commercial enterprises at their locations constantly use their services and revenues through their employees and supporting public, technological, structural, social and political-regulatory infrastructures.

Examples of entrepreneurial design elements of a sustainable design of value creation cycles alongside processes for production, supply chains and applications of produced goods and services are:

1. *Holistic*: Internalisation of the ecological footprint: Company balance sheets, accounting, services and products integrate the follow-up costs of production, use, recycling and return in a holistic way.
2. *Efficiency in material and energy use*: Energy and materials are used as sparingly and efficiently as possible throughout the production process, logistics chains and product life cycle, and are returned to the value chains, reused or recycled.

3. *High product quality*: Produced goods, services and preliminary products combine functional and aesthetic high quality, durability, ease of repair and updating.
4. *Positive social and cultural impact*: manufacturing conditions, working conditions and wage structure, their use and application promote social and cultural development and strengthen cultural commons, for example through working and wage conditions, corporate policy, cultural values, social institutions and fair-trading conditions.

The strategic reorientation of the objectives of companies and investments requires:

1. *Integrated business performance measurement* to enhance publication transparency, including all ecological and cultural effects and risks for investors, companies and issuers of financial market products.
2. *Integrated risk/return analysis and performance measurement* of capital investments, taking into account ecological and socio-cultural criteria, in order to arrive at new, future-oriented asset allocation assessments and strategies.
3. *Information and transparency:* manufacturing conditions, use of raw materials, emissions and compensation as well as usage characteristics and life-cycle costs are communicated transparently and in a way that is rich in information and promotes a sense of responsibility among partners, customers, buyers and consumers.

In total: new methods and tools in bookkeeping, balance reporting and planning in companies' performance and strategic planning is the claim for the future.

The clear identification of specific negative and positive external effects on the sustainability of business activities, on capital investments and on products and production not only serves the purpose of transparency. It is a prerequisite for carrying out adequate risk-return analyses and making sustainability-oriented entrepreneurial investments and asset allocation decisions for financial and portfolio analysis and for capital investment decisions, as well as for optimising entrepreneurial effects. The (slow) process of socio-economic re-cultivation requires changes in almost all economic practices and processes and their social integration. Each economic institution is unique in terms of the natural and social resources it uses, its employee structures, supplier and logistics chains, groups of actors involved and political guidelines, traditional trade relations, market connections and dependencies, requiring tailor-made transformation strategies. Business enterprises and actors that fail to re-structure and position themselves sustainably will be forced out of the market, as is already taking place. It is demonstrated with the attacks against the fossil and nuclear minded centralists of the energy industry, and against backward-thinking automobile corporations as is currently experienced following their strategic announcement, aiming at increasing their sales numbers.

In order to protect and build up the planetary commons through economic alliances and alliances of states, as well as in domestic markets, the necessary reorganisation of economic processes, of activities and offers is connected with enormous entrepreneurial efforts. To this end, the diversity and complementary cooperation of formal and informal institutions across sectors is a precondition and indispensable, as is the optimisation of the social, cultural and political environment. This requires the active communicative participation of all groups and institutions involved in the economic process, such as a well organised civil society, science, the media, but also through direct stakeholder communication with and in municipalities, countries and regions, resident companies and consumer groups. Important instruments for this are:

- *Transdisciplinary multi-stakeholder monitoring*: Monitoring and support of business development and transformation through consultation with civil society organisations, local and regional political bodies and authorities, directly and indirectly involved business enterprises and associations, experts, science, art and culture, etc.
- *Strengthening of locations through multi-stakeholder participation and deliberative and collaborative democracy*: All stakeholder groups affected by economic activities are involved in strategic planning processes, implementation decisions and ongoing business operations at the locations and across sectors. The socio-economic and socio-cultural strengthening of the locations also becomes an economic objective.
- *Democratisation of economic governance and the production bases*: Redesign of the entrepreneurial environment, establishment of capital participation models, types of companies and forms of organisation that strengthen and develop locations socio-economically through their participation structures.

Only in trans-sectoral alliances involving all groups of actors can a sustainable development of the damaged material and immaterial commons be achieved.

METRICS, MEASURABILITY AND EVALUATIONS

The areas of application for the operationalisation of the Sustainability Zeroline extend across all economic actors and their economic activities, production processes and value chains right through to their goods, products and services. It can also be used for the analysis of national economies, state alliances and economic regions up to municipalities and communities and their infrastructures.

The sample indicators for the bio-geo-sphere and anthroposphere differ in their variance, characteristics, quantifiability and qualifiability. They are characterised by their numerous interdependent relationships. As practical as a uniform metric would be means all indicators could be recorded and quantified together in common survey procedures, this would be of little use, because the differences between the indicators cannot be determined without significant qualitative losses. This is particularly obvious for indicators describing anthroposphere impacts. Indicators are, for example, freedom, peace or the arts and are hard to map in quantitative terms. In the bio-geo-sphere this applies to the majority of domains of analysis, because neither biodiversity nor food chains and ecosystems by current methods can be reasonably quantified. However, there is sufficient empirical evidence and methodological diversity for all groups of indicators to be able to assign sufficient evaluation approaches and examples to each indicator. The decision on metrics and quantification options for indicators requires detailed expertise of all those institutions and people who have been dealing with the details of related measuring methods, quantification models, recording systems, and so on in a thematically profound and differentiated manner for many years; experts who are particularly well versed in these areas. This is a point of concern with regard to existing criteria, ratings and evaluation methods, which – with their extensive pool of individual criteria and accounting experience – provide comprehensive connections to existing methodologies and accounting instruments, rankings, ratings and indices. The Sustainability Zeroline does not aim to compete with any existing methods and instruments. It aims to eliminate the lack of a binding and consistent benchmark and reference points for sustainability.

In contrast: The existing methodologies and indicator systems will be supported and strengthened by the Sustainability Zeroline model and it will sharpen the focus on their specific characteristics and advantages. It is only through their reference to a reference brand such as Sustainability Zeroline that the individual evaluations achieve their specific relevance. Their specific meaning can be related to the overall evaluation and classification of the respective business enterprise. The economic analysis becomes more consistent and clearer, the details and individual criteria lead to more depth and strategic guidance. This is an expectation triggered by recent initiatives taken by big economic players, as for example, the current attempts to develop new reporting standards such as the Value Balancing Alliance in Europe (VBA 2020).

AN OUTLOOK TOWARDS THE CREATION OF AN ECONOMY OF SUSTAINABILITY

Clarity in the assessment of our actions in our world, based on a strong common, analytical agreement, allows for a strategically challenging and constructive struggle to achieve what we stand for: a liveable, living world that enables a peaceful and just future in peace and mutual respect for all people. The Sustainability Zeroline is an opportunity to sharpen and strengthen a process of transformation towards a world in which human development and living evolution are not contradictory but mutually re-enforcing.

The Sustainability Zeroline above all is an occasion, a food for thought and a – perhaps consensus-based – starting point for debates and discussions. And this may also be because the theoretical basis of Sustainability Zeroline is so strangely simple and banal that one must wonder what the wild discussions at the hundreds of congresses, conferences, symposia and expert talks today are all about? By this statement, the authors also provide an impetus for the programmatic work of the institutional author of this book: The World Capital Institute (WCI 2020).

NOTE

1. See also Treaty of Point Elliott, 1855, HistoryLink.org (accessed July 2021).

REFERENCES

Baird, C. (2008). Mont Pelerin Society. In Hamowy, Ronald (ed.), *The Encyclopedia of Libertarianism*. Thousand Oaks, CA: SAGE; Cato Institute. pp. 342–343. doi:10.4135/9781412965811.n210. ISBN 978-1412965804

Barea, I. (1966, republished 1992). *Vienna: Legend and Reality*. Chapter on "Biedermeier", pp. 111–188. London: Faber & Faber.

Braungart, M., McDonough, M. (2002). *Cradle to Cradle*. London: Vintage.

Capra, Frithjof (1999). *Lebensnetz: Ein neues Verstaendnis der lebendigen Welt*. Munich: Droemer Knaur, p. 344.

Crutzen, P. J. (2002). Geology of mankind. *Nature* 415 (23). https://doi.org/10.1038/415023a. Quoted in Paul J. Crutzen: *A Pioneer on Atmospheric Chemistry and Climate Change in the Anthropocene* (pp. 211–215). Cham: Springer.

Dahm, D. (2019). *Benchmark Nachhaltigkeit – Sustainability Zeroline. Das Maß für eine zukunftsfähige Ökonomie*. Bielefeld: Transcript Verlag.

Fromm, E. (1976). *Haben oder Sein – Die seelischen Grundlagen einer neuen Gesellschaft*. Munich: dtv Verlagsgesellschaft.

Goydke, T., Koch, G. (eds) (2020). *Economy for the Common Good: A Common Standard for a Pluralist World?* Hamburg: Tradition Publisher.

Hayek, F.A. (1944/1994). *The Road to Serfdom*. Chicago: University of Chicago Press.

Jackson, S. (2019). Humboldt for the Anthropocene, *Science* 365 (6458), 1074–1076, 2019: doi:10.1126/science.aax7212.

Koch, G. (2013). It's too complex. Blog contribution to the 5th Global Peter Drucker Forum, https://www.druckerforum.org/blog/its-too-complex/ (viewed 15 January 2021).

Ostrom, E. (1990). *Governing the Commons: The Evolution of Institutions for Collective Action.* Cambridge: Cambridge University Press.

Si'ahl (1855). https://www.duwamishtribe.org/chief-siahl (viewed July 2021).

Tylor, E.B. (1871). *Primitive Culture.* Cambridge: Cambridge University Press.

Thatcher, M. (1980). Speech at Conservative Party conference, https://www.margaretthatcher.org/document/104431 (viewed July 2021).

UNCED (1992). UN Conference on Environment and Development: Rio Declaration and other documents.

UN SDGs. https//:sdgs.un.org (viewed July 2021).

VBA (2020). Value Balancing Alliance, https://www.value-balancing.com/ (viewed January 2021).

WCI (2020). World Capital Institute, https://www.worldcapitalinstitute.org/ (viewed December 2020).

11. Deep adaptation and collapsology
Jason Monios and Gordon Wilmsmeier

DEEP ADAPTATION ON A CONTINUUM OF RESPONSES

In 2020, climate change and environmental concerns such as extreme weather and biodiversity loss were perceived as the top five threats facing global society in terms of likelihood, "the first time in the survey's history that one category has occupied all five of the top spots" (WEF, 2020: 4). Yet, over decades we have either tacitly accepted climate change or fuelled it with denial and scepticism on the back of well-coordinated campaigns with funding from industry and free-market think tanks (Anderreg, 2010). The current discourse revolves around fine-sounding rhetoric of multilateral and multi-stakeholder coordination and the "win–win" of market-based instruments, even as the annual UN Climate Change Conferences begin to reveal political fractures.

Lately there has been a growth in distracting noise regarding whether collapse is inevitable or just likely, and whether that means a total collapse of human life or of human civilisation, or just individual collapses of certain systems, and whether these are being caused purely by climate change or by other related problems such as resource depletion. Even though we cannot predict the future with certainty, the leading experts on climate change and planetary futures believe that we are currently on course for collapse. Professor Will Steffen believes that we have passed 9 of the 15 known global tipping points (Lenton et al., 2019) and remarked:

> Given the momentum in both the Earth and human systems, and the growing difference between the 'reaction time' needed to steer humanity towards a more sustainable future, and the 'intervention time' left to avert a range of catastrophes in both the physical climate system (e.g. melting of Arctic sea ice) and the biosphere (e.g. loss of the Great Barrier Reef), we are already deep into the trajectory towards collapse. (Moses, 2020)

Professor Joachim Schellnhuber concurs: "I think there is a very big risk that we will just end our civilisation. The human species will survive somehow but we will destroy almost everything we have built up over the last two thousand years. I am pretty sure" (Moses, 2020).

Agreeing that we are on a pathway to at least some level of collapse does not invalidate the importance of mitigation efforts, partly because many actions are relevant for both mitigation and adaptation. For example, switching to renewables is necessary for mitigation as it reduces emissions, but also for adaptation as it reduces dependence on long distance energy supply chains. However, it does mean that we cannot focus only on mitigation. Kenis and Mathijs (2014: 151) argue that "the focus on CO_2 has narrowed the debate to ignore the human-societal root causes and processes of change and led to a focus on technical solutions that remain within the parameters of what currently exists and is convenient. Such discourses have depoliticizing and disempowering consequences." Given that some level of collapse is certain, we should work towards both mitigating its severity as well as adapting to a changed world. But the danger is that we focus only on small adaptations premised on a world not very different to the one we currently inhabit rather than accepting the need for radically changed systems. The current COVID-19 pandemic is instructive, in that the much-touted reduction in greenhouse gases as a result of reduced industry and traffic in April and May 2020 is now only a memory. Governments poured billions into supporting airline and automotive industries rather than using the disruption to push towards decarbonising and reducing transport.

This chapter discusses the recent concepts of collapsology and deep adaptation. Deep adaptation may be placed on a continuum stretching from mitigation to adaptation to transformational adaptation to deep adaptation (Monios and Wilmsmeier, 2020). Mitigation seeks to prevent climate change by reducing carbon emissions, while adaptation aims to respond to climate effects, regardless of whether and to what degree they are mitigated. The distinctions on the continuum arise because "regular" adaptation assumes smaller changes such as increasing coastal defences, whereas the next level of transformational adaptation goes further, envisioning major structural changes such as relocating or abandoning infrastructure but still expecting to maintain current systems. Deep adaptation can be defined as adaptation predicated upon collapse, where current systems collapse in a short timescale in chaotic and unpredictable ways (Bendell, 2018; Read, 2019). Under this scenario, maintaining current practices is likely to become impossible, therefore adaptation to such changes will involve forced and unplanned transitions.

CLIMATE CHANGE, LIMITS TO GROWTH AND NONLINEARITY

One limitation of approaches to climate change is that arguments about predictions of temperature rise and carbon in the atmosphere ignore other triggers of collapse relating to resource depletion (being also of course related to and accelerated by climate change). *The Limits to Growth* (Meadows et al., 1972) analysed the interrelation of key systems, resources and trends, and remains relevant today due to the accuracy of its predictions. The authors concluded that:

> If the present growth trends in world population, industrialisation, pollution, food production, and resource depletion continue unchanged, the limits to growth on this planet will be reached sometime within the next one hundred years. The most probable result will be a rather sudden and uncontrollable decline in both population and industrial capacity. (p. 23)

Turner (2014) plotted recent data against the report's 1972 predictions and showed that their forecasts for all the key indicators are roughly correct. Turner was quoted in a recent article saying: "there's an extremely strong case that we may be in the early stages of a collapse right at the moment" (Moses, 2020). Many of these effects are already becoming evident (cf. the "great acceleration" – Steffen et al., 2004), and will have profound effects on society long before anything that could be defined as a collapse. The key issue is to recognise these events currently happening as part of a trajectory towards collapse requiring major change rather than trying to adapt to each of these events in a piecemeal fashion.

While much focus on climate adaptation revolves around direct impacts of weather and sea level rise, the indirect effects are already being felt. Globally, 145 million people live within one metre above the current sea level (Anthoff et al., 2006) and almost 1 billion people live in low-elevation coastal zones (Neumann et al., 2015), meaning that as regions become difficult to inhabit, massive migrations are to be expected. Already in 2017, 18 million people were made homeless by weather events (Internal Displacement Monitoring Centre, 2018). Other related effects will be droughts (the global demand for fresh water by 2050 is expected to increase by 55%, with 40% of the world's population living in areas of severe water stress – OECD, 2012) and crop failures. The International Organization for Migration estimates 200 million climate refugees by 2050 (IOM, undated). According to the World Bank (2018), climate change will push tens of millions of people to migrate at least within their own countries by 2050. These upheavals will certainly produce localised conflict at least, which will further disrupt global supply chains thus impacting countries all around the world. The disruptions of the port of New York due to Hurricane Sandy and the port of New Orleans from Hurricane Katrina resulted in damage to port-related infrastructure and businesses in the wider area that reached into the billions. If the IPCC is correct that such once-per-century storms will become once-per-year storms by 2050, such levels of disruption and cost must be expected (Monios and Wilmsmeier, 2020).

Another indirect impact frequently overlooked is the opportunity cost of climate adaptation. A UN report recently calculated that by 2050 poorer countries would require at least USD 500 billion per year to adapt (United Nations, 2016) but even rich nations are already facing stark choices. According to Goodell (2018), since 1989 the US federal government has spent USD 2.8 billion purchasing 40,000 homes from residents in endangered areas to allow them to move elsewhere. It is estimated that 300,000 homes in the US housing 550,000 people and worth USD 117.5 billion and 14,000 commercial properties worth USD 18.5 billion are at risk of chronic flooding by 2045 (Union of Concerned Scientists, 2018). Governments cannot afford to rehouse so many citizens nor protect such long coastlines, and the money that is spent will reduce availability for other public spending, contributing to crises in other key areas such as healthcare.

BETWEEN AWARENESS, EMPTINESS AND ACTION: CLIMATE MAINSTREAMING AND POST-POLITICS

Have we finally reached the moment where the reality of anthropogenic climate change is "dislocating" (Methmann, 2010) the modern narrative of fossil-fuel-driven economic growth? Climate doublethink is rife among policy makers who present decarbonisation policies alongside plans for new coal mines. Rather than reforming the current system, we are offered variants such as stakeholder capitalism (Schwab, 2019) and green growth (Wilshusen and MacDonald, 2017). Methmann (2010: 346) describes the phenomenon of "climate mainstreaming", defined as "the integration of climate change into business balance sheets". In this context, climate protection functions in a similar manner to the way sustainable development has been understood as being transformed into an empty signifier. As Methmann (2010: 349) puts it, an assemblage of "rather heterogeneous and contradictory policies – a label for greenwashing par excellence". Wullweber (2015: 80) notes that:

> the basic assumption underlying the concept of empty signifier is that the general interest is not objectively given. Rather, it arises out of a political process in which people seek to determine the positions they perceive as important and appropriate. The general interest exists merely as an imaginary state. However, precisely because such an interest is not given per se but constitutes an empty space; political actors can compete to fill that empty space.

Methmann (2010) argues that this process of mainstreaming makes it possible "to integrate climate protection into the global hegemonic order without changing the basic social structures of the world economy" and "to make fossil-fuel-based growth and free trade appear as part of the solution". Thus, the empty signifier allows organisations to speak of climate protection as a real movement; however, "the reference to climate protection spreads, but climate protection itself changes its meaning and becomes ambiguous. Institutions mostly rephrase existing activities and goals in the terms of climate protection but in reality, almost never change them according to the challenges of global warming" (Methmann, 2010: 346).

A serious problem therefore exists whereby the co-option of climate discourse by economic language prevents climate action. A decade ago, Swyngedouw (2010) asked if the institutions of global economic governance would be able to support dislocated social and economic structures and thus contest the challenges of climate change. The reason for this disconnect is that the real victory of neoliberalism had been to remove politics from economics, or indeed replace the former with the latter (Davies, 2017). Swyngedouw (2007: 13) argues that "environmental issues and their political 'framing' contribute to the making and consolidation of a post-political and post-democratic condition, one that actually forecloses the possibility of a real politics of the environment".

Swyngedouw (2007: 24, 26), drawing on Žižek (1999) and Mouffe (2005), defines the post-political as a "political formation that actually forecloses the political, that prevents the politicization of particulars". Therefore, post-politics "permits the politicization of everything and anything, but only in a non-committal way and as non-conflict. Absolute and irreversible choices are kept away; politics becomes something one can do without making decisions that divide and separate." Swyngedouw argues that both climate deniers and those who argue for radical change are treated equally as fundamentalists. Thus, the only accepted "solutions" are to be found within the neoliberal hegemony and no alternatives that challenge this system are considered.

As a result of the failure of neoliberalism and post-politics to respond to the climate crisis, and the mainstreaming and co-opting of climate discourse through market-based measures, a need has emerged for a radical response. Seemingly up to now climate change impacts have not sufficiently challenged the status quo, at least not of those groups that continue with the same hegemonic discourse, which stands in contrast to the vulnerable parts of society that live, experience and suffer climate change on a daily basis but without sufficient voice. Swyngedouw (2007: 35) argues that:

> To the extent that the current post-political condition, which combines apocalyptic environmental visions with a hegemonic neoliberal view of social ordering,

constitutes one particular fiction (one that in fact forecloses dissent, conflict, and the possibility of a different future), there is an urgent need for different stories and fictions that can be mobilised for realisation. This requires foregrounding and naming different socio-environmental futures, making the new and impossible enter the realm of politics and of democracy, and recognizing conflict, difference, and struggle over the naming and trajectories of these futures.

In this chapter we argue that it is as a response to this need for radical new stories that recent efforts to delineate deep adaptation and collapsology have emerged.

DEEP ADAPTATION AND COLLAPSOLOGY

Transformational adaptation was described above as a step beyond incremental adaptation. It requires fundamental changes to the nature of a system (Park et al., 2012; Mushtaq, 2018), likely to involve a much larger scale of response, actions that are truly new to a system and that transform places and shift locations (Kates et al., 2012). While transformational adaptation does necessitate moving beyond a business-as-usual approach, it nevertheless tends to assume a relatively high degree of continuation of current practices, essentially transforming the system to maintain functionality. Deep adaptation, by contrast, goes beyond these questions to consider the breakdown of functionality (Monios and Wilmsmeier, 2020).

The original paper introducing the concept of "deep adaptation" was published by Bendell in 2018 and has since been downloaded hundreds of thousands of times. A succinct definition was later proposed by Read (2019) as "adaptation premised upon collapse". Bendell (2018: 11) writes that "The evidence before us suggests that we are set for disruptive and uncontrollable levels of climate change, bringing starvation, destruction, migration, disease and war". These changes can include physical changes (pulling back from the coast, closing climate-exposed industrial facilities, planning for food rationing, letting landscapes return to their natural state) as well as cultural shifts, including "giving up expectations for certain types of consumption". The concept is a clear attempt to break with the business-as-usual logic dominating climate discourse, to achieve "an end of the idea that we can either solve or cope with climate change". Some critics have argued that an inevitable collapse cannot be proven and that talking in this way encourages a lack of response to the climate emergency, thus Bendell has updated the original paper to modify the language a little around the point of inevitability. But it seems clear that some level of collapse is inevitable, even if it cannot be predicted accurately, and to ignore this point veers dangerously close to an unfounded belief in technological utopianism. The main authors of the collapsology movement (see below) have recently published a response to critics of deep adaptation arguing that

> the house fire isn't certain, but because you take it seriously (it certainly *can* happen) you act accordingly. And if you act, then it is less likely to happen. In other words, we better take societal collapse for granted to have any chance of avoiding it or, at least, reducing its worst effects. (Servigne et al., 2020)

The key difference between transformational and deep adaptation is the notion of collapse – several of the top climate scientists quoted above agree that it is likely that we are already on the path to collapse. Thus, a parallel can be drawn with the (mostly Francophone) body of work on "collapsology" (Servigne & Stevens, 2015) which has been in existence since before the advent of deep adaptation. This notion of collapse is essential in order to avoid the positive connotations of "transition". Servigne and Stevens (2015) argue that it is inappropriate to use terms such as "transform" or "transition" to describe this level of upheaval that could be on the scale of hundreds of millions of deaths. The term "collapse" does not imply that human civilisation will end but that the upheaval will be on a par with a world war, a collapse of our current hyper-connected systems.

Collapsology has faced criticism as being a gentrification of survivalism, whereas we would say that collapsology is more collegial. It aims to understand the threats and act on them, changing society in such a way as to avoid as far as possible but then live with the consequences. It is not about individualism and running away to a secret bunker. Survivalism is accepted by the media because it fuels more consumption and spending whereas collapsology is mocked because it aims to reduce consumption and transition towards a more sustainable society.

RESPONSES TO VULNERABILITY: RESILIENCE TO REDUNDANCY?

The various impacts expected under climate change reveal the need for interconnected thinking, as we may prepare to adapt to certain risks such as sea level rise by adapting transport infrastructure, but related risks such as changing demand and supply chain ruptures or migration and conflict may equally threaten the economic viability of this infrastructure. We need to think beyond system resilience, which focuses too much on returning to "normal", and consider system change, such as prioritising more local systems of energy and food supply. Even remaining within the resilience paradigm, many of the key resilience recommendations, such as "distributed decision making, modularity, redundancy, ensuring the independence of component interactions" (Linkov et al., 2014) are ignored by decision makers. There is too much focus on making infrastructure stronger and putting planning systems and training in place, but very little focus on transforming systems, which would mean reducing their efficiency. In the race towards efficient systems and economies of scale, we are simply piling up risk and reducing resilience. This model was considered appropriate in the last couple of decades which were more stable politically and environmentally, but now we are in a "Minsky moment": the stability of recent times has led us to take greater risks because we think this stability will continue. Resilience belongs with "regular" adaptation but does not go far enough towards deep adaptation and collapsology.

Manheim (2020) draws on the "vulnerable world" hypothesis of Bostrom (2019) to create the "fragile world" hypothesis: "if technological development continues indefinitely, systemic fragility will increase to the point that the possibility of a shock sufficient for complete collapse approaches certainty". He argues that "collapse risk requires that failures are rare enough that technological growth allows building systems far more complex than are maintainable or replaceable before failure occurs". There is too much tendency currently to focus on individual locations (e.g. cities) or systems (e.g. a national grid) but insufficient focus is given to the interconnectedness of systems.

A recent interview with Professor Joachim Schellnhuber highlighted the importance of redundancy vs efficiency in adapting to climate change. He raised the importance of nonlinearity and the inability to predict outcomes, meaning that we cannot continue to focus on efficiency if we want to create innovative responses to climate change. Inevitably some solutions will fail but the urgency of the climate crisis means that we must employ several different solutions, knowing that only some will succeed:

> efficiency is the enemy of innovation. You have to strand assets, you have to waste capital, because you invest into the wrong things, because you cannot know beforehand. But you also invest in the right things. So, we have to have the courage to squander money. To throw money at things that have potential. It is venture-capital at a global scale we have to muster. We cannot efficiently get ourselves out of this predicament. So, we have to save the world but we have to save it in a muddled way, in a chaotic way, and also in a costly way. That is the bottom line, if you want to do it in an optimal way, you will fail. (Breeze, 2019)

CONCLUSION: TOWARDS DEGROWTH?

The discourses of deep adaptation and collapsology are a response to the notion that "there is no alternative", thus can be seen as an attempt in today's post-political atmosphere to finally articulate a possible alternative to the current hegemonic order. Swyngedouw (2010) has criticised the post-political rhetoric fetishising climate change as the external enemy of humankind.

According to Kenis and Mathijs (2014: 150), the result is that "dissensus is obliterated, the contingency of the present is rendered invisible and alternative voices remain unheard." Deep adaptation and collapsology are working to make climate change politically visible once again. Drawing on Rancière, Kenis and Mathijs (2014) point out that visibility requires a critical mass to be able to genuinely repoliticise the movement. On the other hand, these two concepts "risk being perceived as politicizing for politicization's sake, which can end in a paradoxical situation in which this politicization is not taken seriously, thus undermining its own goal" (Kenis and Mathijs, 2014: 155). This is the risk that leads some authors to criticise deep adaptation and collapsology as "doomists".

We can therefore understand deep adaptation and collapsology as a diagnosis that allows us to re-evaluate what is at stake and provide us with concepts and tools to repoliticise and begin to address the acute threat. Kenis (2019) cites Mouffe (2002: 5) that "the political in its antagonistic dimension cannot be made to disappear simply by denying it, by wishing it away" and argues that the same is true for the post-political. It must be understood in order to change it. In an analogous manner, while critics may argue that deep adaptation and collapsology are "doomism" that drain hope from environmental movements, we would argue the reverse – that it is only by accepting the severity of the current situation and the strong likelihood of at least some level of collapse, that we can accept the need for radical system change. If we accept that some level of collapse and radical change is coming anyway, the most ethical reaction is to try to make these changes voluntarily and in as equitable a manner as possible.

A full programme for deep adaptation and collapsology can be found by consulting the key authors referenced above. However, three key pillars can be highlighted here by way of conclusion. First, decarbonise existing energy use as far as possible. While some moves in this direction are evident, they tend to focus on maintaining current systems such as widespread automobility but merely switching to alternative technologies such as electricity or hydrogen. The problem is that there is insufficient supply to cover our existing needs, such as industry, buildings and transport, 80% of which are currently supplied by fossil fuels, and if future growth is added – currently predicted at 50% growth by 2040 – then potential supply is far short (ETC, 2017). Therefore, the second key pillar is a corollary of the first – the need to plan for degrowth (Kallis et al., 2018), to reduce energy need as far as possible and reduce exposure to vulnerable supply chains. Kallis (2019: 271) remarks that "any kind of economic growth (meaningfully defined), whether capitalist or not, is materially and ecologically unsustainable". Green growth is an illusion that needs to be avoided as there is no evidence of absolute decoupling of emissions and growth (Hickel and Kallis, 2019; Vadén et al., 2020). The third pillar is also related, which is to relocalise as far as possible key systems such as energy, food and water. We need to build in redundancy and modularity. This does not mean an end to globalisation but a reduction and rebalancing. All of these steps currently appear politically difficult, even though they will create jobs and enhance national security which are usually attractive to politicians. Yet the lack of action during the opportunity of COVID-19 underlines this challenge. Thus, by learning from the work on deep adaptation and collapsology, we can appreciate the necessity of undertaking these transitions before they are forced upon us. The question is whether these new movements can overcome the co-opting of climate rhetoric by post-politics.

REFERENCES

Anderreg, W. (2010). The ivory lighthouse: communicating climate change more effectively. *Climatic Change, 101* (3–4), 655–662.

Anthoff, D., Nicholls, R. J., Tol, R. S. J., & Vafeidis, A. T. (2006). *Global and regional exposure to large rises in sea-level: A sensitivity analysis.* (Working paper 96). Tyndall Centre for Climate Change Research.

Bendell, J. (2018). *Deep adaptation: a map for navigating climate tragedy* (IFLAS Occasional Paper 2). IFLAS. http://insight.cumbria.ac.uk/id/eprint/4166/

Bostrom, N. (2019). The vulnerable world hypothesis. *Global Policy, 10* (4), 455–476.

Breeze, N. (2019, 3 January). It's nonlinearity – stupid! *The Ecologist.* https://theecologist.org/2019/jan/03/its-nonlinearity-stupid

Davies, W. (2017). *The Limits of Neoliberalism: Authority, Sovereignty and the Logic of Competition.* Sage.

Energy Transitions Commission (ETC). (2017). *The future of fossil fuels: How to steer fossil fuel use in a transition to a low carbon energy system.* ETC. https://www.energy-transitions.org/publications/the-future-of-fossil-fuels/

Goodell, J. (2018). *The Water Will Come: Rising Seas, Shrinking Cities and the Remaking of The Civilized World.* Black Inc.

Hickel, J., & Kallis, G. (2019). Is green growth possible? *New Political Economy, 25* (4), 469–486.

Internal Displacement Monitoring Centre. (2018). *Global Report on Internal Displacement 2018.* IDMC.

International Organization for Migration (IOM). (undated). *A Complex Nexus.* IOM. https://www.iom.int/complex-nexus#estimates

Kallis, G. (2019). Capitalism, socialism, degrowth: a rejoinder. *Capitalism Nature Socialism, 30*(2), 267–273.

Kallis, G., Kostakis, V., Lange, S., Muraca, B., Paulson, S., & Schmelzer, M. (2018). Research on degrowth: annual review of environment and resources. *Annual Review of Environment and Resources, 43,* 291–316.

Kates, W. R., Travis, R. W., & Wilbanks, J. T. (2012). Transformational adaptation when incremental adaptations to climate change are insufficient. *Proc. Natl. Acad. Sci. (PNAS), 109,* 7156–7161.

Kenis, A. (2019). Post-politics contested: why multiple voices on climate change do not equal politicisation. *Environment and Planning C: Politics and Space, 37* (5), 831–848.

Kenis, A., & Mathijs, E. (2014). Climate change and post-politics: repoliticising the present by imagining the future? *Geoforum, 52*, 148–156.

Lenton, T. M., Rockström, J., Gaffney, O., Rahmstorf, S., Richardson, K., Steffen, W., & Schellnhuber, H. J. (2019). Climate tipping points – too risky to bet against. *Nature, 575*, 592–595.

Linkov, I., Bridges, T., Creutzig, F., Decker, J., Fox-Lent, C., Kröger, W., Lambert, J. H., Levermann, A., Montreuil, B., Nathwani, J., Nyer, R., Renn, O., Scharte, B., Scheffler, A., Schreurs, M., & Thiel-Clemen, T. (2014). Changing the resilience paradigm. *Nature Climate Change, 4* (6), 407–409.

Manheim, D. (2020). The fragile world hypothesis: complexity, fragility, and systemic existential risk. *Futures, 122*, 102570.

Meadows, D. H., Meadows, D. L., Randers, J., & Behrens, W. W. (1972). *The Limits to Growth*. Potomac Associates.

Methmann C. P. (2010). Climate protection as empty signifier: a discourse theoretical perspective on climate mainstreaming in world politics. *Millennium Journal of International Studies, 39* (2), 345–372.

Monios, J., & Wilmsmeier, G. (2020). Deep adaptation to climate change in the maritime transport sector – a new paradigm for maritime economics? *Maritime Policy & Management*. In press.

Moses, A. (2020, 16 September). 'Collapse of civilisation is the most likely outcome': top climate scientists. *Voice of Action*. https://voiceofaction.org/collapse-of-civilisation-is-the-most-likely-outcome-top-climate-scientists

Mouffe, C. (2002). *Politics and Passions. The Stakes of Democracy*. CSD Perspectives.

Mouffe, C. (2005). *The Return of the Political*. Verso.

Mushtaq, S. (2018). Managing climate risks through transformational adaptation: economic and policy implications for key production regions in Australia. *Climate Risk Management, 19*, 48–60.

Neumann, B., Vafeidis, A. T., Zimmermann, J., & Nicholls, R. J. (2015). Future coastal population growth and exposure to sea-level rise and coastal flooding – a global assessment. *PLOS ONE, 10* (3), e0118571.

OECD. (2012). *OECD Environmental Outlook to 2050: The Consequences of Inaction*. OECD.

Park, S. E., Marshall, N. A., Jakku, E., Dowd, A. M., Howden, S. M., Mendhamf, E., & Fleming, A. (2012). Informing adaptation responses to climate change through theories of transformation. *Global Environmental Change, 22*, 115–126.

Read, R. (2019, 8 February). Climate change and deep adaptation. *The Ecologist*. https://theecologist.org/2019/feb/08/climate-change-and-deep-adaptation

Schwab, K. (2019, 2 December). What kind of capitalism do we want? *Project Syndicate*. https://www.project-syndicate.org/commentary/stakeholder-capitalism-new-metrics-by-klaus-schwab-2019-11

Servigne, P., & Stevens, R. (2015). *Comment Tout Peut S'Effronder*. Éditions de Seuil.

Servigne, P., Stevens, R., Chapelle, G., & Rodary, D. (2020, 3 August). Deep adaptation opens up a necessary conversation about the breakdown of civilisation. *Open Democracy*. https://www.opendemocracy.net/en/oureconomy/deep-adaptation-opens-necessary-conversation-about-breakdown-civilisation/?fbclid=IwAR2q-eknkMJlDguzBywtNqXdG7OO1fV3dNaQjXN2waxr5BjjaTkOiQvyTmI

Steffen, W., Sanderson, R. A., Tyson, P. D., Jäger, J., Matson, P. A., Moore III, B., Oldfield, F., Richardson, K., Schellnhuber, H.-J., Turner, B. L., Wasson, R. J. (2004). *Global Change and the Earth System: A Planet Under Pressure*. Springer.

Swyngedouw E. (2007). Impossible "sustainability" and the postpolitical condition. In R. J. Krueger & D. Gibbs (Eds.), *The Sustainable Development Paradox: Urban Political Economy in the United States and Europe* (pp. 13–40). Guilford Press.

Swyngedouw, E. (2010). Apocalypse forever? Post-political populism and the spectre of climate change. *Theory, Culture and Society, 27* (2/3), 213–232.

Turner, G. (2014). *Is Global Collapse Imminent?* (MSSI Research Paper No. 4). Melbourne Sustainable Society Institute, The University of Melbourne.

Union of Concerned Scientists (2018). *Underwater: Rising Seas, Chronic Floods, and the Implications for US Coastal Real Estate.* Union of Concerned Scientists. https://www.ucsusa.org/global-warming/global-warming-impacts/sea-level-rise-chronic-floods-and-us-coastal-real-estate-implications

United Nations. (2016, 10 May). UNEP report: Cost of adapting to climate change could hit $500B per year by 2050. *United Nations.* https://www.un.org/sustainabledevelopment/blog/2016/05/unep-report-cost-of-adapting-to-climate-change-could-hit-500b-per-year-by-2050/

Vadén, T., Lähde, V., Majava, A., Järvensivu, P., Toivanen, T., Hakala, E., & Eronen, J. T. (2020). Decoupling for ecological sustainability: a categorisation and review of research literature, *Environmental Science & Policy, 112,* 236–244.

Wilshusen, P. R., & MacDonald, K. I. (2017). Fields of green: corporate sustainability and the production of economistic environmental governance. *Environment and Planning A, 49* (8), 1824–1845.

World Bank. (2018). *Groundswell: Preparing for Internal Climate Migration.* World Bank.

World Economic Forum (WEF). (2020). *The Global Risks Report 2020.* WEF.

Wullweber, J. (2015). Global politics and empty signifiers: the political construction of high technology, *Critical Policy Studies, 9* (1), 78–96.

Žižek, S. (1999). Carl Schmitt in the age of post-politics. In. C. Mouffe (Ed.), *The Challenge of Carl Schmitt* (pp. 18–37). Verso.

12. Genuine savings and economics for the Anthropocene
Eoin McLaughlin and Cristián Ducoing

INTRODUCTION. GDP IS AN OBSOLETE INDICATOR FOR THE ANTHROPOCENE

Constructed to simply quantify the monetary value of all goods and services entering into market exchange, Gross Domestic Product (GDP) is regarded as the 'invention of the 20th century' (Coyle, 2015; Masood, 2016). However, there is a growing recognition that maximizing year on year growth in GDP is unlikely to be a realistic target for the 21st century due to numerous negative consequences, and that 'sustainable development' has become the key to global survival (Rockström et al., 2009; Steffen et al., 2015). When constructing GDP estimates, factors that have negative impacts on human well-being (e.g., the environmental damages such as pollution) are given equal weight to elements that are beneficial to society. One of the recent IPCC reports includes dire warnings of the dangers of future climate change, which, in the main, has been a direct consequence of following the goal of maximizing GDP (Allen et al., 2019). Unsurprisingly, there is now a growing call, including from some of the most respected levels in the economics profession, for changes to be made to how we measure economic activity, economic development and well-being more generally (Stiglitz et al., 2009; Stiglitz et al., 2018). Economists and economic historians have a responsibility to be at the forefront of a quest to replace their favourite 20th-century metric, for one that is 'fit for 21st century' purpose.

Despite such calls for change, GDP continues to be used, often in policy circles, as the main and often, only, measure of choice to guide decisions designed to increase welfare, in part because it is simple to calculate, but also because it allows country performance comparisons. Politicians can point to GDP growth as a measure of their 'success', both locally and globally.

Theoretically superior alternatives to GDP, as a measure of current and future well-being, however, do exist, for example, Genuine Savings (GS). Although promoted and supported by the World Bank, to date, GS, has not established the knowledge-base of 'sufficient' long-run evidence to displace GDP. Currently, the main sources of world-adjusted net savings estimates are reported in various World Bank Reports (Hamilton & Clemens, 1999; Lange et al., 2018). It appears that, without an overwhelming, body of empirical evidence in support of GS, GDP will continue to be the metric of choice taking us myopically towards a dystopian future.[1]

This chapter summarizes evidence from sustainability indicators for four countries in the period 1870–2000 that is based on the application of a fit-for-purpose methodology. Ongoing environmental and developmental challenges compel us to focus on more comprehensive welfare indicators than GDP, that can address long-term sustainability. A recent article in *PNAS* calls for a greater integration of both economics and sustainable development (Polasky et al., 2019). This is the core concept informing the GS approach. Seeing wealth as the foundation of future income and hence welfare, means that changes in wealth (saving/investment) provide an indication of the feasibility of future, sustainable, development paths. The project proposed here will produce such comprehensive welfare indicators, focusing more on stocks (e.g., wealth), including natural capital, which provide future generations with the capabilities to increase future well-being, rather than on flows (e.g., income), conventionally measured by GDP, which simply measures annual outcomes without recourse to their long-term implications.

An historical focus is necessary to provide evidence as to whether past policies and choices, guided by GDP as a welfare enhancing measure, have maximized (or even increased) well-being, sustainably. Historical data and outcomes provide our only measures to test both the GDP approach and the theoretically superior, GS approach. Historical data will enhance current metrics and provide a deeper understanding of natural capital, human capital, technological change and environmental degradation in the long run, to guide policy for the future. Much of the current work on environmental economics considers uncertain, unknown futures (scenarios or predictions), which may simply never exist. However insights and data from the past can test alternative modelling approaches to inform policymaking in the present and the future. Fenichel et al. (2016) suggest that a better understanding of how past changes influence sustainability can be used to forecast the impact of future changes based upon how these actual changes led to observed outcomes.

If the GS framework is able to inform sustainable social, economic and environmental futures, they should, as a minimum, be able to explain the past. We will therefore analyse whether historical experiences can explain variations in past and current levels of comprehensive/inclusive wealth within and across countries and what future sustainable development prospects would look like. If the GS concept is to complement or replace other indicators, it requires evidence which includes long run estimates for a wide range of countries.

However, there are a lack of standardized methodologies and results across and within countries of GS (Hanley et al., 2015).

A major goal of this chapter is to demonstrate that actions aimed at sustainably increasing current and future well-being should not be based on simply maximizing the growth rate of current GDP. Instead, the 'rules' of GS should be adopted to achieve sustainable development. However, to achieve such goals, theoretical developments and empirical evidence from the GS approach need to be produced and compared to the current, popular paradigm of GDP.

METHODS AND DATA

Genuine savings has become an indicator of *weak sustainability* supported by the World Bank and several scholars have developed their theoretical underpinnings (Ferreira et al., 2008; Hanley et al., 2015; World Bank, 2006). The major differences between weak and strong sustainability relate to assumptions surrounding natural capital. Weak sustainability assumes natural capital can be aggregated, that there is substitutability of different forms of capital and monetary valuation of natural capital. Whereas strong sustainability assumes no substitutability and critical thresholds, there are also issues surrounding monetization and preference is for measurement in physical units.

In order to calculate the long run GS for our four countries sample, we have largely followed the Lange et al. (2018) and World Bank (2006) methodology, as outlined by Bolt et al. (2002) for calculating GS. Effectively this builds on a range of increasingly comprehensive measures of year-on-year changes in total capital over time.[2] The starting point is total wealth, known as comprehensive or inclusive wealth:

$$K = K_P + K_N + K_H + K_S \tag{12.1}$$

Where we distinguish between different sorts of capital. Produced capital includes roads, machines, factories; Natural capital, all 'gifts of nature' (renewable and non-renewable alike), human capital, includes learning, skills, and experience, and social capital, including trust, community links, rule of law and absence of corruption. We also incorporate the value of time to account for technological change. Genuine savings looks at the changes in these various capital stocks.

Genuine savings in turn is the sum of year-on-year changes in these capital stocks, aggregated with appropriate (or approximates thereof) shadow prices.

$$GS = \sum_{i=1}^{N} p_i \dot{K}_i \tag{12.2}$$

Where p_1 is the shadow price (and negative for pollution stocks) for the various capital stocks [i=1...N] and $\dot{K}_i$ is the year-to-year change in the various capital stocks. As K represents capabilities for future well-being, the key insight here is that GS>0 (non-declining K) implies sustainability whereas GS<0 (declining K) implies unsustainable development.[3]

From these foundations in stocks, we have constructed the following indicators to illustrate year-to-year changes in various capital stocks that highlight different aspects of approaches to sustainability:

1. Net Investment = net fixed produced capital formation and overseas investment
2. Green Investment = Net + Δ natural capital
3. Genuine Savings = Green Investment + education expenditure
4. GSTFP = GS + Net Present Value of TFP
5. GScarbon = GS − carbon emissions
6. GSTFPcarbon = GSTFP − carbon emissions.

The first component accounts for depreciation of reproducible capital, the second makes allowances for depreciation of *natural* capital, the third makes adjustments to incorporate the appreciation of human capital, the fourth incorporates the value of technological change. The fifth and sixth account for damages caused by future climate change (implicitly growth today at

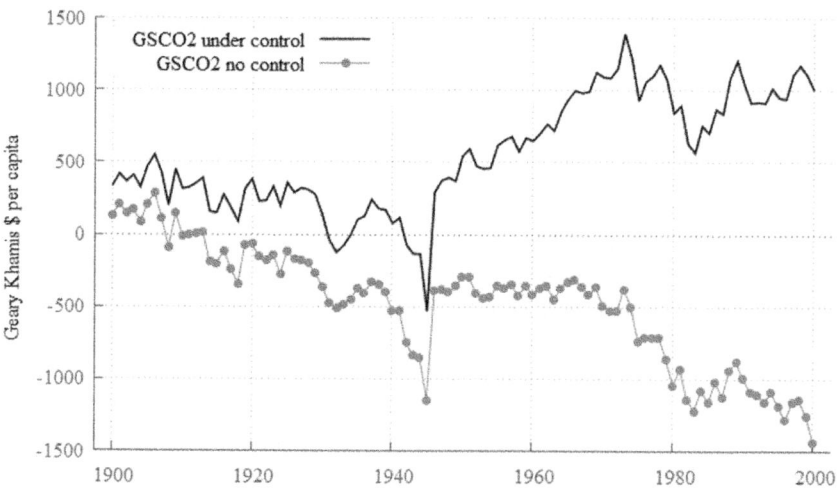

Source: GS data from Blum et al. (2017) and CO_2 from Pezzey and Burke (2014).

Figure 12.1   *Global CO_2 social cost under two scenarios. CO_2 estimated at $131 and $1455*

the expense of the growth future). The incorporation of CO_2 is made taking account of different explicit and implicit assumptions regarding future damage costs. Here we illustrate two social costs of CO_2 estimated following Pezzey & Burke (2014). These two scenarios take into account the capacity of society to *control* the externalities and damage produced by CO_2. The first scenario, estimate the social cost of CO_2 at $131, a higher figure that mainstream models (Nordhaus, 2017). While the second scenario estimates a higher price of $1455. The results in a *hypothetical* Global GS measure are summarized in Figure 12.1.

FOUR COUNTRIES IN THE LONG RUN

How do we measure progress? Much ink has been spilt poring over this question, many published works use GDP as the main indicator of progress as we have seen in the introduction. This narrow focus has led the world. We highlight the historical record of four countries, the United States, United Kingdom, Chile and Australia, to illustrate the risks behind a narrow GDP-led focus. These four countries can be thought of as representatives of each of their respective geographical regions. The United States, as the leader in terms of technological progress over much of the 20th century run, has seen a decoupling between well-being and GDP since the 1970s (Costanza et al., 2014). From Europe, we have chosen the United Kingdom, the origin of the industrial revolution and a long run example of steady growth in its mainstream concept. Australia was the richer country in terms of GDP around the end of the 19th century and it has been included in the *settler economies* group with Argentina, Canada, New Zealand and the United States. Finally, we have included a peculiar Latin American country, Chile. Chile has been in the top five of Latin American income countries, measured as GDP per capita and in the last thirty years, it has become the example by the mainstream policy advisors for developing countries. One of the main features of the economic structure of this country is its high dependence on natural resources (Ducoing et al., 2018).

GDP figures for these countries show a picture of steady growth, a bit least stable in the case of Chile. When we compare with a comprehensive figure such as GS, the picture is quite different. As we can see in Table 12.1, the share of GS during the 20th century did not achieve 10 percent of the yearly income. GS in Britain were just 5.49 percent, being the lower rate of the three developed countries of this sample. Figure 12.2 shows these astonishing differences between a mainstream measure of progress as GDP and adjusted wealth measures, such as GS. The results show that under a weak sustainability approach, these countries are not saving enough for future generations.

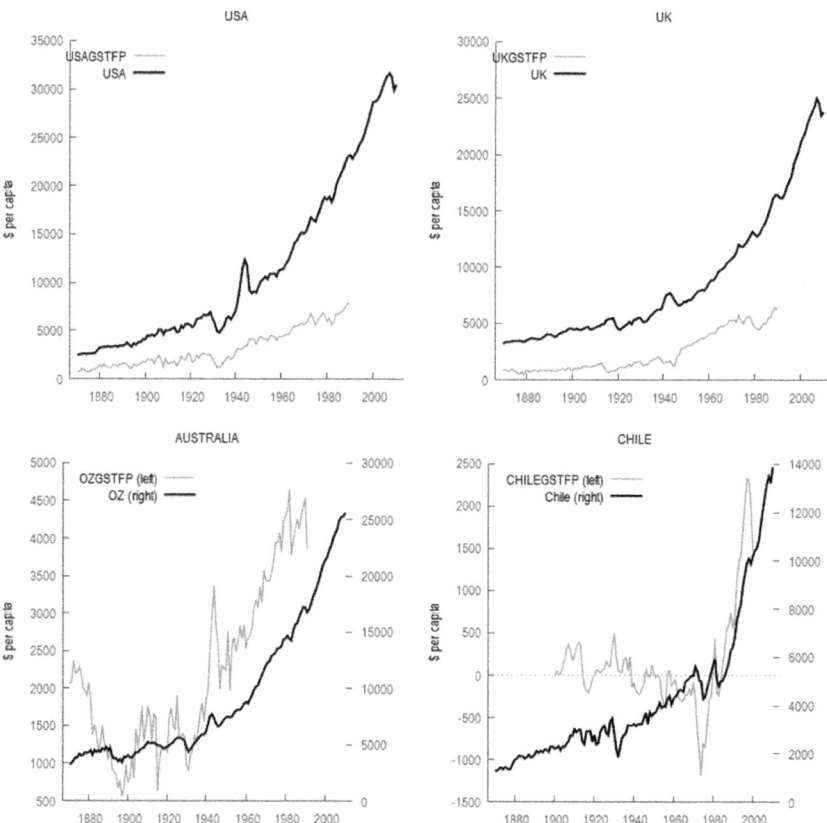

Source: GS estimations from Blum et al. (2017) and GDP from Bolt and Van Zanden (2014).

Figure 12.2   Genuine savings and GDP per capita in USA, UK, Australia and Chile. 1870–2000

CONCLUSIONS

How could we achieve sustainable development? What is the right measure to weigh sustainability? Why are we using GDP instead of alternative indicators? These difficult questions are far beyond the scope of this chapter, but some thoughts can be extracted as preliminary conclusions: The use of GDP as a development indicator has been misleading policy makers for decades. The goal to increase 'growth' at any cost has generated unexpected and harmful externalities that are stretching planet boundaries. In this chapter we have presented a brief report on the effects of measuring countries in an alternative

Table 12.1 *Net investment, green, genuine savings and GSTFP as share of GDP in Australia, Chile, United States and Britain*

	1900–2000			
	Net Investment	Green	GS	GSTFP
	%	%	%	%
Britain	4.63	2.80	5.49	28.58
US	7.07	4.42	8.11	32.62
Australia	7.68	4.33	6.55	24.69
Chile	2.06	-5.64	-3.75	1.82

Source: GS estimations from Blum et al. (2017) and GDP from Bolt and Van Zanden (2014).

development perspective, genuine savings. The United States and United Kingdom, under a weak sustainability model, have maintained savings rates that could be enough to keep welfare (measured as consumption) in the future. However, a big 'if' appears when we include bigger social costs of CO_2 in our estimations. In the case of Australia, as a previous study has shown, the depletion of natural resources has kept the capacity of future consumption, but its savings rates have been steadily lower than other OECD countries, raising concerns in its future capacity to produce welfare (Greasley et al., 2017). The case of Chile becomes more dramatic, because it is the best example of how GDP misleads long-run goals of sustainability for shorter run gains. The income growth of Chile was impressive in the years 1986–1998, and the aggregate improvement welfare can't be denied. However, these gains have proved to be fragile and the last years have seen a constant social unrest caused by environmental and societal harmful externalities of a model based in natural resource extraction. In Chile, the natural capital losses have not been compensated by equal gains in human and physical capital, putting in danger the well-being of future generations.

NOTES

1. An apt analogy here is *heating a house by stripping the floorboards*.
2. An extended version of the methodology used to calculate historical GS can be read from Blum et al. (2017), Hanley et al. (2015) and Lindmark and Acar (2013).
3. For greater elucidation see Hanley et al. (2015) and (Blum et al., 2019).

REFERENCES

Allen, M., Antwi-Agyei, P., Aragon-Durand, F., & Babiker, M. (2019). Technical Summary: Global warming of 1.5° C. An IPCC Special Report on the impacts of global warming of 1.5° C above pre-industrial levels and related global. *Keywan Riahi*. http://pure.iiasa.ac.at/id/eprint/15716/

Blum, M., Ducoing, C., & McLaughlin, E. (2017). A sustainable century? In K. Hamilton and C. Hepburn (eds), *National Wealth: What is Missing, Why it Matters*. Oxford University Press, 89–113

Blum, M., McLaughlin, E., & Hanley, N. (2019). Accounting for sustainable development over the long-run: Lessons from Germany. *German Economic Review*, *20*(4), 410–446. https://doi.org/10.1111/GEER.12148/HTML

Bolt, K., Matete, M., & Clemens, M. (2002). Manual for calculating adjusted net savings. *Environment Department, World Bank*, 1–23. http://siteresources.worldbank.org/INTEEI/1105643-1115814965717/20486606/Savingsmanual2002.pdf

Bolt, J., & Van Zanden, J. L. (2014). The Maddison Project: collaborative research on historical national accounts. *The Economic History Review*, *67*(3), 627–651

Costanza, R., Kubiszewski, I., Giovannini, E., Lovins, H., McGlade, J., Pickett, K. E., Ragnarsdóttir, K. V., Roberts, D., De Vogli, R., & Wilkinson, R. (2014). Development: Time to leave GDP behind. *Nature*, *505*(7483), 283–285. https://doi.org/10.1038/505283a

Coyle, D. (2015). *GDP: A Brief but Affectionate History-Revised and expanded Edition*. https://books.google.es/books?hl=es&lr=&id=t7pKCAAAQBAJ&oi=fnd&pg=PP1&ots=aAI-Nu4b1g&sig=jwRKtoMJVS2StlpUPVBHcDD5kyo

Ducoing, C., Peres-Cajías, J., Badia-Miró, M., Bergquist, A.-K., Contreras, C., Ranestad, K., & Torregrosa, S. (2018). Natural resources curse in the long run? Bolivia, Chile and Peru in the Nordic countries' mirror. *Sustainability (Switzerland)*, *10*(4). https://doi.org/10.3390/su10040965

Fenichel, E. P., Levin, S. A., McCay, B., St. Martin, K., Abbott, J. K., & Pinsky, M. L. (2016). Wealth reallocation and sustainability under climate change. *Nature Climate Change*, *6*(3), 237–244. https://doi.org/10.1038/nclimate2871

Ferreira, S., Hamilton, K., & Vincent, J. R. (2008). Comprehensive wealth and future consumption: Accounting for population growth. *World Bank Economic Review*, *22*(2), 233–248. https://doi.org/10.1093/wber/lhn008

Greasley, D., McLaughlin, E., Hanley, N., & Oxley, L. (2017). Australia: a land of missed opportunities? *Environment and Development Economics*, 1–25. https://doi.org/10.1017/S1355770X17000110

Hamilton, K., & Clemens, M. (1999). Genuine savings rates in developing countries. *The World Bank Economic Review*, *13*(2), 333–356

Hanley, N., Dupuy, L., & Mclaughlin, E. (2015). Genuine savings and sustainability. *Journal of Economic Surveys*, *29*(4), 779–806. https://doi.org/10.1111/joes.12120

Lange, G.-M., Wodon, Q., & Carey, K. (Eds.) (2018). *The Changing Wealth of Nations 2018: Building a Sustainable Future*. The World Bank. https://doi.org/10.1596/978-1-4648-1046-6

Lindmark, M., & Acar, S. (2013). Sustainability in the making? A historical estimate of Swedish sustainable and unsustainable development 1850–2000. *Ecological Economics*, *86*, 176–187

Masood, E. (2016). *The great invention: The story of GDP and the making and unmaking of the modern world*. https://books.google.es/books?hl=es&lr=&id=fE89DAAAQBAJ&oi=fnd&pg=PT5&dq=The+Great+Invention:+The+Story+and+the+Unmaking+of+the+Modern+World&ots=grbpwIGayP&sig=xwut_VK8uewH1Kn6G458G049uSw

Nordhaus, W. D. (2017). Revisiting the social cost of carbon. *Proceedings of the National Academy of Sciences of the United States of America*, *114*(7), 1518–1523. https://doi.org/10.1073/pnas.1609244114

Pezzey, J. C. V., & Burke, P. J. (2014). Towards a more inclusive and precautionary indicator of global sustainability. *Ecological Economics*, *106*, 141–154. https://doi.org/10.1016/j.ecolecon.2014.07.008

Polasky, S., Kling, C. L., Levin, S. A., Carpenter, S. R., Daily, G. C., Ehrlich, P. R., Heal, G. M., & Lubchenco, J. (2019). Role of economics in analyzing the environment and sustainable development. *Proceedings of the National Academy of Sciences of the United States of America*, *116*(12), 5233–5238. https://doi.org/10.1073/pnas.1901616116

Rockström, J., Steffen, W., Noone, K., Persson, Å., Chapin, F. S., Lambin, E. F., Lenton, T. M., Scheffer, M., Folke, C., Schellnhuber, H. J., Nykvist, B., de Wit, C. A., Hughes, T., van der Leeuw, S., Rodhe, H., Sörlin, S., Snyder, P. K., Costanza, R., Svedin, U., ... Foley, J. A. (2009). A safe operating space for humanity. *Nature*, *461*(7263), 472–475. https://doi.org/10.1038/461472a

Steffen, W., Richardson, K., Rockström, J., Cornell, S. E., Fetzer, I., Bennett, E. M., Biggs, R., Carpenter, S. R., De Vries, W., De Wit, C. A., Folke, C., Gerten, D., Heinke, J., Mace, G. M., Persson, L. M., Ramanathan, V., Reyers, B., & Sörlin, S. (2015). Planetary boundaries: Guiding human development on a changing planet. *Science*, *347*(6223). https://doi.org/10.1126/science.1259855

Stiglitz, J. E., Sen, A., & Fitoussi, J.-P. (2009). Report by the Commission on the Measurement of Economic Performance and Social Progress. *Sustainable Development*, *12*, 292. https://doi.org/10.2139/ssrn.1714428

Stiglitz, J., Fitoussi, J., & Durand, M. (2018). *Beyond GDP: Measuring What Counts for Economic and Social Performance.* https://ideas.repec.org/p/spo/wpmain/infohdl2441-4vsqk7docb9nmophtp29pk68cr.html

World Bank. (2006). *Where is the Wealth of Nations?: Measuring Capital for the 21st Century.* The World Bank. https://books.google.com/books?id=4bJQge7WFHAC&pgis=1

PART IV

Justice in the Anthropocene

13. Epistemic injustice
Sabelo J. Ndlovu-Gatsheni

INTRODUCTION

Epistemic injustice speaks to cognitive injustice, that is, deliberate non-recognition of diverse ways through which diverse people make sense of the world and their lives (Santos, 2014, 2018). It is a foundational injustice that is linked to 'coloniality of being,' that is, denial of the very humanity of other people and their reduction to a sub-human status (Maldonado-Torres, 2007). In epistemic justice, one finds the convergence of the 'colour line' and the 'epistemic line' (Du Bois, 1903; Ndlovu-Gatsheni, 2018b). Boaventura de Sousa Santos (2007) introduced the concept of 'abyssal thinking' which is predicated of refusal of 'co-presence' of worlds and peoples, in the process generating epistemicides (the killing of other people's knowledges).

Epistemicides are underwritten by genocides (physical killing of people whose humanity is questioned), linguicides (killing of languages of who are targets for enslavement and colonization and impositions of colonial languages), cultural imperialism (the attacking of cultures of enslaved and colonized peoples), theft of history (denial of possession of history by non-European peoples), and indeed ontolocides (exiling of people from their languages, cultures, histories, and from themselves through alienation, resulting in dehumanization and dismemberment) (Ndlovu-Gatsheni, 2018a, 2018b).

Because epistemic injustices are committed through epistemological colonization which target the minds of the victims, it is one of the most resilient crimes of Euromodernity and its technologies of racism and colonialism. Therefore, the dismantling of the physical empire does not solve the problem of epistemic injustices because they remain deeply etched in the minds of colonized peoples (psyche), in curriculum, institutions, modern systems, and iconography. This is why there is resurgent and insurgent planetary decolonization struggles, which are targeting iconographies of peoples who committed genocides and epistemicides. This planetary decolonization is embodied by the Rhodes Must Fall and Black Lives Matter movements today. These movements are confronting a planetary cognitive empire which has successfully invaded the global mental universe of the people while privileging epistemologies from Europe and North America as well as the theories, concepts and scholarship of white men from Germany, Britain, Italy, France, and United States of America mainly. Thus, at the centre of epistemic injustices are also problems of patriarchy, androcentrism, and sexism of knowledge itself.

This short chapter begins with defining what epistemic injustice is. The second section challenges Gerry Dunne's (2020) notion and claim that Miranda Fricker's book *Epistemic Injustice: Power and the Ethics of Knowing* (2007) is the pioneering work on the subject of epistemic injustices. The chapter highlights how an expansive archive from Africa, Latin America, Asia, Black radical tradition, and feminist scholarship has grappled with the question of epistemic injustices. The third section delves into how the cognitive empire deliberately imposed epistemic injustices and demonstrates how the so-called global economy of knowledge with its uneven intellectual and academic division of labour perpetuates epistemic injustices. The fourth part of the chapter suggests ways of overcoming epistemic injustices. The last part is the conclusion and it highlights the concerns of the current conjuncture in the world of knowledge.

GRAPPLING WITH THE QUESTION OF EPISTEMIC INJUSTICES

It is not correct to credit Miranda Fricker's work *Epistemic Injustice: Power and the Ethics of Knowing* (2007) as the pioneering work on the subject of epistemic injustice. Such seminal works as Claude Ake's *Social Science as Imperialism: The Theory of Political Development* (1979), Ngugi wa Thiong'o's *Decolonizing the Mind: The Politics of Language in African Literature* (1986), Linda T. Smith's *Decolonizing Methodologies: Research and Indigenous Peoples* (1999), Jack Goody's *The Theft of History* (2006), and many others dealt with the question of epistemic injustices long before Fricker's interventions. This argument is not meant to minimize Fricker's contributions to the question of epistemic injustices but to highlight how scholars from Africa in particular experience epistemic injustice in the form of their work not being taken seriously and treated as though it does not exist at all, while privileging those works produced by Europeans and North Americans.

Latin American scholars such as Enrique Dussel (1993, 1995), Walter D. Mignolo (1995), Anibal Quijano (2000), and many others have produced seminal works on the question of epistemic injustices, tracing it back to the very moment of the unfolding of the European Renaissance and Euromodernity and highlighting the often-ignored reality of coloniality of power, knowledge, and being (Maldonado-Torres, 2007). They have meticulously linked the questions of genocides, enslavement, racism, patriarchy, sexism, and heteronormativity as key coordinates of epistemic injustices (Lugones, 2007; Grosfoguel, 2013; 2019).

In South East Asia, the whole project of the Subaltern Studies which emerged in the 1980s sought to address the epistemic injustices inflicted on the poor people whom they termed the subaltern, drawing from the work of Antonio Gramsci (Guha, 1983, 1997). It was from the Subaltern Studies that the famous question of whether the subaltern can speak emerged (see Spivak, 1988). The feminist scholarship in its various iterations and schools has been pioneering in raising issues of epistemic injustices (Collins, 2000). It was feminist scholarship that introduced the concepts of 'intersectionality' in understanding multiple forms of oppression and their differential multiple impacts on women depending on their race, class and ethnicity (see Lorde 1984; Crenshaw, 1995; Lugones, 2003). With specific reference to epistemic injustices in relation to gender, the seminal work *Engendering African Social Sciences* (1997) edited by Imam, Mama and Sow, highlighted the core issues of androcentrism, sexism, and patriarchy in knowledge production and social science.

The black radical tradition with William E. B. Du Bois (1903) as its leading light highlighted the question of epistemic injustices long ago including raising the question of how black people feel about anti-black racism and its epistemic implications. It was from the staple of black radical tradition that Cedric Robinson offered the seminal work *Black Marxism: The Making of the Black Radical Tradition* (1983) critiquing European Marxism for complicity in epistemic injustices. Therefore, it does not make sense to argue that the recent work of Miranda Fricker is pioneering on the subject of epistemic injustices (Dunne, 2020).

Even the Eurocentric idea of Greece as the cradle of knowledge has been vigorously challenged by African scholars such as Cheikh Anta Diop (1954, 1974a, 1974b) who highlighted how Egypt instead of Greece was a seat of knowledge, which was then taken to Greece by those who studied in Egypt. This is a debate that is well-captured in Martin Bernal's *Black Athena: The Afroasiatic Roots of Classical Civilization* (1987) where the very question of civilization is brought into the centre of contestations about epistemic injustices aligned to Eurocentrism. The Afrocentricity School pioneered by scholars such as Diop and today led by Molefi Kete Asante (2007) has been grappling with the question of epistemic injustice for a very long time, in the process recovering African endogenous knowledge that has been pushed to the margins by Eurocentrism. But how were epistemic injustices committed? This question takes us to the concept of the cognitive empire and how it invaded the mental universe of the modern world as it globalized and universalized Eurocentric epistemologies.

THE COGNITIVE EMPIRE AND COMMISSION OF EPISTEMIC INJUSTICES

In his *The End of the Cognitive Empire: The Coming of Age of Epistemologies of the South* (2018) Boaventura de Sousa Santos used the term 'cognitive empire' in the title and insinuated that this empire is coming to an end because of epistemologies of the South, which are 'coming of age.' What did not come out clearly in Santos's analysis is the definition of the cognitive empire and delineation of its key operative techniques as opposed to the traditional physical empire. What is correct in Santos's analysis is that indeed the cognitive empire is currently on the defensive as it is confronting resurgent and insurgent planetary decolonization. However, it can be premature to postulate an end to the cognitive empire. The cognitive empire is the pillar of all empires – the physical and commercial empires. Ngugi wa Thiong'o (1986) called it the 'metaphysical empire,' Robert Gildea (2019) termed it the 'empire of the mind' and Ashis Nandy (1983) depicted it as an 'intimate enemy' (an enemy that resides like a virus in the body and mind of the host). These related but different descriptions and names highlight its amorphous character and invisible operative logics as it commits epistemic injustices across the world.

To gain a deep understanding of the cognitive empire, there is need to reflect on Euromodernity itself because this is where 'the colonizer's model of the world' emerged as an idea and practice (see Blaut, 1993). Euromodernity materialized through colonization of time and human subjectivity. Guminder K. Bhambra (2007) captured this in terms of 'rupture and difference' as the key logics of Euromodernity. Euromodernity's self-representation and indeed autobiography is that of a 'rupture' in time leading to the monopolization of the temporality 'modern' by Europe and designation of all others as 'primitive/ancient' (pre-modern), that is, outside modern time and thus backward. The paradigm of difference enabled colonization of being human itself through techniques of social classification of human population and racial hierarchization in accordance with invented differential ontological densities (Ndlovu-Gatsheni, 2020). Here was born a bio-centric and teleological European account of the human, with the European 'Man' at the centre as the master and owner of the world (Wynter 1995; Ndlovu-Gatsheni, 2013; Ndlovu-Gatsheni, 2018b). This is why Achille Mbembe (2019) posited that colonialism was fundamentally about who owns the earth. This development was always underwritten by a 'colonial turn' in knowledge which advanced only Eurocentric epistemologies as being valid, legitimate, truthful, and universal.

The 'colonial turn' was at one level underpinned by the discourses of Hellenocentrism (all started in Greece), Westernization (the West as the template of a complete human being), Eurocentrism (Europe as the centre of the world), Secularism (de-godding/death of God and rule of European 'Man' armed with scientific knowledge), Periodization (linear conceptions of time marked by ruptures), and Colonialism (conquest and occupation of a world that was deemed to be empty) (Dussel, 2011). At a second level, it was symbolized by the historic Valladolid Debates (1550–1551), in which Bartolome da La Casas and Gines De Sepulveda argued the ontological question of the humanity of the 'natives.' La Casas's position of sub-humans who worshipped the wrong gods, were primitive and could be rescued through conversion to Christianity laid the foundation for anthropology and ethnography whereas Sepulveda's take which dismissed the very humanity of indigenous people formed the basis for scientific racism in knowledge production (Castro, 2007; Saurez-Krabbe, 2016; Ndlovu-Gatsheni, 2020). Achille Mbembe (2001: 288) explained that 'conversion' involved destruction of the previous worlds and forcing the convert to 'give up what she or he believed.'

On the practical unfolding of the cognitive empire, Ngugi wa Thiong'o (1986, 2009) is very insightful. He identified two techniques which made the invasion of the mental universe of the colonized people's world possible. For the first technique, he gave the example of a computer whereby the removal of the hard disk of previous memory and the downloading of a new software results in total change to demonstrate how the cognitive empire works. This means that using the church, school and university; the cognitive empire is actively involved in the process of removal of the hard disk of previous indigenous people's knowledge and memory and is constantly downloading into their minds the software of European knowledge and memory (Ngugi wa Thiong'o, 2009). The consequences amount to what Ali A. Mazrui (1978) termed 'cultural schizophrenia' and Ngugi wa Thiong'o (2012: 39) articulates it very well:

> The colonial process dislocates the traveler's mind from the place he or she already knows to a foreign starting point even with the body still remaining in his or her homeland. It is a continuous alienation from the base, a continuous process of looking at oneself from the outside of self or with the lenses of a stranger. One may end up identifying with the foreign base as the starting point towards self, that is, from another self towards oneself, rather than the local being the starting point, from self to other selves.

The second technique of the cognitive empire is the one that Ngugi wa Thiong'o (1986) likened to a detonation of a 'cultural bomb' at the centre of the universe with far reaching consequences. This is how he put it: 'The effect of the cultural bomb is to annihilate a people's belief in their names, in their languages, in their environment, in their heritage of struggle, in their unity, in their capacities and ultimately in themselves' (Ngugi wa Thiong'o 1986: 15).

A people who experienced this 'cultural bomb' tend to develop complicated consciousness of themselves to the extent of aspiring to be like their colonizers.

Of course, the colonized people across the world rose against the physical empire and rolled out the decolonization struggles of the twentieth century. Those struggles became mainly ranged against the physical empire, which is aimed at dismantlement of direct colonial administrations. Indeed, the direct colonial administrations were rolled back and a modicum of political sovereignty was granted albeit within an unchanged imperial world system and global coloniality. This is why Ramon Grosfoguel (2007) posited that the decolonization of the twentieth century was a myth, as direct colonialism metamorphosed into global coloniality to sustain the colonial matrices of power. Within this context, attainment of a modicum of political sovereignty did not change the epistemic injustices committed by the cognitive empire.

This analysis takes us to the current debates about the global political economy of knowledge in which there is a belief that because of the globalization process, knowledges of the world have coalesced into a commonwealth with no centre and periphery, with no binaries and dichotomies. This debate came out prominently in *Knowledge and Global Power: Making New Sciences in the South* (2019: 8) edited by Fran Collyer, Raewyn Connell, Joan Maia and Robert Morrell where they noted that

> there is a widespread idea that we live in a knowledge society, an information society, or a technological society. Yet in most fields of research, there is also an idea that the disciplines we work in, and the concepts we work with, do not come from any particular place in that society. They are just in the air so to speak.

We just wish the resolution of epistemic injustices would produce a commonwealth of knowledge and better still ecologies of knowledges. We are far from this reality. There are a number of reasons for this. The first is that despite frantic efforts to speak across the North–South divide or the pretenses that these binaries no longer exist, the reality is that Europe and North America continue to enjoy a privileged position in knowledge production. There is still the resilient uneven intellectual and academic division of labour with Africa and the Global South still performing the position of where raw data is hunted for, gathered and extracted in raw form to be processed into theory in Europe and North America. Intellectual and academic dependence remains endemic if not pandemic with scholars of the Global South still compelled by power dynamics to publish in journals and leading presses located in the Global North if they are to gain any recognition as knowledge producers (see Hountondji, 1990, 1997; Ndlovu-Gatsheni, 2018b).

While this situation is indeed being challenged, contested, negotiated, and subverted to the extent that there is increasing talk of partnerships, Paulin

J. Hountondji (1997: 1) insisted that 'The fact bears repeating: in the field of science and technology, Third World countries, especially those in Black Africa, are tied hand and foot to the apron strings of the West.' It was also Hountondji (2002) who adapted Samir Amin's concept of economic extraversion and applied it to the domain of knowledge to reveal that there is a structural condition that sustains the dominance of Europe and North America to the extent that the rest of the majority world (the Global South) finds itself saddled with standards, protocols, methodologies, theories and indeed epistemologies developed in the Global North. This reality takes us to the initiatives and attempts developed by victims of the cognitive empire to resolve epistemic injustices.

TOWARDS RESOLUTION OF EPISTEMIC INJUSTICE

There is no easy way out when it comes to decolonizing thought and dealing with epistemic injustices. Epistemic injustices cannot be resolved without changing the foundational colonial crime of pushing non-European people out of the human family and without deimperialization of the modern world system. This is why one finds such scholars as Sylvia Wynter proposing a new 'counter-cartography of the history of human life' pivoted on 'a relational conception of human existence' and drawing from Frantz Fanon's concept of sociogeny (Wynter, 2001; Erasmus, 2020: 2).

In this counter-cartography of the history of human life, the key aspect is the definitive entry of the descendants of racialized, feminized, enslaved and indeed dehumanized people into the academies across the world vehemently rejecting the sub-human category they have been consigned to (Maldonado-Torres, 2008; Ndlovu-Gatsheni, 2018b). These descendants of dehumanized people are openly challenging all aspects of Eurocentrism and they are declaring for the whole world to know that as human beings they were born into valid and legitimate knowledge systems. The recovery of subjugated knowledge is made possible by the fact that except where absolute genocide was involved, elements of pre-colonization knowledges have survived.

With specific reference to India, Aditya Nigam in his most recent book *Decolonizing Theory: Thinking Across Traditions* (2020) has empirically demonstrated how pre-colonization knowledges have survived and continue to influence the present in the process, even undercutting the Eurocentric impositions of secularism. According to Nigam (2020) the pre-colonization knowledges and spiritualties which he termed 'non-synchronous synchronicities' have never obediently succumbed to Euromodernity, Eurocentrism and colonization.

The first important move in dealing with epistemic injustices for the victims is to undertake the painstaking process of self-introspection with a view to confront at individual and collective level the problematic consciousness imposed by the cognitive empire. It was this consciousness that Du Bois (1903) depicted as cascading from two souls, two thoughts, two unreconciled strivings, and two warring ideas in one black body. This process is necessary but very difficult because the intellectuals and academics from the Global South are produced by the modern world university system, which itself is a culprit in the commission of epistemic injustices and sustenance of global colonial matrices of power. The issue here is the agenda of dealing with miseducation while working towards re-education. This is what has been rendered by feminist movements and indigenous people's movements as learning to unlearn in order to relearn (see Ndlovu-Gatsheni, 2018b). While there are numerous pedagogies of learning new knowledge, there is a scarcity of those that help in unlearning what was imposed by the cognitive empire. They have to be developed in struggle as the victims of the cognitive empire grapple with abandoning what was meant for colonization.

Within the academies and universities that claim to be advancing the agenda of resolution of epistemic injustices there are many steps that have to be taken. The first is that of re-provincializing Europe while de-provincializing those areas like Africa, Asia, Caribbean and Latin America that were marginalized and peripherized. This agenda has to build on the concepts of 'moving the centre' and shifting the biography and geography of knowledge so as to make sure that the definitive entry of the descendants of the enslaved, racialized, feminized, and dehumanized substantially stir the toxic pond of knowledge in such a way that openings emerge to ecologies of knowledges (Ngugi wa Thiong'o, 1993; Santos, 2007; Ndlovu-Gatsheni, 2018b). A number of moves have to be practically taken within the academy to deparochialize social theory; diversify and pluralize the syllabus and the curriculum; shift and deliberately digress from what has been given as the canon; decentring of exhausted and irrelevant knowledge; and dehierarchization of hierarchies in epistemology so as to open spaces for previously marginalized and excluded voices (Ndlovu-Gatsheni, 2018b).

CONCLUSIONS: THE CURRENT CONJUNCTURE AND STRUGGLES FOR EPISTEMIC FREEDOM

The current conjuncture is characterized by resurgent and insurgent planetary decolonization of the twenty-first century which is ranged against all crimes of the cognitive empire and its physical manifestations through killing black people and the general anti-black politics and practices. The decolonization that is upon the world now invites the victims of the cognitive empire to engage the current times from their own vantage points and their own terms including with Europeans and their knowledge, drawing from the rich epistemologies of the Global South.

There is a big window of opportunity because the Eurocentric epistemologies that have dominated the world since the colonial encounters of the fifteenth century have become exhausted. They are experiencing a terminal epistemic crisis. The modern world is in desperate need for another knowledge. Such events as the 2008 global financial crisis and the current COVID-19 pandemic are some of the signs of an epistemic and indeed civilizational crisis. Europe and North America has to reckon with its responsibility for epistemic injustices as part of the search for another knowledge and indeed as they venture into the future beyond the COVID-19 pandemic.

This is possible because the current conjuncture is dominated by profound dissatisfaction with the dominant and hegemonic knowledge. This fundamentally means that epistemic injustices have to be confronted as the world looks forward to epistemic reconstitution involving embracement of ecologies of knowledges, thinking across traditions and indeed adoption of mosaic epistemologies that makes convivial scholarship and epistemic freedom possible.

REFERENCES

Ake, C. 1979. *Social Science as Imperialism: The Theory of Political Development*. University of Ibadan Press.

Asante, M. K. 2007. *An Afrocentric Manifesto: Toward an African Renaissance*. Polity Press.

Bernal, M. 1987. *Black Athena: The Afroasiatic Roots of Classical Civilization*. Rutgers University Press.

Bhambra, G. K. 2007. *Rethinking Modernity: Postcolonialism and the Sociological Imagination*. Palgrave Macmillan.

Blaut, J. M. 1993. *The Colonizer's Model of The World: Geographical Diffusion and Eurocentric History*. The Gilford Press.

Castro, D. 2007. *Another Face of Empire: Bartolome de La Casas, Indigenous Rights and Ecclesiastical Imperialism*. Duke University Press.

Collins, P. H. 2000. *Black Feminist Thought: Knowledge, Consciousness, and the Politics of Empowerment*. Routledge.

Collyer, F., Connell, R., Maia, J., and Morrell, R. (eds) 2019. *Knowledge and Global Power: Making New Sciences in the South*. Wits University Press.

Crenshaw, K. 1995. 'Mapping the Margins: Intersectionality, Identity Politics and Violence against Women of Colour.' In K. Crenshaw., N. Gotanda., G. Peller., and K. Thomas (eds), *Critical Theory*. The New Press, pp. 283–313.

Diop, C. A. 1954. *Nation, Negres at Culture*. Presence Africaine.

Diop, C. A. 1974a. *The African Origins of Civilization: Myth and Reality*. Lawrence Hill Books.

Diop, C. A. 1974b. *Precolonial Black Africa*. Lawrence Hill Books.

Du Bois, W. E. B. 1903. *The Souls of Black Folk*. Dover Publications.

Dunne, G. 2020. 'Epistemic Injustice.' In M. A. Peters (ed.), *Encyclopedia of Educational Philosophy and Theory*. Springer, pp. 3–8.

Dussel, E. 1993. 'Eurocentrism and Modernity (Introduction to the Frankfurt Lectures).' *Boundary 2*, 20(3), pp. 65–76.

Dussel, E. 1995. *The Invention of the Americas: Eclipse of 'the Other' and the Myth of Modernity*. Continuum.

Dussel, E. 2011. *Politics of Liberation: A Critical World History*. SCM Press.

Erasmus, Z. 2020. 'Sylvia Wynter's Theory of the Human: Counter-, Not Post-Humanist.' *Theory, Culture & Society*. DOI: 10.1177/0263276420936333, pp. 1–19.

Fricker, M. 2007. *Epistemic Injustice: Power and the Ethics of Knowing*. Oxford University Press.

Gildea, R. 2019. *Empires of the Mind: The Colonial Past and the Politics of the Present*. Cambridge University Press.

Goody, J. 2006. *The Theft of History*. Cambridge University Press.

Grosfoguel, R. 2007. 'The Epistemic Decolonial Turn: Beyond Political-Economy Paradigms.' *Cultural Studies*, 21(2–3), pp. 211–223.

Grosfoguel, R. 2013. 'The Structure of Knowledge in Westernized Universities: Epistemic Racism/Sexism and the Four Genocides/Epistemicides of the Long 16th Century.' *Human Architecture: Journal of the Sociology of Self-Knowledge*, XI (1), pp. 73–90.

Grosfoguel, R. 2019. 'What is Racism? Zone of Being and Zone of Non-Being in the Work of Frantz Fanon and Boaventura de Sousa Santos.' In J. Cupples and R. Grosfoguel (eds), *Unsettling Eurocentrism in the Westernized University*. Routledge, pp. 264–273.

Guha, R. 1983. *Elementary Aspects of Peasant Insurgency in Colonial India*. Duke University Press.

Guha, R. 1997. *Dominance without Hegemony: History and Power in Colonial India*. Harvard University Press.

Hountondji, P. (ed.) 1997. *Endogenous Knowledge: Research Trails*. CODESRIA Books.

Hountondji, P. J. 1990. 'Scientific Dependence in Africa Today.' *Research in African Literatures*, 21(3), pp. 5–15.

Hountondji, P. J. 2002. *The Struggle for Meaning: Reflections on Philosophy, Culture, and Democracy in Africa*. Athens: Ohio University Research in International Studies Africa Series No. 78.

Imam, A. M., Mama, A., and Sow, F. (eds) 1997. *Engendering African Social Sciences*. CODESRIA Book Series.

Lorde, A. 1984. *Sister Outside: Essays and Speeches*. Cross Press.

Lugones, M. 2003. *Pilgrimages/Peregrinajes: Theorizing Coalitions against Multiple Oppressions*. Rowman & Littlefield.

Lugones, M. 2007. 'Heterosexualism and the Colonial/Modern Gender System.' *Hypatia*, 22(1), pp. 186–209.

Maldonado-Torres, N. 2007. 'On the Coloniality of Being: Contributions to the Development of a Concept.' *Cultural Studies*, 21(2–3), pp. 240–270.

Maldonado-Torres, N. 2008. *Against War: Views from the Underside of Modernity*. Duke University Press.

Mazrui, A. A. 1978. *Political Values and the Educated Class in Africa*. University of California Press.

Mbembe, A. 2001. *On the Postcolony*. University of California Press.

Mbembe, A. 2019. *Necropolitics*. Duke University Press.

Mignolo, W. D. 1995. *The Dark Side of the Renaissance: Literacy, Territoriality and Colonization*. The University of Michigan Press.

Nandy, A. 1983. *The Intimate Enemy: Loss and Recovery of Self under Colonialism.* Oxford University Press.

Ndlovu-Gatsheni, S. J. 2013. *Empire, Global Coloniality and African Subjectivity.* Berghahn Books.

Ndlovu-Gatsheni, S. J. 2018a. 'Metaphysical Empire, Linguicides and Cultural Imperialism.' *The English Academy Review: A Journal of English Studies*, 35(2), pp. 96–115.

Ndlovu-Gatsheni, S. J. 2018b. *Epistemic Freedom in Africa: Deprovincialization and Decolonization.* Routledge.

Ndlovu-Gatsheni, S. J. 2020. *Decolonization, Development and Knowledge in Africa: Turning Over A Leaf.* Routledge.

Ngugi wa Thiong'o. 1986. *Decolonizing the Mind: The Politics of Language in African Literature.* James Currey.

Ngugi wa Thiong'o. 1993. *Moving the Centre: Struggles for Cultural Freedom.* James Currey.

Ngugi wa Thiong'o. 2009. *Something Torn and New: An African Renaissance.* Basic Civitas Books.

Ngugi wa Thiong'o. 2012. *Globalectics: Theory and The Politics of Knowing.* Columbia University Press.

Nigam, A. 2020. *Decolonizing Theory: Thinking Across Traditions.* Bloomsbury India.

Quijano, A. 2000. 'Coloniality of Power, Eurocentrism, and Latin America.' *Nepantla: Views from the South*, 1(3), pp. 533–579.

Robinson, C. J. 1983. *Black Marxism: The Making of Black Radical Tradition.* The University of North Carolina Press.

Santos, B. de S. 2007. Beyond abyssal thinking: From global lines to ecologies of knowledges. *Binghamton University Review*, 30(1), 45–89.

Santos, B. de S. 2014. *Epistemologies of the South: Justice against Epistemicide.* Paradigm Publishers.

Santos, B. de S. 2018. *The End of the Cognitive Empire: The Coming of Age of Epistemologies of the South.* Duke University Press.

Saurez-Krabbe. J. 2016. *Race, Rights and Rebels: Alternatives to Human Rights and Development from the Global South.* Rowman & Littlefield.

Smith, L. T. 1999. *Decolonizing Methodologies: Research and Indigenous Peoples.* Zed Books and Otago University Press.

Spivak, G. C. 1988. 'Can the Subaltern Speak?' In C. Nelson and L. Grossberg (eds), *Marxism and Interpretations of Culture.* Palgrave Macmillan, pp. 271–313.

Wynter, S. 1995. '1492: A New World View.' In V. L. Hyatt, and Rex Nettleford (eds), *Race, Discourse, and the Origin of the Americas: A New World View.* Smithsonian Institution Press, pp. 68–89.

Wynter, S. 2001. 'Toward the Sociogenic Principle: Fanon, Identity, the Puzzle of Conscious Experience, and What It is Like to be "Black".' In M. F. Duran-Cogan and A. Gomez-Moriana (eds), *National Identities and Sociopolitical Changes in Latin America.* Routledge, pp. 78–95.

14. The urgency for epistemic and political climate justice

Jacobo Ocharan, Velina Petrova and Irene Guijt

INTRODUCTION

The halls of the Madrid Convention Center during the UN Framework Convention of Climate Change (UNFCCC) Conference of the Parties (COP 25) in December 2019 were full of many thousands of dark-suited bureaucrats. The exceptions were the few hundred indigenous and local population and the youth groups wearing flamboyantly colourful clothes. The diversity they brought to the conference was not only in attire but more significantly in the solutions they proposed to fight the climate crisis. Unfortunately, their 'badges' did not allow them to enter the rooms where the negotiations were happening – they were mere 'observers.' On day 10 of the conference, it was clear that the negotiations were, yet again, not taking into account the claims made by these civil society groups. So they organized a peaceful gathering of hundreds, in front of one of the negotiation rooms, to show their deep discontent and despair. Almost 100 protesters, including some from Oxfam, were evicted from the Convention Center and stripped of their badges (Leonard 2019). The justification: they had not respected the basic rules of diplomacy required for any UN Summit. There is a deep and bitter irony to being neither invited to the table, nor allowed to disturb those special and knowledgeable enough to sit around it. The ideas and claims of those who had been evicted, sometimes expressed in chants or drama, were not considered valid for decision-making on the very climate policies that affect them profoundly every day. This is not a story of individual protest or any specific organization; this is the story of systemic and systematic exclusion of those people most affected by climate change from the conversations at the climate crisis solution table.

Climate breakdown has arguably become the most widely recognized and debated threat to a healthy future for everyone. Yet despite global protests and the Paris Agreement calling for urgent action and economic transformation, greenhouse gas emissions continue to track the highest emission trajectory, which will see the planet exceed – on an increasingly consistent basis – an average 1.5 degrees Celsius of extra warming between 2030 and mid-century, and exceed 2 degrees about 10 years later. But as warming is uneven, many parts of the inhabited Earth will see higher temperatures on a regular basis before then. To avert the most catastrophic effects of climate change, it is now widely understood that societies must quickly cease to use fossil fuels and act more boldly to adapt to climate chaos.

However, some groups face greater risks and consequences. Climate change worsens and deepens inequalities of all kinds – the most vulnerable people experiencing impacts more acutely in their daily lives and livelihoods than those who have the economic basis or political options that give access to alternative locations, incomes or food and water. These impacts reduce people's resilience, making them ever more vulnerable to the stresses and shocks that the climate emergency brings. Inequality fundamentally holds back advances against the climate crisis by excluding the contributions of the most affected people, or their civic groups, youth leaders, and advocates. The neo-colonial approach of climate politics, based on climate science from a specific body of knowledge rooted in Western epistemologies, will not be enough to fight the climate crisis.

This chapter describes how climate justice and inequality are currently framed. It highlights that these framings emerge from two injustices that deserve more attention: epistemic and political injustice. Climate change policies are rooted in a particular body of knowledge that ignores invaluable complementary understandings. The global climate justice endeavour is held back further by the fundamental political injustices of climate change policy processes and economic systems. This chapter makes the case for tackling systemic structural inequality of how the climate crisis is being tackled by reducing the inequalities of knowledge and political power.

An Evolving Understanding about Climate Justice and Inequality: Is Inequality Part of the Problem?

The link between climate change and inequality is now well recognized by policy makers, science, practitioners and affected populations. The main argument, central to climate justice, is that structural and underlying social and economic inequalities determine differential impacts of climate disaster on people and communities. In turn, the disproportionate effects of climate change on the most vulnerable increases inequality. This inequality is not an ideological argument; it has been well-substantiated: "The rapidly accelerating growth in total emissions – and the attendant rise in climate change risks and damage – has categorically not occurred to benefit the poorer half of the world's population" (Kartha et al., 2020; see Box 14.1, Figure 14.1).

Nearly half of the total growth in absolute emissions was due to the richest 10%, with the richest 5% alone contributing over a third (37%) (see Figure 14.1). The remaining half was due almost entirely to the contribution of the middle 40% of the global income distribution (the next eight ventiles). The impact of the poorest half (the bottom ten ventiles) of the world's population was practically negligible.

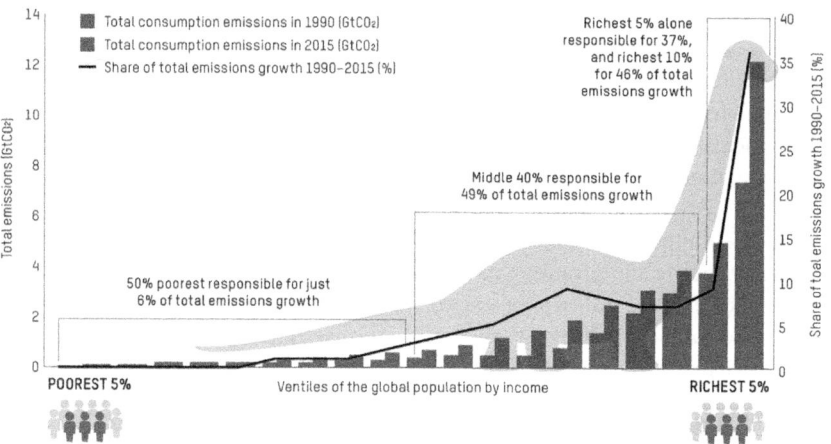

Source: Kartha et al. (2020).

Figure 14.1　The carbon inequality 'dinosaur' of emissions growth from 1990 to 2015

The world population is arranged in 20 groups (ventiles) by income, from the poorest 5% on the left to the richest 5% on the right. The line shows each ventile's increase in per capita emissions (as a percentage of its 1990 per capita emissions), while the bars show each ventile's increase in total emissions (as a percentage of total global emissions increase).

Unless global and national structural inequalities are tackled, climate change will continue to have a disproportionate effect on the most vulnerable in society, and will, in turn, contribute to increasing inequalities within and between countries. There is a long tail to this crisis: "Even under moderately progressive scenarios of socio-economic development, we show that very little atmospheric space is left for the world's poorest households" (Kartha et al., 2020). While there has been substantial debate on the carbon impact of economic growth in so-called "emerging economies" over the past 20 to 30 years, more attention must be paid to the continuing outsized impact of the minority of the world's richest citizens, wherever they reside, and the continuing developmental needs of the world's poorest citizens. Even as renewable technologies become a viable part of the energy future, the global carbon budget remains a precious natural resource. Socio-economic and climate policies must be designed to ensure its most equitable use.

Until quite recently, climate inequality and, in particular, its inverse – climate justice – were not well-established concepts. Climate justice is a young, twenty-first century idea that highlights the political and ethical nature of the climate crisis and its impact. It unequivocally links the breaching of climate change boundaries with the breaching of the rights of affected people. The concept challenged the status quo in two ways. First, it called out the false neutrality of the term 'climate change' that focused on environmental or biophysical processes, ignoring the role of political and economic systems that gave rise to the climate crisis. Second, it challenged simplistic environmental movements that have, at times, quite profoundly failed to recognize the unequal distributive impacts of environmental degradation on people.

To date, there is no agreed definition of climate justice. Its multiple facets can be described in terms of four aspects of justice and three forms (see Box 14.1). Harlan et al. provide a useful summary of how climate justice must be upheld based on four key principles:

1. Equity in distributing the burdens and sharing the benefits of climate change in communities and among nations.
2. Social and political processes that recognize currently or previously marginalized groups as rightful participants in the governance and management of climate change.
3. Freedom of peoples to make choices that maximize their capabilities to survive now and in the future; and
4. Rebuilding damaged historical relationships between parties, correcting past wrongs against humanity, and restoring the Earth. (Harlan et al., 2015)

BOX 14.1 UNPACKING CLIMATE JUSTICE DIMENSIONS AND TYPES (MAYNE 2020)

Dimensions of justice

1. *Intergenerational* justice i.e., between generations, notably ensuring a safe climate for future generations
2. *International* justice between countries, including equal distribution of the remaining carbon budget between countries, and providing financial, practical and technical support to lower-income countries and

groups of people who have contributed the least to climate change, by high-income countries who cause the most climate damage
3. *Intranational* justice within countries, i.e., pro-poor carbon mitigation and adaptation policies and programmes that reduce poverty and inequality, with respect for people's and, in particular, women's rights

Types of justice

1. *Distributional justice* – relating to the distribution of responsibilities, capacities, costs and benefits of climate action
2. *Procedural* justice – relating to who is at the decision-making table
3. *Recognition justice* – recognizing and taking action to address structural constraints
4. *Corrective justice* – providing remedies for past injustices

Although climate justice as a term was coined around the time of the 6th Climate Justice Summit in 2000 in the Hague, it took over a decade for policy makers to recognize the links between climate change and inequality and to adopt a climate justice perspective. In 2015, article 2.2 of the Paris Agreement (UNFCCC, 2015) first reflected the concept of equity of responsibilities among member states. While the simple acknowledgement fell well short of demands made for climate justice by civil society and many member states, at least there was a principle around which to anchor the urgency of challenging inequality to fight the climate crisis. Since the Paris Agreement, recognition of inequality as central to the climate emergency has accelerated not only in key international agreements and reports (IPCC, 2019) but also in research and the wider discourse.

Notwithstanding the growing understandings and uses of 'climate justice,' critics continue to exert their voice. Their argument centres on two practical problems. First, they point out the complexity of finding solutions if climate change and socio-economic inequalities are to be tackled simultaneously. Second, pointed out is the inherent trade-off between the two goals that will lead to a weakening of efforts to fight the climate crisis (Klinsky et al., 2016).

HOW DOES INEQUALITY PLAY OUT IN CLIMATE JUSTICE SOLUTIONS?

Given the many forms in which climate justice is understood, it is unsurprising that responses have been diverse, each addressing a certain aspect of inequality and often blind to other aspects.

Response 1. Ensure Everyone Decides on Mitigation and Adaptation Action

At the broadest level, inequality exists between the global North and the global South that plays out in two ways. First, an unequal dialogue exists creating a simplistic narrative that the global North should mainly mitigate what causes the climate crisis with the global South needing mainly to adapt to it. But it is fundamentally unequal if those affected by climate change cannot play a key part in the discussion of how the planet must mitigate. Therefore, approaches to climate justice from this perspective have focused on the disproportionate effects of the climate crisis (Islam & Winkel, 2017) and adaptation efforts. Such solutions emphasize fairer distribution, for example, of carbon emission rights, or participation in governance and capacity building. This line of thought focuses on how inequality is a cause of unjust climate change *adaptation*, but it does not include *mitigation*. More participatory discussion on mitigation would redirect thinking towards the predatory economy that underpins the crisis.

A second aspect of mitigation/adaptation solutions has emerged from practice. Vulnerability-oriented work focused on understanding risk to disaster as the intersection of hazard and vulnerability. It has evolved towards a resilience orientation, focusing strongly on strengthening the capacities of affected people and societies to absorb, adapt and transform. This is problematic if inequality is absent – as those not causing the problem are being asked to fix it. More recently, resilience has been more explicitly framed around a consideration of inequality:

> ...if building resilience is now on the agenda of national governments, donors, aid agencies and civil society, this must go beyond the dry, technical fixes that have dominated the discussion so far. Building skills and capacity must go alongside tackling the inequality and injustice that make poor women and men more vulnerable in the first place. (Hillier & Castillo, 2013)

Response 2. Redress National Inequalities

The second way in which inequality plays out in solutions to the climate crisis is between countries. One arena of debate and solutions concerns carbon emissions. Carbon inequality describes the inequality between national responsibility for greenhouse gas emissions. The consumption patterns of high-income countries, households and individuals make them the biggest polluters while suffering fewer impacts (Kartha et al., 2020). The solution has centred on mitigation – central to much research, discussion and negotiations. These efforts focus on a fair distribution of the global carbon budget between the countries, based on the criteria of equity while also taking into account historical responsibilities (Alcaraz et al., 2018). This will bring together the 'right' to equal per-capita emissions pondering what individuals in each country have emitted up to now.

Another country-focused solution effort focuses on rebalancing the loss-damages disequilibrium. In the research and discussion on national responsibilities, it is for the cost of adaptation and the equity of paying for the damage done. The establishment of the Warsaw International Mechanism for Loss and Damage in 2013 was heralded as a major victory for developing nations that are particularly vulnerable to climate change impacts and their NGO allies. But two years later, the Paris Agreement made the explicit exclusion of claims for compensation or assertions of liability (Vanhala & Hastbaek, 2016).

Response 3. Tackle Intersecting Individual Inequalities

Climate justice solutions have also focused on redressing the inequalities that lead certain groups to experience worse impacts. In many contexts, women often suffer disproportionately from climate change impacts compared to men (Habtezion, 2016). Gendered roles and norms mean many women have less access to and control over assets to help them cope with shocks, with climate impacts forcing them to walk longer distances to find water, wait longer at pumps and pay more for water. Disaster fatality rates are much higher for women than for men (ibid.). In some contexts, disasters induced by climate change may increase child marriage as parents seek to cope economically (Girls Not Brides, 2018; McLeod et al., 2020). Some social groups, such as indigenous people, people of colour, low-income women or female-headed households, and youth low-income communities and countries face a quadruple injustice (Mayne 2020):

- Contribute the least to causing it, yet suffer the worst impacts, without adequate compensation from those who caused it.
- Pay the most, as a proportion of income or time, for climate change mitigation policies and programmes, e.g. via carbon taxes or levies on fuel bills or use of their own resources.
- Benefit the least from carbon mitigation policies and programmes due to inequitable design and structural constraints.
- Are less able to participate in decision-making around mitigation and adaptation responses due to exclusive decision-making processes and structural constraints.

Climate justice solutions that focus on the individual include providing equal access to benefits for those on the margins, working on greater mobility and accessibility to key resources, promoting people's democratic rights and participation, and investing in climate change cooperative efforts that are locally led and locally driven (Acakpo-Addra et al., 2017).

As this section described, the climate crisis and inequality nexus has led to many discourses and arenas of action. However, two areas where inequalities persist surround whose knowledge counts and whose voice is shaping policies. Epistemic and political injustice remain key blockers in climate justice.

BREAKING THROUGH THE ENTRENCHING EFFECTS OF EPISTEMIC AND POLITICAL INJUSTICE

To date, the most affected population and their experiences have been excluded from climate policy decision-making. Those most affected, often local and indigenous populations, are not co-creating solutions. Epistemic inequality and political injustice are key to unlocking this exclusionary effect. The current enlightened despotism of 'everything for the people, nothing by the people', even if well-intended, is holding the world back from finding solutions for the climate crisis.

Epistemic Justice: What is the Case for Epistemic Justice?

Where there is epistemic justice, different kinds of knowledge contribute fairly to critical decisions. In the world of climate justice, however, epistemic injustice prevails. Epistemic injustice extends beyond elevating one body of knowledge over another, such as scientific knowledge over experiential knowledge. It includes ignoring knowledge sources due to their identity, for example not accepting an Arabic farmer's experience due to their skin colour. It means ignoring some knowledges because their vocabularies or values do not fit with those that are privileged. It also, most fundamentally, means ignoring or devaluing what sort of questions about climate change and humans' relationship to their natural environment are asked, and the various ways of knowing with which they can be answered.

All communities facing climate change have generated knowledge to understand and cope with it. For centuries, local knowledge processes have (re)shaped cultures, resource management and livelihoods. As climate change accelerates, learning processes accelerate to cope and adjust – but within the limits of what is locally knowable and disconnected from the principal causes of the crisis. On the other side of the knowledge coin, scientists have been observing and documenting climate changes and their impacts for decades with the first Intergovernmental Panel on Climate Change (IPCC) report published in 1990. In this scientific endeavour, local knowledge on climate change phenomena has not been taken seriously or pursued systematically.

It is only recently that initiatives are underway to draw on local knowledge based on the idea of 'complementarity' (Reyes-García et al., 2016). While existing climate models are effective at providing global information on climate change, their ability to detect impacts at local scale are deeply questioned. The myriad of uncertainties generated by climate change has highlighted the need for more detailed local observations. Local knowledge that generates granular observations of change can partner with mainstream science, either, at the least, serving to corroborate what academic science detects, or filling the knowledge gaps where science fails.

The idea of knowledge complementarity has led to efforts to value what is generated locally beyond simply corroborating mainstream science observations: "Most of the scientist-led qualitative research does not include locally led efforts documenting climate change impacts. A network coordinating the scalability of place-based research on climate change impacts is needed to bring Indigenous and local knowledge into global research and policy agendas" (Reyes-Garcia et al., 2019, p. 1). One such network is the Local Indicators of Climate Change Impacts, or LICCI (https://licci.eu/). Its members argue that global data are too coarse and thus that local climate impacts, particularly on socio-economic systems, remain poorly understood. LICCI gathers local data with qualitative and qualitative models to improve the understanding of how the climate crisis is affecting local socio-economic systems and uses insights for creating more appropriate adaptation and mitigation policies.

WHAT'S HOLDING BACK EPISTEMIC JUSTICE?

Disconnecting climate science from local climate knowledge is exacerbating non-inclusive climate policies. Three epistemic problems exist: (1) an inaccessible science, (2) the scientific rejection of local knowledge, and (3) the technical impossibility of including local knowledges.

Climate science has created a body of knowledge about the climate crisis that is abstract, a specialized area of scientific knowledge that has no cultural meaning. This body of knowledge is inaccessible to precisely those people who should be using the evidence to influence decision-making and governmental action to fight the climate crisis, whether global and local (Harlan et al., 2015). One criticism of dominant scientific representations of climate change is the separation of scientific facts, statistics and models from the very geographies and timescales that people can understand and act on:

> Durable representations of the environment do not arise from scientific activity alone...but are sustained by shared normative and cultural understandings of the

world as it ought to be. When it comes to nature, human societies seem to demand not only objectively claimed matters of fact but also subjectively appreciated facts that matter. Environmental knowledge achieves robustness through continual interaction – or conversation – between fact-finding and meaning-making. (Jasanoff, 2010, p. 248)

Local and indigenous people have been largely portrayed as victims of climate change with limited agency to know and respond. This misrepresentation pervades all aspects of society, including science and policymaking. Research on printed media representation of indigenous people and climate change in four Anglo-Saxon countries over a 20-year period found that:

Indigenous knowledge was mainly documented where it easily corroborates scientific knowledge, or when the impacts it identifies are sociocultural, and thus beyond the purview of "scientific" research. In such interpretations, complex knowledge systems are reduced to simple observations, valuable because they originate from regions where scientific data is sparse or confirm scientific findings. A focus on Indigenous belief systems, cosmologies, and alternative ways of knowing and interpreting climate change, are largely absent from the articles reviewed. (Belfer et al., 2017, p. 67)

Quantitative rigor, or the lack of it, is a key technical requirement that leads climate science to avoid incorporating local knowledge. Despite some efforts around epistemic complementarity, the majority of climate scientists and policy makers believe that indigenous knowledge lacks quantitative rigor as it is transmitted verbally over generations living in a particular environment (Alexander et al., 2011). Another technical difficulty is diversity. Many thousands of 'local knowledges' are in a stand-off against one dominant paradigm of Western scientific knowledge. In the rare cases where climate scientists consider some form of local knowledge, only a select few are included. But possibly the greatest obstacle comes from the perception that local knowledge should be folded into climate scientific understandings, instead of a more open-minded approach of mutual co-learning.

Local knowledge production is profoundly different from mainstream climate science, with no interlocutors to bridge this divide. If such different interest groups and knowledges are to communicate and integrate as equals, ways need to be created to codify across this knowledge divide. The greater the inequality in a society, the more difficult the communication and ability to connect across groups (UNDP 2019).

Stubborn barriers exist that hinder local knowledges from informing climate decision-making. Cartesian-based science is appropriate when uncertainty is low, controllability is high and the rate of change slow. It is based on assumptions of equilibrium and controllability so conditions of high uncertainty and low controllability, such as climate, call for additional ways of knowing.

Global environmental, economic and social problems do not occur in isolation but are interconnected in unexpected ways. This understanding of the interdependent nature of the social-ecological system is integrated in the cosmovision of many societies. Using a social-ecological system as the unit of analysis requires a shift in conventional governance practice, eliminating the divide between those governing and those being governed – and eliminating artificial disciplinary divides (Berkes, 2017). But there is also a political inequality that hinders the influence of local knowledge – mainstream science has few incentives to make this effort of bridging worlds as it prevails.

POLITICAL JUSTICE

What is the Case for Political Justice?

Political justice is, in the general case, about ensuring that people's right to inform and influence processes is not undermined. This means observing the basic accountabilities we expect of representational politics; ensuring those are not serving corporate interests disproportionately over the collective will of peoples; answering questions of who assesses climate impacts and assigns the limited carbon budget; and ensuring that those social groups and countries most affected by climate change are meaningfully represented and heard at the table of climate negotiations and international agreements. In other cases, political justice is about purposefully and proactively bringing the most frequently ignored voices and experiences to the decision-making table of climate change policy – usually those of women, indigenous people, people with disabilities, and many others. In the most extreme cases, it is about questioning the dominant economic system that kills people. In 2019, 212 land and environmental defenders were murdered, the highest record in a single year (Global Witness, 2020). Every year, more people are killed defending the environment than are soldiers from the United Kingdom and Australia on overseas deployments in war zones combined (Butt et al., 2019). During the last 15 years, recorded deaths related to conflict over natural resources have increased from two per week to four per week. The rule of law and corruption indices are closely linked to patterns of killings (Butt et al., 2019).

What's Holding Back Political Justice?

In January 2019, the Lancet Commission reported on the global syndemic of obesity, undernutrition and climate change – three interlinked pandemics of ill health occurring simultaneously. The Lancet Commission Report (*The Lancet*, 2019) identifies three fundamental drivers: "Underpinning all of these are weak political governance systems, the unchallenged economic pursuit of

GDP growth, and the powerful commercial engineering of overconsumption" (ibid., p. 1). All fingers are pointing to the destruction caused by the dominant neoliberal economic model (Raworth, 2017; Trebeck & Williams, 2019). These drivers and the economic system they have created have generated enormous global economic wealth, albeit highly unequally distributed. However, it is now becoming clearly unsustainable and actively dangerous for the planet and all its inhabitants.

Consumption economy and extractive industries. The drivers generate and maintain consumption that demands materials and resources. This demand is met through economic growth and extraction, processing and manufacturing, trade and waste disposal. The three most problematic sectors are: (1) extractive industries including fossil fuels and minerals, and accompanying infrastructure including dams to meet energy demands; (2) industrial agriculture for cereals, animals, and plantations; and (3) criminal activities, including drugs, wildlife and human trafficking and the sex industry. From 2000, global material extraction for food, fuel and minerals has accelerated and grown at a faster rate than global GDP, with a diminishing rate of return as resources become more expensive to extract and environmental costs harder to ignore (Watts, 2019).

The focus on changing people's behaviour to use less paper or plastic bags ignores the reality that it is this economic model that is at fault. Just 20 fossil fuel companies have contributed to 35% of all energy-related carbon dioxide and methane worldwide since 1965:

> The great tragedy of the climate crisis is that seven and a half billion people must pay the price – in the form of a degraded planet – so that a couple of dozen polluting interests can continue to make record profits. It is a great moral failing of our political system that we have allowed this to happen. (Mann, cited in Taylor & Watts, 2019)

Going digital is no panacea. In 2018, data centres used at least 1% of electricity consumption (Masanet & Lei, 2020), while the annual Bitcoin carbon footprint is comparable to that of New Zealand (de Vries, 2020; Digiconomist.net, 2020).

The shareholder-first business model provides a strong incentive for big business to relegate the urgency of the climate crisis to a second priority (High Pay Centre & TUC, 2019). In 2018, British Petroleum spent 14 times as much on their shareholders as it invested in low carbon activity; Shell spent at least 11 times more on their shareholders (ibid., p. 13). To avoid catastrophic climate change, the vast majority of proven fossil fuel reserves would need to be left in the ground. Yet between 2020 and 2029, the UK's two largest energy companies will spend $220 billion developing new oil and gas fields (ibid.). In

2018, Shell, the UK's largest energy company pledged to invest $1–2 billion in 'new energy,' while BP, the second largest, invested $500 million in low carbon activity energy at the same time as they paid their shareholders a total of £25 billion (ibid.). This is not heeding the growing calls for a 'better future' in a COVID-changed world.

Political capture keeps the consumption machine churning and shareholder-first practices thriving. It has led to today's grotesque inequality, where the rich 1% crave and perpetuate an extractive economy that is causing the climate crisis. Corruption props up this political economy, blocking access of those seeking change. Political capture maintains the extractive economy that concentrates wealth inequalities. In the process actors are excluded from climate politics, blocking access of the most affected to take action, or being part of any meaningful change; marginalizing and making invisible leadership that can go from local to global with new approaches to the problem; undermining or blocking South–South alliances and movements that can bring new approaches and leaderships to the global stage.

In addition, political capture plays out within the borders of many countries. Major barriers often exist to communication and collaboration between national and sub-national levels. These are caused by power imbalances across governance levels that reflect broader institutional differences between federal and decentralized systems of government. Ironically, the same division seen internationally with the global North shaping mitigation and the global South responsible for adaptation is reproduced at national level. National level actors focus on mitigation policy arenas, while local actors dominate in the adaptation policy domain (Di Gregorio et al., 2019).

Breaking Through the Impasse

The climate crisis is also a socio-economic and a political crisis. The causes and drivers of all crises are one and the same, generating extreme wealth for some alongside entrenching the injustices of poverty for many hundreds of millions. The urgency for action on climate change is widely recognised, even if action is woefully deficient. Global carbon emissions must reach net-zero by 2050 in order for humanity to have a reasonable chance of restraining temperature rise to 'well below' 2 degrees centigrade and to have any chance at all of restraining temperatures to 1.5 degrees centigrade, the stretch goal in the Paris Agreement. To do this, emissions must be halved in the next 10 years or there will be no viable downward trajectory to reach net zero by 2050.

Glimmers of hope exist. The massive surge in global concern and mobilization cannot be denied, and with it a growing recognition for intergenerational justice. Growing movements, including Extinction Rebellion and the rise of youth climate activism, show a shift from closed door policy debates to everyday concerns for every person. Cases of practical changes go well beyond mini islands of success (Mayne & Guijt 2020). The European Union's goal of net-zero by 2050 and China's carbon neutrality by 2060 are accompanied by technological breakthroughs making solar and onshore wind power the cheapest new sources of electricity for two-thirds of the world's population.

These seismic changes are still too slow, with responses to the climate crisis dominated by technocratic solutions that are not implemented urgently enough, while debating next steps and implementation gaps through a global governance system that excludes those with the most to lose. Working towards greater epistemic and political justice means breaking down the inequalities that shape whose knowledge counts and whose power matters in the collective fight to end the climate crisis.

REFERENCES

Acakpo-Addra, E.S. et al. (2017). *Gender Just Climate Solutions*. The Women and Gender Constituency (WGC). https://wedo.org/wp-content/uploads/2019/06/WGC-Solutions-Publication-COP23-ENG-Final-.pdf

Alcaraz, O., Buenestado, P., Escribano, B., Sureda, B., Turon, A., & Xercavins, J. (2018). Distributing the Global Carbon Budget with Climate Justice Criteria. *Climatic Change*, 149, 131–145. https://doi.org/10.1007/s10584-018-2224-0

Alexander, C., Bynum, N., Johnson, E., King, U., Mustonen, T., Neofotis, P., Oettle, N., Rosenzweig, C., Sakakibara, C., Shadrin, V., Vicarelli, M., Waterhouse, J., & Weeks, B. (2011). Linking Indigenous and Scientific Knowledge of Climate Change. *BioScience*, 61(6), https://doi.org/10.1525/bio.2011.61.6.10

Belfer, E., Ford, J.D., & Maillet, M. (2017). Representation of Indigenous Peoples in Climate Change Reporting. *Climatic Change*, 145, 57–70. https://doi.org/10.1007/s10584-017-2076-z

Berkes, F. (2017). Environmental Governance for the Anthropocene? Social-Ecological Systems, Resilience, and Collaborative Learning. *Sustainability*, 9(7), 1232. https://doi.org/10.3390/su9071232

Butt, N., Lambrick, F., Menton, M., & Renwick, A. (2019). The Supply Chain of Violence. *Nature Sustainability*, 2, 742–747. https://www.nature.com/articles/s41893-019-0349-4

de Vries, A. (2020). Bitcoin's Energy Consumption is Underestimated: A Market Dynamics Approach. *Energy Research & Social Science*, 70. https://doi.org/10.1016/j.erss.2020.101721

Digiconomist.net. (2020). Bitcoin Energy Consumption Index. https://digiconomist.net/bitcoin-energy-consumption

Di Gregorio, M., Fatorelli, L., Paavola, J., Locatelli, B., Pramova, E., Nurrochmat, D.R., May, P.H., Brockhaus, M., Sari, I.M., & Kusumadewi, S.D. (2019). Multi-Level Governance and Power in Climate Change Policy Networks. *Global Environmental Change*, 54, 64–77. https://doi.org/10.1016/j.gloenvcha.2018.10.003

Girls Not Brides (2018). *Child Marriage in Humanitarian Settings*. https://www.girlsnotbrides.org/wp-content/uploads/2016/05/Child-marriage-in-humanitarian-settings.pdf

Global Witness. (2020). Defending Tomorrow. https://www.globalwitness.org/en/campaigns/environmental-activists/defending-tomorrow/

Habtezion, S. (2016). Gender and Climate Change: Gender, Climate Change and Food Security. United Nations Development Programme. https://www.undp.org/content/dam/undp/library/gender/Gender%20and%20Environment/UNDP%20Gender,%20CC%20and%20Food%20Security%20Policy%20Brief%203-WEB.pdf

Harlan, S., Pellow, D., Roberts, J., Bell, S., Holt, W. & Nagel, J. (2015). Climate Justice and Inequality. In R. Dunlap & R. Brulle (Eds.), *Climate Change and Society* (pp. 127–163). American Sociological Association. 10.1093/acprof:oso/9780199356102.003.0005

High Pay Centre & TUC. (2019). *How the Shareholder-First Business Model Contributes to Poverty, Inequality and Climate Change.* https://www.tuc.org.uk/sites/default/files/2019-11/Shareholder%20Returns%20report.pdf

Hillier, D., & Castillo, G.E. (2013). *No Accident: Resilience and the Inequality of Risks.* Oxfam GB. https://www-cdn.oxfam.org/s3fs-public/file_attachments/bp172-no-accident-resilience-inequality-of-risk-210513-en_1_0.pdf

IPCC. (2019). Global Warming of 1.5°C. United Nations Intergovernmental Panel on Climate Change. https://www.ipcc.ch/sr15/

Islam, S.N. & Winkel, J. (2017). *Climate Change and Social Inequality.* United Nations Department of Economic and Social Affairs. https://www.un.org/development/desa/publications/working-paper/wp152

Jasanoff, S. (2010). A New Climate for Society. *Theory, Culture & Society*, 27(2–3), 233–253. https://doi.org/10.1177/0263276409361497

Kartha, S., Kemp-Benedict, E., Ghosh, E., Nazareth, A., & Gore, T. (2020). *The Carbon Inequality Era.* 2020 Stockholm Environment Institute & Oxfam. https://www.sei.org/wp-content/uploads/2020/09/research-report-carbon-inequality-era.pdf

Klinsky, Sonia, Timmons Roberts, Saleemul Huq, Chukwumerije Okereke, Peter Newell, Peter Dauvergne, Karen O'Brien, Heike Schroeder, Petra Tschakert, Jennifer Clapp, Margaret Keck, Frank Biermann, Diana Liverman, Joyeeta Gupta, Atiq Rahman, Dirk Messner, David Pellow, Steffen Bauer (2016) Why equity is fundamental in climate change policy research, *Global Environmental Change*, http://www.sciencedirect.com/science/article/pii/S0959378016301285

Leonard, N. (2019, 11 December). *Protesters kicked out of COP25, stripped of badges.* 350.org. https://350.org/civil-society-kicked-out-of-cop25/

Masanet, E. & Lei, N. (2020, 17 March). *How Much Energy Do Data Centers Really Use?* Energy Innovation. https://energyinnovation.org/2020/03/17/how-much-energy-do-data-centers-really-use/

Mayne, R. (2020). Climate Justice: Achieving a Fair and Fast Transition/Climate Justice. Oxfam internal document.

Mayne, R. & I. Guijt. (2020). *Inspiring Radically Better Futures: Evidence and Hope for Impact at Scale in a Time of Crisis.* Oxfam Research Report.

McLeod, C., Barr, H., & Rall, K. (2020). Does Climate Change Increase the Risk of Child Marriage? A Look at What We Know – And What We Don't – With Lessons from Bangladesh & Mozambique. *Columbia Journal of Gender and Law*, 38(1), 96–145. https://doi.org/10.7916/cjgl.v38i1.4604

Raworth, K. (2017). *Doughnut Economics: Seven Ways to Think Like a 21st-Century.* Random House Business.

Reyes-García, V., Fernández-Llamazares, Á., Guèze, M., Garcés, A., Mallo, M., Vila-Gómez, M., & Vilaseca, M. (2016). Local Indicators of Climate Change: The Potential Contribution of Local Knowledge to Climate Research. *Wiley Interdiscip.*

Rev. Clim. Change, 7(1), 109–124. doi: 10.1002/wcc.374. PMID: 27642368; PMCID: PMC5023048

Reyes-García, V., García-del-Amo, D., Benye, P., Fernández-Llamazares, A., Gravani, K., Junqueira, A.B., Labeyrie, V., Li, X., Matias, D.M.S., McAlvay, A., Mortyn, P.G., Porcuna-Ferrer, A., Schlingmann, A., & Soleymani-Fard, R. (2019). A Collaborative Approach to Bring Insights from Local Observations of Climate Change Impacts into Global Climate Change Research. Current Opinion in Environmental Sustainability, 39, 1–8. https://www.sciencedirect.com/science/article/abs/pii/S1877343518301295

Taylor, M. & Watts, J. (2019, 9 October). Revealed: The 20 Firms Behind a Third of all Carbon Emissions. *The Guardian*. https://www.theguardian.com/environment/2019/oct/09/revealed-20-firms-third-carbon-emissions

The Lancet. (2019). The Global Syndemic of Obesity, Undernutrition, and Climate Change: The Lancet Commission report. https://www.thelancet.com/commissions/global-syndemic

Trebeck, K. & Williams, J. (2019). *The Economics of Arrival: Ideas for a Grown-Up Economy*. Polity Press.

UNDP. (2019). Human Development Report 2019: Beyond Income, Beyond Averages, Beyond Today: Inequalities in Human Development in the 21st Century. United Nations. http://hdr.undp.org/en/content/human-development-report-2019

UNFCCC. (2015). Paris Agreement. United Nations Framework Convention on Climate Change. https://unfccc.int/sites/default/files/english_paris_agreement.pdf

Vanhala, S. & Hestbaek, C. (2016). Framing Climate Change Loss and Damage in the UNFCCC Negotiations. *Global Environmental Politics*, 16(4), 111–129. https://doi.org/10.1162/GLEP_a_00379

Watts, J. (2019, 12 March). Resource Extraction Responsible for Half World's Carbon Emissions. *The Guardian*. https://www.theguardian.com/environment/2019/mar/12/resource-extraction-carbon-emissions-biodiversity-loss

15. Towards global environmental governance
Julia M. Puaschunder

THEORY

Climate Justice Within Countries

In order to finance climate change mitigation and adaptation efforts, a diversified taxation scheme is proposed. To find a fair and just distribution of the burden of climate change, a taxation mix of (1) consumption tax, (2) progressive tax and (3) inheritance tax is recommended. Consumption tax can curb harmful emissions and directly nudge behavior towards sustainability. Yet to place a fair share of the burden of climate change mitigation upon society, these taxes have to be adjusted to the individual disposable income in order to not charge low-income households more heavily. Retroactive taxation of past wealth accumulation at the expense of environmental damage can be enacted through inheritance tax of the corporate sector. Industries should be taxed, when a merger or acquisition or a board member change occurs, in order to reap benefits from past wealth accumulation that potentially caused carbon emissions.

Climate Justice Between Countries

Following the introduction of the gains from climate change (Puaschunder, 2017a), the Climate Justice in the 21st century endeavor proposes a model to distribute the benefits of a warming earth in a fair way. Based on legal subsumptions and ethical imperatives, argumentations of those countries having better means of protection and conservation of a stable climate, lead to the pledge of climate change winners having to bear a higher weight of climate stabilization efforts (Rawls, 1971). The climatorial imperative advocates for the need for fairness in the distribution of the global earth benefits among nations based on Kant's (1783/1993) categorical imperative to only engage in actions one wants to experience being done to oneself. Rawls' veil of ignorance (1971) advocates for analyzing ethical dilemmas without taking into consideration whether the individual decision maker is a beneficiary or a sufferer of costs and damages. From this angle, all climate change winners and losers come to the conclusion to avert the warming of the earth. The climatorial imperative thereby advocates for redistributing climate gains and losses, which is argued philosophically and ethically to alleviate climate inequality (Puaschunder, 2017a, 2017b, 2017c, 2020).

MODEL

A macroeconomic cost-benefit analysis aids to find the optimum solution on how to distribute climate change benefits and burden within society and over time. Based on the optimal temperature for Gross Domestic Production (GDP) measured on the pillars of agriculture, industry and service sector productivity, the optimal temperature condition for economic productivity can be derived per country.

RESULTS

Given data of the average temperature per country around the world as well as climate projections of the year 2100 under a business-as-usual path, the world is found to macro-economically benefit from climate change more until 2100 than lose (Puaschunder, 2016a, 2016b, 2016c, 2016d). Climate change winners and losers are distributed highly unequally around the world (see https://blogs.cuit.columbia.edu/jmp2265/). Winning and losing from a warming earth is significantly positively correlated with self-reported CO_2 emissions, leading to the conclusion that the countries with the longest time horizon regarding a warming earth lack motivation to mitigate global climate change. Detected climate-induced migration streams and financial flows manifest that different parts of the earth are affected differently by a warming earth.

Based on a 187 country-strong dataset, a significantly positive inflow of migrants was found into the climate change winner countries (Puaschunder, forthcoming a). A statistically significant correlation highlights a positive Foreign Direct Investment (FDI) inflow into the territories that have more time ahead towards the temporal peak condition for GDP production (Puaschunder, 2020). No significant remittances flow to climate change loser countries is found. The results underline the need to redistribute the gains from climate change to offset losses incurred from global warming and demand for a recognition of climate refugees under the Geneva Convention.

Having found that there are gains from a warming earth demands transfers of benefits into areas of the world that will be primarily losing from climate change (Chichilnisky, 1996, 2016; Chichilnisky et al., 1998; Chichilnisky &

Heal, 2000). Having shed light on the gains of a warming earth allows for the redistribution of climate change benefits to those areas of the world that will be economically losing from a warming earth. In the implementation, climate change bonds but also taxation strategies are recommended (Chichilnisky, 1996, 2016). In order to avoid governmental expenditure on climate change hindering economic growth (Chichilnisky, 2007, 2010, 2016); the 'Climate in the 21st century' idea offers a new way of funding climate change mitigation and adaptation policies. A broad-based climate stability bonds-and-taxation mix could fund a transition to renewable energy (Puaschunder, 2020; *World Bank 2015 Report*, 2015). In order to finance climate change abatement, a climate bonds financing mix could subsidize the current world industry for transitioning to green solutions. Sharing the costs of climate stabilization between and across generations is a Pareto-optimal strategy to immediately instigate climate action without curbing today's economic growth potentials (Chichilnisky et al., 1998; Chichilnisky & Heal, 2000; Chichilnisky & Sheeran, 2018).

DISCUSSION AND IMPLICATIONS

Climate Justice Over Time: Tax-and-Bonds Transfer Strategy

As for redistributing the gains of a warming globe in order to offset for losses incurred by global warming, a climate change bonds-and-tax finance strategy is proposed to bear the burden of climate change in a right, just and fair way within society, around the globe and over time (Puaschunder, 2017a, 2017b, 2017c). In climate change winner countries weighted by GDP per capita, taxation should become the main climate stability financialization strategy. Foremost, the industries winning from a warming climate should be taxed. Regarding concrete climate taxation strategies, a carbon tax on top of the existing tax system should be used to reduce the burden of climate change and encourage economic growth through subsidies. Within a country, high- and low-income households should face the same burden of climate stabilization adjusted for their disposable income. Finding the optimum balance between consumption tax adjusted for disposable income through a progressive tax scheme will foster tax compliance in the sustainability domain.

Governments in global warming loser countries weighted by GDP per capita should receive tax transfers in the present from the winning countries. The climate change loser countries should also borrow by loans or issuing of bonds to be paid back by future generations. Taxing future generations is justified as future generations avoid higher costs of climate change long-term damages and environmental irreversible lock-ins. Overall this tax-and-transfer mitigation policy thus appears as a Pareto-improving fair solution across the world and among different generations.

Tax-and-bonds transfers could be used to incentivize industry actors for choosing clean energy. The revenues raised from taxation and bonds would thereby be allocated to subsidize corporations choosing clean energy. This market incentive could shift the general race-to-the-bottom regarding price cutting behavior and choosing dirty, cheap energy to a race-to-the-top hunt for subsidies for going into clean energy and production.

Concluding, climate change winning countries are advised to use taxation of the gains in sectors to raise revenues to offset the losses incurred by climate change. Climate change losers should issue bonds to be paid back by taxing future generations. Climate justice within a country should also pay tribute to the fact that low- and high-income households share the same burden proportional to their dispensable income, for instance enabled through a progressive carbon taxation. Those who caused climate change could be regulated to bear a higher cost through carbon tax in combination with retroactive billing through a corporate inheritance tax to reap benefits from past wealth accumulation that contributed to global warming.

GREEN NEW DEAL

Historical Foundation

The New Deal was historically a bonds financing strategy of the United States of America during the years 1932 to 1939. In total, around 15 to 35 billion USD were spent on a series of development programs that funded public work projects, financial reform and regulation efforts on economic development. US President Franklin D. Roosevelt's overarching goal of the project was to relief, reform and recover from the Great Depression.

The now newly enacted Green New Deal (GND) advocates for a co-use of carbon tax and green bonds in order to stimulate economic growth. Based on the foundations of Modern Monetary Theory, the GND targets at vitalizing the economy by a transition into renewable energy and sustainable growth. The Green New Deal (GND) serves as a market solution to implement global environmental governance as "the sum of the many ways individuals and institutions, public and private, manage their common affairs." The GND thereby combines Roosevelt's economic approach with modern ideas such as renewable energy and resource efficiency.

Framework

The Green New Deal group operates within the framework of the United Nations Environment Programme (UNEP) since 2008 to create jobs in green industries, thus boosting the world economy and curbing climate change at the same time. In 2019 over 600 organizations submitted a letter to Congress declaring support for policies to reduce greenhouse gas emissions. This includes ending fossil fuel extraction and subsidies, transitioning to 100% clean renewable energy by 2035, expanding public transportation, and strict emission reductions rather than reliance on carbon emission trading.

Since 2019 Senator Edward Markey and Representative Alexandria Ocasio-Cortez have pushed for transitioning the United States to use 100% renewable, zero-emission energy sources, including investment into electric cars and high-speed rail systems, and implementing the social cost of carbon that was part of the Obama administration's plans for addressing climate change within 10 years. Besides increasing state-sponsored jobs, this GND is also aimed to improve vulnerable communities via universal health care, increased minimum wages and preventing monopolies. A 10-year national mobilization targets at work security and working conditions by high-quality health care, affordable housing, economic security, access to clean water, air, healthy food and nature, education, clean, renewable, zero-emission energy, repairing of infrastructure, energy efficient smart power grids, upgraded living conditions, pollution elimination, clean manufacturing and positive work collaborations.

In January 2019, a letter signed by 626 organizations in support of a GND was sent to all members of Congress. It called for measures such as an expansion of the Clean Air Act, a ban on crude oil exports and fossil fuel subsidies and leasing and a phase-out of all gasoline-powered vehicles by 2040. The letter also opposed market-based mechanisms and technology options such as carbon and emissions trading and offsets. Various proposals for a GND have been made internationally, for instance in Australia, Canada and Europe.

Economic Foundations

Economic theories that back the GND include John Maynard Keynes' spending multiplier effect (1936), which captures the ratio of a change in national income to any autonomous change in spending – such as private investment spending, consumer spending, government spending, or spending by foreigners on the country's exports that causes it.

Joseph Stiglitz famously advocated for the GND by making the point that money will always be there and available but unmanageable environmental conditions may impose irreversible lock-ins and tipping points of no return.

Also, Jeffrey Sachs supports the idea of financial overspending for the sake of avoiding irreversible tipping points and environmental lock-ins. Money will always be there and is fungible, whereas environmental resources are depletable and irreplaceably destroyable.

IMPLEMENTATION

Global Environmental Governance features different means ranging from formal institutions (major global conferences and treaties), legal regimes, informal arrangements, intergovernmental relationships, nongovernmental organizations, global capital markets and multinational corporations (Puaschunder, 2020). Intergenerational equity attention aids to stabilize the climate. The current generations face high taxes and expenses. Future generations benefit from these investments for the future. With the right financialization strategy, these costs can be borne by future generations after the climate has been stabilized and is favorable for the humankind to come (Puaschunder 2018, 2019a, 2019b). Green bonds would be able to enact this intergenerationally harmonious solution. These financialization strategies are common in the public sector, for instance the New York water distribution is built on this bonds principle. With financial means that raised money via bonds, lakes could be built in mountains near New York. Now when water is consumed, the consumers pay off previous expenses.

As for monetary and credit policies, the importance of monetary policy in support of climate policy is visible in inflation targeting as a proper policy. Yet, adaptation, the provision of climate disasters, and the recovery are often producing bottlenecks causing higher inflation rates. So, targeting the inflation rate to move down inflation rates does not seem to be the appropriate policy if one has negative shocks on the supply side.

Different attempts exist from a fiscal space perspective. The public sector and governing institutions play a central role in overcoming free-rider problems and initiated market opportunities associated with externalities like climate change. Mitigation and adaptation policies and disaster risk prevention and recoveries may be supported by fiscal policy. Proposed financing tools include (long) maturity bonds – such as discussed in Sachs (2014), Orlov, Rovenskaya, Puaschunder and Semmler (2019) and Braga, Fischermann and Semmler (2020).

To peg emissions to tax payments, such as a carbon tax, appears simple and fair. Around the globe, about 14% of CO_2 emissions are subject to taxation. But most of these taxation efforts are only a few cents or dollars per CO_2 ton of emissions. Climate effects are only predicted for around 40 USD and increasingly doubling the taxation after an introductory phase successively. So far, Sweden has been quite successful with this: Since 1991 the CO_2 tax has been raised to 130 USD and carbon emissions dropped for about one-fourth while the economy could still grow.

Some researchers stress the importance of preventive actions and of policy buffers, designed to enhance resilience to shocks. Insurance policies to ease borrowing constraints, greater reserves, and reserve fund accumulation is suggested. Low-income countries and regions have limited access to issuing climate bonds and exercise little borrowing power. Besides tax increases, risk pooling through self-insurance or some collective insurance schemes, grants from donors, and buildup of financial buffers and disaster funds for contingencies are recommended.

Departing from their central focus on monetary and economic stability (e.g., legal tender and setting the interest rate to achieve market stabilization), central banks have recently gained interest in aiding on the financialization of climate change mitigation and adaptation. Around the globe, emissions trading covered around 20% of the global CO_2 emissions in about 40 countries of the world and over 20 cities, municipalities and provinces of the world ranging from China to the EU.

Innovation efforts and financialization foster technological innovations, which are usually a result of a mix of private and public activities. The public sector can set frameworks and incentives, to support inventions through R&D and de-risk of innovation through public support and subsidies and setting incentives. Public actions – such as tax and subsidies – could enable the transition to a low carbon economy and contribute to a faster transformation of the energy system toward a less carbon-based energy provision.

Engaged portfolio managers can impact on resource consumption choices and thereby implicitly determine energy prices. In an integrated economy, oil price fluctuations are causing disturbances in many industries. Portfolio and hedge fund managers strive for reducing risks to the overall portfolio, in the short and the long run. Renewable energy appears to be chosen for socio-psychological motives and with this is less dependence on market performance. This makes renewables more crisis-stable as investors stay with this market option during downturns, as it is chosen more based on personal values than financial performance and profit motives. Investment options based on renewable energy can reduce the risks and political dependencies on commodities associated with non-renewables.

Green Bonds allow to raise funds for green innovations. Solar power and wind turbines, eco-friendly infrastructure and more research and development in clean energy and green technology are all investments for climate change. Addressing market changes and the financialization of climate justice are estimated to comprise of 5–7% of the contemporary world GDP, accounting for 5–6 billion USD. Green bonds could fund all these endeavors.

Environmental pricing reform is the process of adjusting market prices to include environmental costs and benefits. A negative externality exists where a market price omits environmental costs. Then rational (self-interested) economic decisions can lead to environmental harm, as well as to economic distortions and inefficiencies. Environmental pricing reform is a market-based or economic instrument for environmental protection. Examples include green tax-shifting (ecotaxation), tradeable pollution permits, or the creation of markets for ecological services. "Ecological fiscal reform" differs in more narrowly dealing with fiscal (i.e., tax) policies as opposed to using non-fiscal regulations to achieve the government's environmental goals.

Absorbing CO_2 and forestation focuses on carbon-negative market solutions as CO_2 can be absorbed from the atmosphere. Examples of this are carbon-absorbing forests, green rooftops in cities, carbon-negative clothing through fungus-wear but also the absorption of CO_2 from the atmosphere by machinery and windmills as well as premia to stop deforestation. Another ground-breaking innovation could be decentralized energy grids that are run on blockchain approaches. Thereby single households could generate energy, for instance via solar panels on the rooftop or isolated heating devices. Immediately as the energy is generated, the individual household could either use the energy or distribute energy to close neighbors in a grid. This point-to-point solution between closer distributors and decentralized energy sharing could revolutionize the dependency on a few energy providers.

Behavioral insights offer the possibility of behavioral changes. In most recent decades, affluent people in high-income countries have defined environmental conscientiousness as a luxury good. High-end consumers around the world then have proven interest in goods that do not cause CO_2 emissions. They travel and shop environmentally conscientiously with respect for the wider community and are investing to fund social and environmental causes in their local communities. Behavioral insight – hence the behavioral economics application onto global governance – proves in many powerful laboratory and field experiments the power of behavioral nudges and winks on consumer choices with less money incentives. Nudges, the behavioral means to change people's choices based on their emotions, status and other environmental and social conditions, have proven to be powerful and easily implementable sources to educate and change people's behavior without direct enforcement (Puaschunder forthcoming a, b).

Sustainable tourism is the concept of visiting somewhere as a tourist and trying to make a positive impact on the environment, society, and economy. Tourism can involve primary transportation to the general location, local transportation, accommodation, entertainment, recreation, nourishment and shopping. It can be related to travel for leisure, business and visiting friends and relatives. There is now broad consensus that tourism development should be sustainable.

FUTURE RESEARCH ENDEAVORS

Future research may address the redistribution of climate change gains and losses (1) temperature range variations' economic impact, (2) commodity price estimates based on scarcity, (3) economic peak temperature for production re-estimates. (1) Temperature range estimates should be refined and connected to economic output. Does the economic output of countries with a vast temperature range based on latitude and altitude differ from countries with cyclical temperature changes? (2) Contemporary attention to global warming is assumed to affect commodity and beverage prices hyperbolically at extinction.

With the novel Coronavirus (COVID-19) spreading around the world from the beginning of 2020 on, calls are made that the medicine of the future should prevent diseases instead of just treating their consequences. In the novel Coronavirus crisis, prevention and general, holistic medicine determine whether COVID-19 puts patients on a severe or just mild symptom trajectory. Obesity, but also the general status of the immune system are decisive in whether the Coronavirus becomes a danger for the individual. The COVID-19 crisis is therefore an important accelerator for necessary, fundamental changes in the health system, which also results in ecological impacts as a healthy diet which is usually less carbon intensive.

REFERENCES

Braga, J.P., Fischermann, T. & Semmler, W. (2020). Ökonomie und Klimapolitik: So könnte es gehen. *Die Zeit, 11*, 5, March 10, 2020.
Chichilnisky, G. (1996). *Development and global finance: The case for an international bank for environmental settlements*. United Nations Development Programme, Office of Development Studies.
Chichilnisky, G. (2007). *The economics of global environment: Catastrophic risks in theory and policy*. Springer International.
Chichilnisky, G. (2010). *The economics of climate change*. Edward Elgar Publishing.
Chichilnisky, G. (2016). Reversing climate change. *Global Policy* retrieved online at https://www.globalpolicyjournal.com/blog/01/09/2016/reversing-climate-change-interview-graciela-chichilnisky
Chichilnisky, G. & Heal, G. (2000). *Environmental markets: Equity and efficiency*. Columbia University Press.
Chichilnisky, G., Heal, G. & Vercelli, A. (1998). *Sustainability: Dynamics and uncertainty*. Kluwer.
Chichilnisky, G. & Sheeran, K. (2018). *Handbook on the economics of climate change*. Edward Elgar Publishing.
Kant, I. (1783/1993). *Grounding for the metaphysics of morals*. Hackett.

Keynes, J.M. (1936). *The general theory of employment, interest and money.* Macmillan.
Orlov, S., Rovenskaya, E., Puaschunder, J.M. & Semmler, W. (2019). *Green bonds, transition to a low-carbon economy, and intergenerational fairness: Evidence from an extended DICE model.* International Institute for Applied Systems Analysis Working Paper WP-18-001, 2018. IIASA.
Puaschunder, J.M. (2016a). Intergenerational climate change burden sharing: An economics of climate stability research agenda proposal. *Global Journal of Management and Business Research: Economics and Commerce, 16,* 3, 31–38.
Puaschunder, J.M. (2016b). *Mapping Climate Justice.* Proceedings of the 2016 Young Scientists Summer Program Conference, International Institute for Applied Systems Analysis (IIASA).
Puaschunder, J.M. (2016c). On eternal equity in the fin-de-millénaire: Rethinking capitalism for intergenerational justice. *Journal of Leadership, Accountability and Ethics, 13,* 2, 11–24.
Puaschunder, J.M. (2016d). The call for global responsible intergenerational leadership in the corporate world: The quest for an integration of intergenerational equity in contemporary Corporate Social Responsibility (CSR) models. In D. Jamali (Ed.), *Comparative perspectives in global corporate social responsibility* (pp. 275–288). IGI Global Advances in Business Strategy and Competitive Advantage Book Series.
Puaschunder, J.M. (2017a). *Climate in the 21st century: A macroeconomic model of fair global warming benefits distribution to grant climate justice around the world and over time.* Proceedings of the 8th International RAIS Conference on Social Sciences and Humanities organized by Research Association for Interdisciplinary Studies (RAIS) at Georgetown University, Washington, D.C., United States, March 26–27, pp. 205–243.
Puaschunder, J. M. (2017b). Mapping climate in the 21st century. *Development, 59,* 3, 211–216.
Puaschunder, J.M. (2017c). The climatorial imperative. *Agriculture Research and Technology, 7,* 4, 1–2.
Puaschunder, J.M. (2018). *Intergenerational responsibility in the 21st century.* Vernon.
Puaschunder, J.M. (2019a). *Corporate and financial intergenerational leadership.* Cambridge Scholars Publishing.
Puaschunder, J.M. (2019b). *Intergenerational equity: Corporate and financial leadership.* Edward Elgar Publishing.
Puaschunder, J.M. (2020). *Governance and climate justice: global south and developing nations.* Palgrave Macmillan. Springer Nature.
Puaschunder, J.M. (forthcoming a). *Behavioral economics and finance: Nudging and winking to make better choices.* Springer Nature.
Puaschunder, J.M. (forthcoming b). *Verhaltensökonomie und Verhaltensfinanzökonomie: Ein Vergleich europäischer und nordamerikanischer Modelle.* Springer Gabler.
Rawls, J. (1971). *A theory of justice.* Harvard University Press.
Sachs, J.D. (2014). Climate change and intergenerational well-being. In L. Bernard & W. Semmler (Eds.), *The Oxford handbook of the macroeconomics of global warming* (pp. 248–259). Oxford University Press.
World Bank 2015 Report (2015). Washington, D.C.: World Bank.

16. Transition agendas: going beyond consumerism?

Boris Manov and Asen Balabanov

INTRODUCTION

We cannot predict the future. At least that is what one of the most influential liberal thinkers of the twentieth century claims in his book *The Poverty of Historicism* (Popper, 1957, p. 1). Inspired by the near-bloodless fall of the communist regimes in the early 1990s, Francis Fukuyama went even further, declaring the oncoming time "the end of history" (Fukuyama, 1989, pp. 3–18). With a series of more or less unforeseen cataclysms, history has decided to remind itself and shake even the seemingly most stable economic and social systems. It would be enough to mention the financial crisis of 2008, followed by the refugee wave of 2015, and more recently, today's pandemic crisis, consequences of which are yet to be felt and examined. Even though, environmentalists from all around the world argue, that, compared to the already observed implications of failing to meet the challenges of climate change and depletion of natural resources, the consequences of these cataclysms are like child's play, these crises still gave us some very valuable lessons. And perhaps the most valuable of these is that modern civilization is *one*. The "thickening" (Chardin, 1964/2003, p. 274) of humanity has led to the fact that there is no promised land, ideology or sphere of influence on the planet that can remain unaffected by such cataclysms. That is why today, in all parts of the planet Earth, one of the main questions is – can we continue living like this? Not only a scientific issue, but also a spiritual one, can we blindly put our faith into our technical progress or God's providence, or whether we should take matters into our own hands. And whether we agree with Karl Popper or not, scientists, thinkers, politicians, are all looking for solutions to today's global challenges in the future. Because otherwise we may not have one.

If there is positive news in the turbulences of recent decades, that is that in them more and more analysts are finding signs of a radical change for humanity. We are witnessing significant transformations not in one, but in in each sphere of human activity – economy, culture, society, psychology, politics, and so on, which gives grounds to speak even of a New Age in history. There is a variety of concepts describing and explaining this major transition. Some speak of an informational age, others of a planetary civilization, those who emphasize the importance of the human factor, call it the Anthropocene. And others find signs of the predicted decades ago upcoming of a reasonable layer around the earth – the so-called *Noosphere*. While one part welcomes the changes that are taking place, another argues that they do not significantly improve the state of human civilization, and a third even condemns them as a regression of social relations. For instance, the huge leap in technological development is indisputable, but does it lead to a corresponding social and cultural progress? Devices and machines have been created that can wipe thousands of human beings off the face of the earth in a matter of seconds, but, at least so far, no device has been invented that can feed or heal more than a few at the same time. Despite the introduction of increasingly sustainable and renewable technologies, the global economic system still relies on wasting too many natural resources to thrive and grow successfully. Resources that take too long to recover, and in some cases are even unrecoverable.

TRANSITION PATHS

This chapter will not consider environmentalist's dystopian predictions of the decline and demise of human civilization as a result of the climate crisis, nuclear war as a result of a struggle for key resources, or for world domination. It is based on the premise that taking into account the serious threats facing humanity is a *"conditio sine qua non"* in order to take appropriate counter-measures. But that those who are declared doomed, often resign or lose their human face. Therefore, with the understanding that the dilemma of progress in human history is not only a matter of scientific data and interpretation, but also a matter of belief, this chapter explores three possible paths for the progressive development of our modern civilization, focusing on their political aspects:

1. Smooth transition by the method of piece-meal social engineering (Lozev, 2018, para. 40), proposed by Karl Popper;
2. Gradual harmonic transformation on an evolutionary basis;
3. Accelerated drastic and dramatic changes – revolution.

Each of these three scenarios for the future has its advantages and disadvantages, the clarification of which will allow us to answer the question of which would be the most productive as a response to today's global challenges.

The *first one* may be considered to be the dominant view in the political systems of liberal Western societies. In addition, the unsuccessful attempts to put communist ideology into practice in the former Eastern bloc reinforced the positions of proponents of this approach in both West and East. In essence, this is the idea of undertaking small changes in social systems on a *trial-and-error* basis, with everything that works and leads to improvement being preserved, and what is unproductive being rejected. At least ideally. The advantages of this scenario are that it does not lead to dramatic social upheavals, because the changes made are insignificant, and the effect of failure of these changes is correspondingly insignificant. Also, the basic framework of society remains unchanged, which should ensure that the changes do not lead to the degeneration of the political regime into a dictatorship, to the violation of fundamental human rights, violence and coercion against civil society in the name of, for example, a utopian better future. If we should imagine the realization of such a scenario in practice today, it means the gradual and voluntary imposition of liberal political and economic ideology around the world, the slow and gradual replacement of energy-intensive sources and productions with renewable energy sources and efficient production, the gradual reduction of discarded waste and its comprehensive recycling, the gradual reduction of greenhouse gas emissions through milestones, reduction of deforestation, nature pollution, and so on. Last but not least, confidence in politicians and political systems, that they will not succumb to the pressure of *capital* when it is to the detriment of the public interest, that effective tools will be found for public and civil control, as well as priority and targeted financing of scientific innovation research that would gradually solve or at least limit the main human problems, such as hunger, inequality, water scarcity, diseases and pandemics, and so on.

It is in this context that Karl Popper emphasizes public control over politics, stating that "political power and its control is everything – economic power should not be allowed to take precedence over political power" (Lozev, 2018, para. 40). And this is exactly where the weak spots of such a political engineering in the transition to a Planetary Civilization are. If stability or preservation of the current model of civilization is sought, this would be the ideal approach. But if the opinion of the majority of scientists is taken into account and it is assumed that there are generally two alternatives – our modern civilization has to change or we will disappear – then this would be a losing strategy. Or at least one that is no longer up to date. Because we might just not have that much time anymore.

One of the concepts of the new age, *the Anthropocene theory*, claims scientific data clearly showing, that as a result of human activity, balance in the environment, on which the existence of our civilization actually depends, is seriously disturbed and four of the nine planetary boundaries have already been crossed – climate change, loss of biosphere integrity, land-system change, and altered biogeochemical cycles (phosphorus and nitrogen) (Welcome to the Anthropocene, n.d.).

And if that would be true, then it is time to abandon the conservative approach. Apart from the very slow pace of change, another significant problem of piece-meal social engineering is something Popper strongly emphasized – dominance of economic power over political power. The rapid growth of social inequality at a global scale since the 1990s, the concentration of world wealth in the hands of an ever lesser part of the world's population is a serious threat to democracy in modern societies, including those with well-established democratic practices. Under the pressure of the global financial elite, huge budget funds continue to be spent as a priority in areas that in no way contribute to solving the world's human problems, like for instance the *arms race*. On the opposite side, no one seriously believes that by 2030 even a minimum of the Global Goals (https://www.globalgoals.org/) set by the UN in 2015 can be achieved, as there is little or no budget whatsoever allocated.

Here we come to one of the key aspects of the second approach, that of the purpose. Obviously, the idea of piece-meal social engineering categorically rejects any utopian theories, aiming at a better universal future, as vain dreams. Yet our historical development unequivocally shows that things, considered centuries ago as utopian are now part of people's daily lives. For example, flying. Or the internet. Philosophically, can something be achieved without having the faintest idea of what the future should look like? Such an idea, undoubtedly utopian for its time, was developed by two scientists of quite different origins and profiles at the beginning of the twentieth century, as an example of how universal scientific knowledge does not divide, but unifies – the noosphere theory. The noosphere is a philosophical concept developed and popularized by the biogeochemist Vladimir Vernadsky, and the French philosopher and Jesuit priest Pierre Teilhard de Chardin. Vernadsky defined the *noosphere* as the new state of the biosphere and described as the planetary "sphere of reason". The noosphere represents the highest stage of biospheric development, its defining factor being the development of humankind's rational activities (Noosphere, 2021). The concept was formulated at a time when almost no one foresaw the future environmental problems of mankind, at a time when the paradigm of man as the crown of evolution and master of nature dominated, with the goal of "creating such a social order, which achieves harmony within society and harmony of the interaction of society and the biosphere", "establishing a new type of human relationship with the biosphere – a relationship of equality, harmony and unity" (Manov, 2014, pp. 302–303). Both the natural scientist and the Jesuit-palaeontologist saw the unification of human reason or scientific knowledge as main tools for achieving such a harmonious system. Pointed in the right direction, it would have the potential to save humanity from its major ills such as famine, war, violent death, disease, and so on, not in the foreseeable future, or in the afterlife, but today here on Earth. Again, they both declared the onset of the noosphere as inevitable, categorically rejecting fears of the impending doom of mankind, one justifying the irreversibility of the evolution and the other believing in God's providence, but both relying also on human rational choice. Vernadsky wrote that "scientific knowledge, manifesting itself as a geological force creating the noosphere, cannot lead to results that contradicts the geological process of which it is a creation" (Vernadsky, 1938/2009, p. 14). While analyzing Chardin's work, Manov focuses on the author's strong statement, that "divine providence cannot allow *Creation*, namely Man, for one reason or another, cease to exist, to die, it will only change its beingness, the way and forms of its manifestation" (Manov, 2014, p. 327).

Applied to the present analysis, taking such an approach would in the first place mean rethinking the legendary ideas of the French Revolution of *Freedom, Equality and Brotherhood* (Chardin, 1964/2003, pp. 223–224). *Freedom* meaning the creation of conditions for the liberation of the full potential and creative possibilities of Humanity, for instance through the introduction of a Universal Basic Income. *Equality* means the right of every person to participate according to his qualities and strengths in the elaboration of this harmonious social order, for example, by voting in a referendum, but also by participating directly in government. And under *Brotherhood* is understood the abolition of all racial, sexual, ethnic, religious, etc. boundaries and giving major priority to the only unifying "Man".

In fact, both Vernadsky and Chardin avoid talking about rule or government but emphasize the idea of a scientific think tank that critics of the Noosphere theory would probably compare to Plato's ideas of government by philosophers. The idea of an impartial and apolitical think tank by scientists is undoubtedly attractive, but it can be cited as one of the few vulnerable points of the noosphere doctrine. Scientists are human, and we think *nothing human is alien to them*, so the belief in pure reason and democracy of scientific knowledge may seem politically naive. Another less productive point is the belief that the occurrence of the Noosphere is inevitable. If this is indeed the case, why should society make any effort to change?

Although nowhere explicitly stated, a careful reading reveals that many of the ideas of one of the possibilities for a third progressive path of humanity, the Resource-Based Civilization Model of Jacques Fresco and his followers, naturally continue the theory of the noosphere. The notions of a harmonious coexistence of man and nature, the insistence that state borders of the planet are artificial, that natural resources are universal property and heritage, that all people are equal, regardless of gender, origin, religion, and so on, claims that scientific and technological progress has reached a level at which it can solve all the problems of mankind. Naturally, there are also differences, mainly in the point that Fresco does not believe in the inevitability of his project called "Venus". Or that God or Evolution will protect us from self-destruction and so he insists that there is no time to postpone change. Reforms are already overdue. Jacques Fresco progresses the idea of an *impartial scientific think tank* with his vision of decision making about how to allocate resources among the inhabitants of future smart cities, by supercomputers based on a series of calculations. Which is one of the problematic moments in his theory. Yes, probably today's and tomorrow's supercomputers could easily calculate the daily needs of a city's inhabitants. But would these same residents agree to be controlled by computers? What if the machines fail, revolt or wage war on their creators?

Moreover, the acceptance of the Venus Project as human future's goal is connected with the realization of truly revolutionary changes in the way of our society's existence. Eliminating money, globalizing national resources, abolishing nation states and their institutions, abandoning the consumer economy and the idea of owning a home, car, villa, yacht in general, and instead sharing everything everyone needs. With everyone else alive on Earth.

Like Vernadsky and Chardin, Fresco believes that people have not yet developed their creative abilities, and that this can only happen if we are free from dependence on money. As long as we need to sell our life time for money to survive, we will be forced to accept activities that do not meet our abilities and needs.

One of the great merits of the Venus project is the acceptance of the coming automation of production processes not as a problem, but as a blessing for humanity. Since machines can replace humans in performing monotonous or labour-intensive processes, this would enable people to free themselves from the routine of dull work or from hard and dangerous physical labour, and instead devote themselves to creative and social activities, to their families and their loved ones. Saving resources through the production of sustainable products, sharing vehicles and holiday homes, sustainable production and the use of energy efficient sources would allow us to live in harmony with nature and keep our planet clean and green for future generations.

Too utopian to ever be realized? The last crisis that hit the planet in 2020, the Coronavirus pandemic led to not just negative effects. Reduced air, sea and land traffic, reduced consumption of fuels and goods have had a huge beneficial effect on the environment. Not only that, the pandemic has shown that when there is imminent threat, society is able to take drastic measures in its way of life. But the pandemic has also shown that when danger passes, lessons that are learned are quickly forgotten and unfortunately even the most harmful practices are restored.

CONCLUSION

As expected, human history and development continues, with our current civilization facing unprecedented in their scale, significance and frequency, all-encompassing challenges that even threaten its physical survival. The course of human development depends on how we will be able to respond to these challenges. Progress is still possible, but not certain. The conservative resistant forces that still largely drive this social transformation, both economically and politically, are not to be underestimated. If their vision for the end of these changes is imposed, and not visions like those of the creators of theories about the noosphere or the resource-based civilization, it is very likely for all of us, that there will be not a coming era of prosperity, liberation of human possibilities and their creative development, but an era of spiritual poverty and a robotic voluntary slavery.

On the basis of the outlined, albeit in general terms, models for a planetary future, can it be determined which of them would be the most productive? And why not all three at once? Perhaps the right question is not whether modern man can continue to live this way, because it is obvious that we can. At least for a while. But should we? If too many people think something in the system is wrong, then it is high time to change the system. Or at least to improve it. In the words of Jacques Fresco, "we must stop constantly fighting for human rights and equal justice in an unjust system, and start building a society where equal rights are an integral part of the design" (Fresco, 2002, p. 31).

The main prerequisite for this is the actions of political elites. In their struggle for power today, these elites find it difficult to accept the ideas of the scientific communities. Probably fearing their place in the world hierarchy to be endangered. Maybe because they do not have answers to global challenges or do not accept the ones offered by modern science. Or because they represent someone's corporate interests, not those of the people who put them at the top of politics and empower them with the task of looking for new ways to develop our civilization. And this would mean that the existing political system of representative democracy is outdated, that it is necessary to look for new forms of political organization of society, including in traditional "well" functioning, respecting fundamental human rights and freedoms, democratic societies.

As shown in the exposition, a main precondition for this is the reduction of global inequality and dominance of capital over politics, as well as unification of science and practice. It seems to be the only way leading to the belief of human evolution still going on, and its direction towards the harmony of theories such as the Noosphere, can become a reality. And that perhaps God has a plan, and it is not for us to exterminate ourselves and almost everything living on the planet we call home.

REFERENCES

Chardin, P.T. (1964). *The Future of Man.* Harper & Row. Published in Bulgaria 2003.
Fresco, J. (2002). *The Best That Money Can't Buy: Beyond Politics, Poverty, & War.* Global Cyber-Visions.
Fukuyama, F. (1989). "The End of History?". *The National Interest* (16): 3–18.
Lozev, K. (2018). *About Marx and Marxism according to Karl Popper.* NotaBene-bg.org. http://notabene-bg.org/read.php?id=690
Noosphere. (2021, 7 January). In *Wikipedia.* https://en.wikipedia.org/w/index.php?title=Noosphere&oldid=998976048
Manov, B. (2014). *Paradigms and Doctrines of the European Political Thought. Theory and History of Political Historiography.* SWU University Press.
Popper, K. (1957). *The Poverty of Historicism.* Beacon Press.
Vernadsky, V.I. (1938). *Scientific Thought as a Planetary Phenomenon.* Nongovernmental Ecological V.I. Vernadsky Foundation, 1997. Published in Bulgaria 2009.
Welcome to the Anthropocene (n.d.). *Planetary boundaries.* http://www.anthropocene.info/planetary-boundaries.php

PART V

Knowledge systems for the Anthropocene

17. Scientific knowledge for the Anthropocene

Marc Zimmer

ECONOMIES AND SOCIETIES AS COMPLEX ECOSYSTEMS

Since its formation, a total of 108 billion humans have lived on earth. A fifteenth of all these people are alive and kicking right now. Science and technology are the reason why the numbers have undergone such robust growth.

Thanks to the development of weapons, protective houses, medicine, fertilizer and so on, humans have moved from being an occasional snack for predators to being at the top of the food chain. We have overcome constraints imposed by nature and evolution. There are no predators – outside of ourselves – to limit our growth. Humans have become, in the worst sense of the word, the dictators of nature. We, thanks to science, can impose our will on nature and are in charge of our own destiny. We are not doing a great job. Currently we are losing species 1,000 times faster than the natural rate of extinction. Humans represent just 0.01% of all living things by mass, yet during our time on earth we have caused the loss of 83% of all wild mammals and half of plants (Bar-On, Phillips, & Milo, 2018). Although we, thanks to science, have become the masters of nature, ultimately it is the biosphere that will prevail showing us limits of humankind's mastery.

Natural selection has been around for 4 billion years. Now we are no longer governed by natural selection. Not only are we in charge of our own evolution but we have also changed how nature evolves. Ever since the continents moved apart delicate ecosystems have been kept apart by mountain ranges and oceans. Increased human movement has made the borders between ecosystems more porous, we both inadvertently and on purpose introduced new species into ecosystems they could never have visited themselves without hitching a ride on our cars, ships and planes. In some way we have created a whole new global pseudo-ecosystem (Harari, 2016). This has consequences – for example epidemiologists have been predicting and preparing for a viral outbreak like COVID-19 for many years. Live animal markets, which can act as viral mixing bowls, increased world travel, and population density have made a pandemic such as this one pretty much inevitable. Despite this, we weren't ready. At the same time, we also, thanks to modern science (CRISPR, gene drives etc.) have the ability to control the evolution of other species. Not since cyanobacteria has a single taxonomic group been so in charge.

Scientific discoveries have not only enabled the expansion of the human population, but they are also the key to expanding economies in the post-industrialization world. To keep the financial systems growing they need to come faster and faster. At the same time scientific discoveries will have to curtail the environmental and ecological degradation associated with the growth. These expectations are unrealistic, science cannot deliver all this. They are a sign of a broken system. Expecting science to deliver us from the consequences of our overconsumption is just as dangerous as failing to acknowledge its great potential.

Our economies and societies are complex ecosystems constrained by thermodynamics, and from thermodynamics we know that all this growth requires energy and generates waste (entropy). At best science may be a band aid on an unsustainable growth-based economy, but it won't be the cure. According to Philip Ball, former *Nature* editor "Creating a true science of sustainability is arguably the most important objective for the coming century; without it, not an awful lot else matters. There is nothing inevitable about our presence in the universe" (Al-Khalili, 2018, p. 22).

Throughout the last century children had better lives than their parents, this cannot continue forever, something has to give – the economic system built on growth, the environment, our energy consumption and/or our eating habits. There is no denying that an increase in scientific knowledge will increase the quality and length of our lives, but it will also bring increasing environmental and ethical problems, and it will widen the gap between the have and the have nots. New developments in science won't just improve transport (cars, trains), how we communicate (phones, internet) and eat (fertilizers) like they did in previous generations, instead they have the potential to improve our bodies (CRISPR) and minds (optogenetics). In *Homo Deus: A Brief History of Tomorrow*, Yuval Harari (2016) argues that this increased inequity will be compounded by the fact that in the past medicine used to be aimed at healing the sick, while in the future medicine will increasingly be designed to improve the healthy. Treating the sick is an egalitarian process, while upgrading the healthy will increase existing inequities by giving an edge to the rich. "People want superior memories, above-average intelligence and first-class sexual abilities. If some form of upgrade becomes so cheap and common that everyone enjoys it, it will simply be considered the new baseline, which the next generation of treatments will strive to surpass" (Harari, 2016, p. 203). For the first time in history the rich will not only have material benefits, but they will also be able to have genetic improvements. As in the past they will have better lives than the poor, however thanks to new techniques like CRISPR they may actually be better too. Today the richest 100 people own more than the poorest 4 billion. In the future this financial inequality may lead to biological inequalities and increased geopolitical *tensions* (Harari, 2018).

Exponential versus Linear Growth

Authors and thinkers like Ray Kurzweil and Bruno Giussani suggest that science and technology are growing exponentially, while the structures of our society (government, education, economy etc.) are designed for predictable linear increases that are dysfunctional in today's exponential growth (Giussani, 2018). This is why our nation state system can't deal with the challenges of modern science. The challenges presented by modern science (CRISPR, gene drives and artificial intelligence) are much larger than those brought about during the industrial revolution (steam engines and electricity). The world is not designed to deal with these challenges. There is always a country marketing itself as a place to do research that is banned at home. And it just takes one country pursuing a high-risk, high-profit path for all the other countries to follow – in the nation state/growth economy they have to follow in order to prevent themselves from falling behind.

Many governments, including the US, control research by not funding certain areas of research. This doesn't work when other foundations or start-up companies fund the work. It also fails when the techniques and materials being used are cheap, as is the case with CRISPR, and government funding isn't needed. In the absence of clearer guidelines or regulations scientists have to rely on themselves, on their own scientific norms. This doesn't work too well because academia is intensively competitive, "the drivers are about getting grants and publications, and not necessarily about being responsible citizens," says Filippa Lentzos from Kings College London, who studies biological threats (in: Yong, 2018). High profile results matter. Additionally, to prevent their competitors from knowing what they are doing and to save themselves from being scooped, scientists keep their experiments under wraps until they are ready to publish. At which point it is often too late.

Machine learning *exploits* the (often hidden) information contained in vast data sets to solve questions of interest. In *AI Superpowers: China, Silicon Valley, and the New World Order*, Kai Fu Lee writes that artificial intelligence, which encompasses machine learning, has surpassed the intellectual development stage, where its growth is dependent on a few extremely talented intellectuals making breakthroughs (Lee, 2018). Instead, it is in the application phase. AI is a little like electricity after Edison and Westinghouse brought it to our homes and inventors searched for new uses of the power source. Researchers in small start-ups and the big mega companies such as Baidu, Alphabet (Google's parent company) and Facebook are looking for new applications of artificial intelligence (AI).

AI and Computational Chemistry

Artificial intelligence has been in the news a lot due to its use in autonomous cars, face recognition and fake news. If it hasn't changed the way we live yet, it soon will. But will it change the way science is done? I am a computational chemist. My students and I use programs developed and refined by many researchers over decades of work. They are very good at calculating three-dimensional changes that occur when we make small changes to known structures of molecules. However, a big challenge remains in computational chemistry, namely, how to simulate protein folding. Misfolded proteins are most likely responsible for Alzheimer's, Parkinson's, cystic fibrosis and Huntington's disease.

Every two years the top computational chemists test the abilities of their programs to predict the folding of proteins and compete in the Critical Assessment of Structure Prediction (CASP) competition. In the competition teams are given the linear sequence of amino acids for proteins for which the 3D shape is known but hasn't been yet published, they then have to compute how those sequences would fold. In December 2018 98 teams competed in the 13th CASP competition in Cancun, Mexico. As you have probably guessed the competition was won (barely) by the rookie AlphaFold team, the Google machine learning entry.

Two years later, on Monday, November 30 it was announced that AlphaFold2 won the 2020 competition by a healthy margin. It whipped its competitors, and its predictions were comparable to the best existing experimental results. Soon AlphaFold2 and its progeny will be the method of choice to determine structures before resorting to techniques that require painstaking laborious work on expensive instrumentation.

We don't exactly know what the program is doing and why it uses certain correlations, but we do know it works. Besides helping us predict the structures of important proteins, understanding AlphaFold's "thinking" will also help us gain new insights into the mechanism of protein folding.

One of the reasons for AlphaFold2's success is that it could use the Protein Database, which has over 170,000 experimentally determined 3D structures, to train itself to calculate the correctly folded structures of proteins. The potential impact of AlphaFold can be appreciated if one compares the number of all published protein structures (~170,000) with the 180 million DNA sequences, protein sequences deposited in Universal Protein Database (UniProt). Like the microscope has opened a new world invisible to the human eye, so AlphaFold can help us sort through treasure troves of DNA sequences hunting for new proteins with unique structures and functions (Zimmer, 2020).

One of the most common fears expressed about AI is that it will lead to large-scale unemployment. AlphaFold still has a significant way to go before it can consistently and successfully predict protein folding. However, once it has matured and the program can simulate protein folding, computational chemists will be integrally involved in improving the programs, trying to understand the underlying correlations used, and applying the program to solve important problems such as the protein misfolding associated with many neurodegenerative diseases. AlphaFold and its offspring will certainly change the way computational chemists work, but it won't make them redundant.

Other areas won't be as fortunate. In the past robots were able to replace humans doing manual labor, with AI our cognitive skills are also being challenged. According to an analysis by investment bank Morgan Stanley, autonomous trucks have the potential to save the industry $160 billion annually in labor, fuel and productivity. They would also improve the safety and decrease the environmental impact of trucking. The report concludes that, "Longhaul freight delivery is one of the most obvious and compelling areas for the application of autonomous and semi-autonomous driving technology," and that it "will be adopted far faster in the cargo markets than in passenger markets" (iTECH, 2014). Once again China is leading in this area of AI applications. They have already started building highways designed for autonomous trucks (Lee, 2018).

Biosensors in our Bodies Sending Data

In recent years at least 60% of my incoming introductory chemistry class have been pre-med students. Many don't survive organic chemistry. In the US all pre-med students have to take organic chemistry, not because organic chemistry is useful in the practice of medicine, but because the skills needed to master organic chemistry are very similar to those used in medicine. In organic chemistry you have to memorize many functional groups and reactions before combining them in new ways to make a previously unseen product. Similarly, medical diagnosis relies on gathering all the patient's unique symptoms and finding the underlying cause. I think you can see where I am going with this, medical doctors collect all the patient's blood work, symptoms and so on, and then go to their vast, memorized database of known maladies to find the cause and then proceed to treat the symptoms – a procedure perfectly suited to machine learning. I am sure that by the time the students in my current introductory chemistry class complete their residency and start looking for jobs in the medical field AI will have changed the way medicine is practiced. We, the insured and relatively well-off patients will have small biosensors permanently embedded in our bodies. They will continuously monitor levels of key proteins, metabolites, minerals and so on, and send this information to a central computer – it will process our vitals. Based on the data gathered from millions of patients the AI system will probably know that we are sick even before we experience the symptoms ourselves, it will prescribe a treatment, monitor its efficacy and adjust the dosage appropriately. If needed the AI system will send the patient to the doctor.

In the US researchers have developed algorithms that are at least as good as humans at interpreting X-rays, mammograms and photos of skin cancers. While in China Baidu has developed an AI-powered diagnostic program, RXThinking, to compensate for the fact that most well-trained doctors in China are concentrated in the wealthiest cities, resulting in large swaths of rural areas having too few doctors. RXThinking acts as an automated triage station. Given a set of symptoms the algorithm makes a series of diagnoses with a confidence rating for the diagnosis, which is then given to the doctor. The program uses over 400 million medical records and is continually updated with the latest medical publications. At this point the program is used to lighten the work load of doctors in rural areas and to provide the same high quality diagnoses offered in expensive hospitals in the wealthy cities (Lee, 2018). But I am sure that will change as health insurances and medical aid schemes attempt to reduce costs.

AI diagnostics are making their way into the US market too. In late 2018 Amazon announced a collaboration with the Fred Hutchinson Cancer Research Center to evaluate "millions of clinical notes to extract and index medical conditions." They will use the data to train their machine learning medical diagnostics algorithm that already compares favorably with all other published diagnostic programs (Diamandis, 2019). CVS has also partnered with an AI company to bring computerized diagnostics to its 1100 minute-clinics, where the program will diagnose and recommend the best care and treatment for their customers. Over the last five years AI healthcare start-ups have raised more money than any AI other sector.

Because of AI's reliance on data, data has become a valued and desired asset. Everyone is collecting it. Ten years ago, investors struggled to figure out Google, Facebook and WeChat's business model. What were they making or selling? Today it's obvious – data, lots of it and we are willing to give it to them for nothing, just for the privilege of using their products. Now we are stuck, if we don't want to give these companies our data that means we have to forsake using Facebook, Gmail and Google. Data is the new knowledge.

Like science, or maybe because it is part of science, AI is progressing so fast it is very difficult to foresee all the consequences and to provide guidelines for its development.

THE ROLE OF SCIENCE IN STOPPING CLIMATE CHANGE

Climate change is the biggest challenge facing our generation. And I don't see science solving this problem. I grew up in South Africa. My childhood dream was to be a game warden. A dream that was shattered when I failed botany in my freshman year (it clashed with anti-apartheid protest meetings at lunch). I am often amazed by my friends who have gone on to become game wardens. They believe that climate change is real and that it is caused by fossil fuel combustion; however, they are not particularly concerned. They have misplaced confidence in science. Like many others, my old university friends believe that scientists will find a painless solution. I don't think that will happen. Scientists cannot tell us what to do about climate change. Drastic social solutions are required. Science will help us develop and improve electric autonomous cars, renewable energy sources, meat alternatives, carbon dioxide sequestration and so on. But that is not enough; it is going to require a change in the way we live. We will have to build smaller houses, travel less, eat less red meats and throttle back our consumption. Bigger, more significant changes than the ones brought about by the COVID-19 pandemic.

To make the changes required to prevent catastrophic climate change we are going to have to change the minds of both climate change believers who have too much confidence in science and climate change deniers who don't trust climate scientists – two very different audiences. That is going to require a lot of faith in science and scientists in a time when there is not much faith going around and truth is negotiable.

I am not the only one concerned. The decline in trust "terrifies" former MIT president Susan Hockfield, she says we have no way of making it into the future if we don't trust scientists to be expert in their fields. Hockfield said on the podcast Recode – Decode with Kara Swisher:

> We have to insist on an understanding that there are people who understand areas better than we do. I don't pretend to be an engineer. I don't pretend to be a physicist. If the physicists at MIT tell me that they've figured out gravitational waves, I'm going to trust them more than I'm going to trust myself to imagine whether or not there are gravitational waves. (Patel, 2020)

She added that she understands "that people might debate the fine points of climate change, but the fact is that the best science indicates that we're in trouble" (Johnson, 2019).

In May 2019 New Zealand became the first country in the World to announce a national budget that placed the "well-being" of its citizens over economic growth. According to Bill McKibben this would be a sign of a mature economy, an economy that has reached its adult size and no longer needs to grow. To stop climate change we will all have to trust the science enough to realize that we are all grown up and that it is time to stop growing. We don't need new malls, housing developments, bigger cars and so on. We have to accept that our teen years are over and that we have to live responsibly and, in the interest of our communities, place limits on our behaviors. It may not be the American way, but perhaps machine learning, and AI might help us move to a "well-being" economy. If the advances in these fields result in the feared reductions in jobs (e.g., of truck drivers, medical doctors, service industries), perhaps we can "make lemonade out of lemons" by redistributing responsibilities, reducing our working hours and focusing on upkeep and sustainability overgrowth – that is move from a growth economy to a mature economy.

Interestingly, although China is much more populous and in a much steeper growing phase than the US, I think because it is an autocratic state it will be able to make the changes required to halt climate changes easier and faster than the US. Julian Savulescu, an Australian philosopher and bioethicist at the University of Oxford, echoes these thoughts, he thinks democracies can't solve climate change, "for in order to do so a majority of their voters must support the adoption of substantial restrictions on their excessively consumerist lifestyle, and there is no indication they would be willing to make such sacrifices" (Persson & Savulescu, 2014, p. 2).

KNOWLEDGE IS USELESS WITHOUT TRUST IN EXPERTS

We all need experts. Scientists will be happy to tell you that journalists, consumers and policy makers don't understand science and need expert advice. But they themselves struggle to see that they also need input from experts. We scientists need to learn how to increase the diversity of our field, consider the ethics of what we are doing and how to reach out to non-scientists and explain our work.

The knowledge gap between experts (be they scientists, medical doctors or economists) and non-experts is increasing and perversely our reliance on these experts is decreasing. This is due in part to the fact that we often have a misguided overconfidence in our knowledge, something Leonid Rozenblit and Frank Keil, psychologists at Yale University termed the illusion of explanatory depth. According to them, "most people feel they understand the world with far greater detail, coherence, and depth than they really do" (Rozenblit & Keil, 2002, p. 521) and consequently have less use for experts than they should.

A little more than a century ago Richard Burdon Haldane wrote a report for the British Parliament in which he, a career politician, advocated that scientists should be in charge of deciding who should receive government research funding, that politicians should stay out of funding deliberations and that in making science policy they should listen to the advice of experts. His ideas would come to be known as the Haldane Principle and it would take the upheaval of the Second World War for them to take root.

In the Second World War scientific technologies, such as radar and the atomic bomb, were central to the allies' victory. Following the war, scientists advocated for a separation between science policy and politics, arguing if science could win wars it could also help to hold peace. They were successful and many nations adopted the Haldane Principle. By 2015 US federal agencies had about 1,000 advisory panels relying on the expertise of roughly 60,000 members. They advise government agencies on everything from foodborne illness, drug safety to air and water pollution. They were there to provide fact-based advice and evidence. After all no one can reasonably expect our elected representatives to master every topic – they have to rely on the expertise of professionals.

However, that was about to change, on June 14, 2019 President Trump signed an executive order to reduce all advisory boards by one-third. A rather telling move that will have long-term consequences (Goldman, 2019). In a comment published in *Nature* in 2018 Ehsan Masood worries that an expanding global network of populist political movements are deriding independent scholarship and are erasing the protections granted by the Haldane Principle. Masood calls out Trump and the UK Brexiteers and worries that, "Today, from Istanbul to Islamabad, from Rome to Rio de Janeiro, a parade of authoritarian leaders is advancing policies that fly in the face of evidence – on energy, emissions, the environment, economics, immigration and more. Worse, these leaders are demanding that academics march to the beat of their drums" (Masood, 2018, p. 620). Scientists are no longer advising politicians and it is politics that are driving science policy. The consequences of their disdain of experts have become painfully obvious in the US and Brazilian response to the COVID-19 pandemic.

It is important to remember that scientists use experiments to test their theories and hypotheses. No map shows the way to prove or disprove a theory; the scientific process is not linear. Neil Gershenfeld, director of MIT's Center for Bits and Atoms, writes that "to find something that's not already on the map, you need to leave the road and wander in the woods besides it." He feels that nonscientists do not recognize that "science appears to be goal-directed only after the fact. While it's unfolding, it is more like a chaotic dance of improvisation, than a victory march" (Gershenfeld, 2018). Uncertainty is an inherent and unremovable component of scientific experimentation. It is not a weakness; it is a strength.

SHOULD THERE BE LIMITS ON SOME SCIENCE: IF SO, WHO GETS TO SET THEM?

In 1926 J.B.S. Haldane (the British statesman Richard Burdon Haldane was his uncle), wrote an essay entitled "On Being the Right Size." Drop a mouse from a building and it survives, "a rat is killed, a man is broken, a horse splashes." Gravity limits our size; it is the enemy of the large. Elephants have massive hearts to pump oxygen-bearing blood through their bodies and they have thick sturdy bones to remain upright. The larger the organism the higher its complexity – just compare a flea with a giraffe. But complexity only gets you so far, to get larger you have to change the conditions, such as going from a terrestrial to marine environment. Paul Saffo, professor of mechanical engineering at Stanford University, argues that Haldane's observations apply to more than just organisms, "Everything from airplanes to institutions has an intrinsic right size, which we ignore at our peril" (Brockman, 2018, p. 328). What about science? What is its right size?

Science and technology have enabled robust growth for humans. We have overcome constraints imposed by nature and evolution, and now have no predators – outside of ourselves – to limit our growth. We can impose our will on nature and are in charge of our own destiny.

But can we let that growth continue unchecked? Many scientific breakthroughs will benefit humanity, but their advantages are often offset by potential abuses. These abuses need to be limited when they are dangerous and when they exceed a commonly established ethical boundary. "The corollary of progress is risk" (Goldin & Mariathasan, 2014, p. 204). I am particularly concerned about developments in CRISPR, optogenetics and artificial intelligence. How do we limit them? Especially when the limits may curtail our competitiveness.

With CRISPR, especially when it is associated with a gene drive, we will soon reach the point where for the first time in history one person or one experiment can potentially wipe out a whole species, including humans. How do we regulate these dangers? How are we going to regulate the weaponization of CRISPR gene drives when CRISPR and gene drives are used in labs all over the world and the production of supercharged viruses and bacteria will be relatively cheap and will require minimal infrastructure? Perhaps this is a case where we need to limit both the practice of science and the distribution of scientific knowledge.

When ethical dilemmas like the sudden ability to genetically modify human embryos arise, there are no standard procedures in place to deal with them. Instead, ad hoc solutions are found, like the international committee convened by the U.S. National Academy of Sciences (NAS) and the National Academy of Medicine in Washington, D.C, that in February 2017 concluded that human embryo editing may only be permitted "for compelling reasons and under strict oversight."

To date all global science regulations have come from scientists themselves, and they have been reactive, responding to advances that were too dangerous to ignore. However, the time has come to introduce a trusted, non-partisan advisory group to proactively devise safety and ethical guidelines that can be used by scientists, legislators and funding agencies across the world. There are organizations like the World Commission on the Ethics of Scientific Knowledge and Technology that was set up by UNESCO that could be the permanent solution, but they lack the prominence, resources and the legitimacy required.

THE FUTURE OF SCIENCE IS TOO BRIGHT TO JUST WEAR SHADES

In my opinion science itself is healthy. Nature's secrets are steadily being unpacked, our understanding of the universe is expanding, and we are continuing to utilize our knowledge. But, and this is a big but, I think the future of science is too bright to just wear shades.[1] Our scientific abilities and knowledge are increasing at a faster and faster rate, as a consequence science is outgrowing its supporting structures. Governments, funding agencies, philosophers and scientists themselves are struggling to find ways to make the most of the power of science, of CRISPR and deep learning while keeping its potential abuses at bay. With increasing scientific ability, the question "Are we being good ancestors?" posed by Dr. Jonas Salk, the polio vaccine inventor, becomes ever more relevant. We are in charge of our own evolution and the health of the planet – that requires a lot of responsibility and a paradigm shift both within science as well as towards science.

If we are able to find a way to appropriately regulate future discoveries and knowledge, communicate with the public effectively, and continue to hold ourselves accountable then science has nothing to worry about – but only time will tell us if we'll be able to accomplish all this.

NOTE

1. "The Future is so Bright – I gotta wear shades" Timbuk 3, 1986.

REFERENCES

Al-Khalili, J. (2018). *What the future looks like: scientists predict the next great discoveries and reveal how today's breakthroughs are already shaping our world.* New York: The Experiment.

Bar-On, Y. M., Phillips, R., & Milo, R. (2018). The biomass distribution on Earth. *Proceedings of the National Academy of Sciences.*

Brockman, J. (2018). *This Idea is Brilliant.* New York: Harper Perennial.

Diamandis, P. H. (2019). AI is rapidly augmenting healthcare and longevity. *Singularity Hub, February, 15* (https://singularityhub.com/2019/02/15/how-ai-is-rapidly-augmenting-healthcare-and-longevity/#sm.00000549n197vffd5w6d1pos4wxyj).

Gershenfeld, N. (2018). Ansatz. In J. Brockman (ed.), *This Idea Is Brilliant: Lost, Overlooked, and Underappreciated Scientific Concepts Everyone Should Know.* New York: Harper Perennial, p. 305.

Giussani, B. (2018). Exponential. In J. Brockman (Ed.), *This Idea Is Brilliant – Lost, Overlooked, and Underappreciated Scientific Concepts Everyone Should Know.* New York: Harper Perennial, p. 115.

Goldin, I., & Mariathasan, M. (2014). *The Butterfly Defect: How Globalization Creates Systemic Risks, and What to Do about It.* Princeton University Press.

Goldman, G. T. (2019). Trump's plan would make government stupid – cuts to science advisory panels for federal agencies will haunt the United States long after the current administration finishes, says Gretchen T. Goldman. *Nature, 570,* 417.

Harari, Y. N. (2016) *Homo Deus : A Brief History of Tomorrow* (First U.S. edition. ed.). New York: Random House.

Harari, Y. N. (2018). *21 Lessons for the 21st Century* (First edition. ed.). New York: Spiegel & Grau.

iTECH (2014). iTECH: Trucking May Save $168 Billion Annually With Driverless Vehicles, Report Concludes. *Transport Topics, January 1* (https://www.ttnews.com/articles/itech-trucking-may-save-168-billion-annually-driverless-vehicles-report-concludes).

Johnson, E. (2019). The decline of trust in science "terrifies" former MIT president Susan Hockfield. *Vox – Recode, May 31* (https://www.vox.com/recode/2019/5/31/18646556/susan-hockfield-mit-science-politics-climate-change-living-machines-book-kara-swisher-decode-podcast).

Lee, K.-F. (2018). *AI Superpowers : China, Silicon Valley, and the New World Order.* Boston: Houghton Mifflin Harcourt.

Masood, E. (2018). Why academic freedom is needed more than ever. *Nature, 563,* 620.

Patel, N. (2020). Recode Decode: Susan Hockfield's talk with Recode's Kara Swisher about her new book: The Age of living machines: how biology will build the next technology revolution. Apple Podcast.

Persson, I., & Savulescu, J. (2014). *Unfit for the Future: The Need for Moral Enhancement.* Oxford: Oxford University Press.

Rozenblit, L., & Keil, F. (2002). The misunderstood limits of folk science: an illusion of explanatory depth. *Cognitive Science, 26*(5), 521–562. doi:10.1207/s15516709cog2605_1.

Yong, E. (2018). A controversial virus study reveals a critical flaw in how science is done. After researchers resurrected a long-dead pox, some critics argue that it's too easy for scientists to make decisions of global consequence. *The Atlantic, October 4.*

Zimmer, M. (2020). *The State of Science: What the Future Holds and the Scientists Making It Happen.* New York: Prometheus, Rowman & Littlefield.

18. The sciences of knowledge
Francisco Javier Carrillo

INTRODUCTION

If the Anthropocene is the single most challenging reality ever confronted by the human species, and knowledge the most substantial leverage to human action, then 'Knowledge for the Anthropocene' becomes a prominent significant issue of our days. This chapter explores where we stand in our understanding of knowledge as an object of scientific inquiry. A brief account is provided of each of the mainstream 'Sciences of Knowledge.' Given the size constraints of the chapter, the more nuanced concepts, issues and authors are just pointed out through references. This chapter is complemented by Chapter 28, where contributions from other scientific fields not yet institutionalized as independent fields, as well as other forms of knowledge studies are also reviewed.

ANTHROPOLOGY OF KNOWLEDGE

Like all sciences of the human do, Anthropology deals with knowledge (Crick, 1982, p. 287). Fredrik Barth set the subfield's coordinates as a comparative perspective on human knowledge across cultures. He and his colleagues identified three constants: a substantive corpus of assertions, a range of media of representation and a social organization (Barth, 2002). Barth leans on a structuralist approach towards the social construction of reality. In doing so, such approach unfolds into the wider scope of Cognitive Anthropology, where knowledge socialization as well as cultural innovation and generational transfer are dealt with on the basis of cognitive psychology theories and methods. An overview of this approach can be obtained from the 2015 special issue of *Social Anthropology* on 'The Cognitive Challenge' (e.g., Regnier & Astuti, 2015, pp. 131–134). Soren Klausen sets the main challenge for Anthropology of Knowledge as "finding an appropriate level of description" by achieving a balance between a proper epistemological perspective and locally situated practices (2020, p. 204).

HISTORY OF KNOWLEDGE

Amongst the sciences of knowledge, History has probably the longest tradition, with deep roots in the History of Science. Pioneer figures such as George Sarton (see Stimson, 1962), John Bernal (e.g., 1971), and Gerald Holton (e.g., 1988) established History of Science and Technology as a discipline. When the focus widened to include other forms of understanding such as ancient scholarship and native knowledge as well as other social agents, settings and practices for knowledge creation and transfer, the History of Knowledge emerged (Lässig, 2016). To exemplify the momentum currently achieved by this field, let us mention three substantial programs. First, Burke's (2015) socio-cultural account of institutions and processes that gave rise to current techno-determinism. Second, Renn's (2020) compilation of studies conducted at the Max Planck Institute for the History of Science, woven by a cognitive-evolutionary narrative. Third, the Lund Center for the History of Knowledge that draws on a wider array of explanatory objects and processes to provide a contemporary perspective on knowledge as an encompassing cultural phenomenon (Heidenblad, Nilsson & Östling, 2020). Other initiatives such as the *Competence Centre History of Knowledge* (ZGW) at Zurich, the *Journal for the History of Knowledge* and the book series of *Studies in the History of Knowledge* from the University of Amsterdam, testify to the vitality of this field.

BIOLOGY AND KNOWLEDGE

Whereas in this case there is no discipline in the same institutional sense as in other branches of Knowledge Sciences, there have been a number of developments across biological sciences that deserve attention. Genetics deals with inheritability issues of thinking and verbal behavior and the epigenetic factors determining their expression (Hermo, Giráldez, & Saura, 2014). Neuroscience, besides dealing with the anatomy and physiology of the nervous system at different levels of analysis (e.g., molecular biology and cytology), also deals with the resulting functions of development and behavior. Besides mapping specific neural correlates of behavior, "… the last frontier… is to understand the biological basis of consciousness and the mental processes by which we perceive, act, learn and remember" (Kandel et al., 2012, p. 5). While Behavioral Neuroscience is the application of biological principles to the study of behavioral processes, including learning, memory, language, thinking and decision making, Cognitive Neuroscience deals with the biological processes underlying cognition. Jean Piaget, a biologist by training, explored in his 1967 *Biology and Knowledge* the relationship between neurogenesis and psychogenesis (Parker, Langer & Milbrath, 2014) and developed his *Genetic Epistemology*, a structuralist view of cognitive and linguistic development that remains influential. In exploring the wider perspectives for Neuroscience before Climate Change, Catherine Malabou makes a compelling case for an integrated Science of Knowledge for the Anthropocene: "Could it be that new histories of mentalities, which could bring together the geological, biological and cultural dimensions of historical (non) awareness, may open a new chapter for Anthropocene studies?" (Malabou, 2017, p. 52).

PSYCHOLOGY OF KNOWLEDGE

Alternative accounts of science have been developed by different Psychology schools, including structuralist and personality psychologies (Osbeck et al., 2010; Feist & Gorman, 2012). However, two dedicated research programs spanning over several decades are of special significance to knowledge sciences, both satisfying the empirical requirement of "actually engaging the subjects in a specifically scientific task" (Fuller, 2019, Ch. 6). The first is the Behavioral Economics program, of high relevance to environmental choice; the second the Knowledge Systems program particularly relevant to an integrated science of knowledge. Both, although having originated within behavioral research, have influenced the broader field of Knowledge Management, Knowledge Economy and Knowledge Society.

Behavioral Economics has had a major impact on Economic Theory and Policy Development. Prospect Theory, proposed by Daniel Kahneman and Amos Tversky in 1979 exhibited the sub-optimality of human choices, empirically challenging the behavioral axioms of neoclassical microeconomics. Simon (1982) set the ground for a *Bounded Rationality* framework where human information processing is limited in both knowledge and computational capacity. Nobel laureate Kahneman (2011) achieved a major influence on psychology and economics, where his views on judgment and decision-making have prompted substantial research. One of the most visible fields of applied behavioral economics is *Nudging*, a key element in contemporary policy design (Thaler & Sustein, 2008).

Knowledge Systems is a more recent movement but consistently developed over several decades. For disambiguation, the use of the term Knowledge Systems in Knowledge Management and Knowledge Based Development is distinct from that used in software engineering (Stefik, 2014). Knowledge Systems are thus defined as value arrays aligning the relevant attributes of knowledge objects, agents and contexts (Carrillo et al., 2019). Knowledge systems include "the practices, routines, structures, mindsets, values and cultures affecting what and how knowledge is produced and used, and by whom" (Fazey et al., 2020, p. 101724).

Knowledge systems research has roots in the application of Experimental Analysis of Behavior to the understanding and development of scientific practice (Carrillo, 1983a; Morales & Carrillo, 2008a, 2008b) and an agenda for Empirical Epistemology (Carrillo, 1983b, 1983c). Through 1983–1986, the UK's *Assessment of Performance Unit* – tasked with developing methods of assessing and monitoring school achievement (Johnson, 1989) – became a rich empirical ground to study the acquisition of scientific patterns of behavior. From 1989 through 2018, the Center for Knowledge Systems (CKS) at Tecnológico de Monterrey was the basis for an extensive program on Knowledge Management (KM), Intellectual Capital (IC) and Knowledge-Based Development (KBD). The CKS combined graduate teaching and research work on KM and IC with an intense consultancy practice for both private (Carrillo & Galvis-Lista, 2014) and public organizations (Carrillo, 2016). The demanding consultancy requirements helped develop a KM and IC process framework and a Capital Systems model that allowed model testing in practical situations. Initiatives such as the World Capital Institute, the Knowledge Cities movement (Edvardsson, Yigitcanlar, and Pancholi, 2016), the international Most Admired City Awards, the *International Journal of Knowledge-Based Development*, and finally a KBD agenda including the ongoing Knowledge for Anthropocene program emerged from this context (Carrillo, 1997; 2001).

Recently, a study involving 175 authors undertook a critical overview of the prospected capacity of existing global knowledge systems to cope with major challenges such as climate change (Fazey et al., 2020). Realizing the gross misalignment between the current knowledge establishment and the challenges of the Anthropocene, they envision the attributes of 'knowledge system for life'. The Second Order Science framework they applied is described in Chapter 27 of this volume by Anthony Hodgson, while the envisioned knowledge systems attributes are referred in Chapter 28.

The conjunction of Behavioral Economics and Knowledge Systems may prove a fruitful addition to a Science of Knowledge. Understanding human response to the climate crisis seems key to tackling it effectively. Shoshana Zuboff has courageously exposed in *The Age of Surveillance Capitalism* the uncanny capacity for prediction and control exercised by the oligopoly of digital companies controlling the internet behavioral surplus. She echoes Karl Polanyi's warning about the vulnerability of both climate and human behavior if left unchecked at the hands of capital: "The commodity fiction disregarded the fact that leaving the fate of soil and people to the market would be tantamount to annihilating them" (Zuboff, 2019, pp. 345–346). Unless the link between climate change and behavior control is understood and exposed, the core questions raised by Zuboff about surveillance capitalism remain unanswered: *who knows? who decides? and who decides who decides?*

KNOWLEDGE ECONOMY

Early calls on the surging significance of intangible value came in the works of Fritz Machlup (1962), Taichi Sakaiya (1991), Paul Romer (1990), and Peter Drucker (1994). Unlike other sciences of knowledge, studies on the knowledge economy did not evolve mainly from economic studies of Science. While there is work on the Economics of Science (Stephan, 1996; Audretsch et al., 2002), it has remained a subfield of the Economics of Knowledge, that is, the analysis of science and knowledge at large as factors of production. Distinctive studies on the Economy of Knowledge as a major social and cultural transformation draw from two drives. First, the reaction to the advent of the Knowledge Society and Culture at the dawn of the second millennium. Second, the thrust from Knowledge and Intellectual Capital Management as well as from Knowledge-Based Development to be further discussed below.

The application of Economic Science to understanding the rise of Knowledge Societies is two-pronged. The two core issues are the characterization of *'knowledge-based'* (1) and the resulting *value dynamics* (2) (Carrillo, 2014). The rest of this section attempts to frame these two issues. The closest to a consensus is the realization of an unprecedented increase in the relative weight of intangible capital in Post-Industrial Society and its boosting effect on productivity (e.g., Powell & Snellman, 2004, p. 201). Besides that, three progressive approaches or *generations* in understanding what 'knowledge-based' means have been recognized (see, e.g., Carrillo, 2001; Laszlo & Laszlo, 2002; Rowley, 2003; Garcia, 2008; Dang & Umemoto, 2009; Wang, 2009; Martínez, 2010; Batra, 2012; Edvarsson, Yigitcanlar & Pancholi, 2016; Arsenijević et al., 2017; Fachinelli et al., 2017; Lajul, 2018; Yigitcanlar & Inkinen, 2019). These generations are characterized in terms of the progressive addition of three necessary and sufficient conditions for knowledge events to occur: *first*, an object-centred infrastructure for increasing the stock of knowledge, aiming at higher-value and technology-intensive production; *second*, an object plus agent-centred platform for facilitating knowledge flows, aiming at human capital development and knowledge-transfer processes; and *third*, an object plus agent plus context strategy balancing traditional and intangible capital systems, aiming at total system welfare. Of these, the first or *instrumental* view is still the dominant logic and the most widely advocated in organizational, urban and regional development plans (e.g., Neef et al., 1998; Barkhordari, Fattahi & Azimi, 2019). The second or *incremental* view has been gaining ground, mainly in the perspective of international agencies, as it establishes feedback between economic growth and human capital (e.g., Bindé, 2005; Faggian, Modrego & McCann, 2019). The third or *disruptive* view is still marginal as it involves a qualitative shift in economic culture that might be prompted by the emergency conditions of the Anthropocene (von Mutius, 2005; Dang & Umemoto, 2009; Van Wezemael, 2012).

Shifting from the first and second to the third view involves changing the limits, elements and rules of economics as a formal system. This means that the space of possibilities becomes qualitatively different (e.g., knowledge goods are non-rival and non-excludable, meaning that access to and consumption of a good by an agent does not prevent simultaneous access and consumption by other agents). Acknowledgement of the socio-behavioral base of economics involves a change from experiencing physical realities, to experiencing represented ones. Value production does not rest mainly on the physical attributes of matter, but in the symbolic and economic attributes of representations. This major transition requires a new understanding of the relation between knowledge and value. Rather than material objects generating a sensory or instrumental record, the representations and interpretations of these objects dominate human experience. The underlying phenomenological evolution implies shifting emphasis from the realm of things to the realm of representation of things, while maintaining their interdeterminacy and functional continuum. Such a substitution process is at the core of psychological life, knowledge-based behavior and culture (Carrillo, 1998). In this natural transposition lies the association of meanings from which semiotics emerges and the association of values from which economics unfolds.

The Knowledge Economy has to do with the viability of human civilization. It has to do with the capacity to achieve a balance between access to, use, and disposal of energy and matter for sustaining life (Smith & Max-Neef, 2011) as much as it has to do with our capacity for learning to coexist in a more-than-human world. It implies the possibility of agreeing upon a manageable set of collective preferences (Hodgson, 2012). Hence, the 'knowledge-based' attribute refers to an economic, political and cultural order, placing as much emphasis on intangible values as on material and monetary ones.

The knowledge context dimension distinctive of the third or *disruptive* view of knowledge provides it with economic relevance (value) and cultural significance (meaning). In Chapter 28, knowledge context and the actual happening of *knowing* as a connection between object, agent and context that enables justified explanation and effective action is explored. Such exploration builds on time-honoured concepts such as *phronesis, lebenswelt* and *savoir*, leading to the more recent ideas of *vernacular knowledge* and *actionable knowledge* (see Chapter 28).

Knowledge Capital deals with value dimensions so far excluded from traditional economics and institutions. The point at which knowledge and civilization redefine each other has more to do with human potential for consciousness and cultural evolution. Here lies the distinctive leverage of

KBD: the transformation of culture on the basis of a human behavior upgrade (Carrillo, 2006; Hodgson, 2012).

SOCIOLOGY OF KNOWLEDGE

Sociology of knowledge, in turn, studies the relationship between human symbolic behavior and its social context. This field has its roots in the Sociology of Science. Prominent amongst pioneers are Emile Durkheim (1858–1917), Max Weber (1864–1920) and Robert Merton (1910–2003), who described the terms of relation between science and society. Durkheim sought in the early 1900s to understand the social origins of language and logical thought, anticipating recent dilemmas on the social construction of knowledge and the relevance of collective consciousness (Durkheim & Swain, 2008). Weber's classic *Science as a Vocation* 1918 lecture (Weber, 1946) anticipated the reflexive turn in the sciences of knowledge. Robert Merton outlined four ideal regulators of scientific behavior: *universalism, communism, disinterestedness and organized skepticism* (Merton, 1973, Part III) becoming for decades the hallmark for a *scientific ethos*.

A major stream of sociological approach to knowledge mentioned above is Social Constructionism. Social Studies of Science were marked by a theory of knowledge stating that our understanding of the world is socially constructed rather than individually grasped from an independent reality. While social constructionism can be traced back in the history of thought, it is Max Scheler's 1924 *Problems of a Sociology of Knowledge* (Scheler, 2012) where a number of 'axioms' on the cultural genesis of public knowledge are set. Karl Mannheim, in turn, stressed the roles that ideologies and utopias play as social interpretative frameworks. But Peter Berger and Thomas Luckmann's publication of *The Social Construction of Reality: A Treatise in the Sociology of Knowledge* (1966) consolidated social constructionism. Of particular relevance is the concept of *Social Stock of Knowledge*. This concept acknowledges the existence and importance of modes of knowing beyond scientific and theoretical knowledge, just as significant for everyday life, including values, beliefs, proverbs, myths, institutions, routines, customs, conventions, roles, division of labor, practices, procedures, and so forth. Michel Foucault went further to treat reason and knowledge themselves as social constructions, for example, in his examination of 'madness' and 'illness'. Another of Foucault's influential contributions was his analysis of the interdependence between knowledge and power. He claimed that knowledge builds discourses that rule our lives through a dominant logic.

Social constructionism has evolved to assimilate substantial inputs from Marxism, feminism, de-colonization and other critical accounts of institutionalized knowledge, such as in Doyle McCarthy's *Knowledge as Culture*

([1996] 2005). Subtitled *The New Sociology of Knowledge*, this book sets to "reestablish at the forefronts of the sociology of knowledge the problem of the functions of knowledge in public life and politics" (pp. 2–3). The inter-determination between knowledge and society and the resulting power dynamics lead to considering the political dimension of such interplay.

Rather than a strictly disciplinary field, social aspects of science have evolved into a wider, more interdisciplinary field known as 'Science and Technology Studies' or 'Science, Technology and Society Studies' to be reviewed in Chapter 28. This very dynamic field now encompasses most of the areas here depicted as separate sciences of knowledge, particularly within the social sciences and the humanities, but also linked with technology and engineering (Felt et al., 2017).

POLITICAL SCIENCE AND KNOWLEDGE POLICY

Nico Stehr and others have dissected the political-economic dynamics of the knowledge society (Adolf, 2018). In 'The Power of Scientific Knowledge' Grundmann and Stehr look at the complexity of policymaking throughout knowledge production and distribution processes, anticipating the political struggles of a knowledge for the Anthropocene. In it, they come across a core realization for knowledge policies: "… knowledge gains in *distinction* on the basis of its ability to change reality" (Grundmann & Stehr, 2012, p. 22), where changing reality does not just mean transforming the environment, but any significant level of behavioral change that positively improves the space of possibilities. That is the sense in which the World Capital Institute regards knowledge as the most important leverage to human action (including, thinking, imagination and other forms of verbal behavior). Contrary to an extended view, knowledge is not merely the capacity to act per se, instead it is a state shift resulting from understanding a value context and making appropriate decisions within it.

Science policy studies have advanced in the direction this chapter is pointing, that is, integrating the cultural, political and economic drivers of knowledge policy choice and implementation (Lane et al., 2011). More specifically, the field of Science Policy and Innovation Studies deals with understanding and re-framing the received processes through which knowledge transitions from research to actionable knowledge (Simon et al., 2019). The fact that actionable knowledge involves much more than a superior research capacity and knowledge base infrastructure is exemplified by the spectacular failure of the United States in coping with the COVID-19 pandemic: the country that still has the largest installed capacity in R&D, being the hardest hit country in the world regarding the total number of COVID-19 cases and deaths (as of February 2021).

Social capacity to act has been impaired by the evolution of the very knowledge system that was supposed to leverage development. The current knowledge ecosystem is subjugated by the capital structure through a tight network of privatization, financing, commercialization, standardization, rankings and incentives. Such tight regime results in a globalized, meta-national scheme of intellectual homogenization (Schmidt-Wellenburg & Bernhard, 2020). From a perspective of 'Next Generation Science Policy' creative knowledge construction can only be promoted through influence rather than power, through three 'designs': meta-governance; concertation and assemblage; and capability and capacity building (Kuhlmann & Rip, 2019, p. 21). Like others, Karl Mannheim's later attempts to avoid the cul-de-sac of relativism found an exit through Pragmatist philosophy.

KNOWLEDGE AND INTELLECTUAL CAPITAL MANAGEMENT

More recently, the field of Knowledge and Intellectual Capital Management (KM&IC) at the organizational level and its progression into Knowledge-Based Development (KBD) at the macro level entered the scene. The latter found expression predominantly at city level (Urban-KBD) through the better-known category of Knowledge Cities.

Unlike most specialized fields of knowledge studies, KM&IC and KBD are technical fields dealing with problem-solving and solutions delivery. Hence, the knowledge base of these fields stems mostly from procedural schemata subject to the test of delivering solutions. Not surprisingly, the early stages of KM&IC in the late 1980s were mainly applied activities with emerging models and practices largely developed by mid- and large-size consultancy bureaus and only internally communicated. Increasingly, the very process of managing KM expertise required more sophisticated models, methods and, above all, empirical testing of the relative efficacy of alternative practices. As KM&IC were entering academia by the mid-1990s, the center of gravity gradually shifted from consultancy firms to graduate programs and research centers. In 1997, the *Journal of Knowledge Management* (JKM), the pioneer and to date the highest-ranking journal in the field started publication. Now, over two dozen specialized journals populate the field (Serenko & Bontis, 2017), while an active ecosystem of professional practice, technical development, empirical testing, systems modeling and theory building acts as communicating vessels.

The knowledge systems perspective presented earlier in this chapter draws from a theoretical-practical environment shaped in the former context. In 1992, the abovementioned Center for Knowledge Systems was created. According to an early report, this was the first dedicated R&D unit identifiable in the KM&IC world (Skyrme & Amidon, 1997). The CKS team was fortunate to experience over its 25 years of existence the combination of an intense consultancy practice (achieving financial independence) and a just as intense research work. At a ranking exercise for the JKM 20th anniversary, the CKS was second amongst the 50 most "productive and influential institutions publishing in the JKM" (Gaviria-Marin, Merigo, & Popa, 2018). The CKS soon engaged in the development of a graduate KM curriculum that served as a basis for a number of programs through several countries.

The former account exemplifies the benefit for the study of knowledge systems provided by an environment that is not only conducive to research and reflection, but also intensely driven by expected outcomes from applied projects (Antonacopoulou, 2007; Lambe, 2014). Renn's (2020) account of the multi-layered evolutionary processes of knowledge construction throughout history implies that in looking for the best fit knowledge for the Anthropocene we should resource not only at the institutionalized sciences of knowledge. Beyond these, we should also look at all other relevant forms of knowledge, particularly those that will be required by and constructed within the social space of most individuals in the upcoming circumstances. Such domains of knowledge have been anticipated in the classical idea of *phronesis*, as well as in the more recent concepts of *Social Stock of Knowledge* (Berger & Luckmann, 1966) and *Vernacular Knowledge* (Illich, 1981). Rather than mere procedural knowledge we are looking for the sets of contextualizable, actionable knowledge that might be best fit for foreseeably extreme and chaotic scenarios (Kirchhoff, Lemos, & Dessai, 2013; Mach et al., 2020).

It might help that we capitalize on the existing knowledge about the design of knowledge systems (Bennet & Bennet, 2014; Fazey et al., 2020). Basically, these involve operationalizing the attributes of knowledge objects, agents and contexts, so as to achieve the maximum possible alignment between those attributes (Carrillo et al., 2019). Once counting with a strategic map, alignment between attributes can be systematically improved through KM&IC processes (Carrillo & Galvis-Lista, 2014).

CONCLUSION

Knowledge as an object of study has been approached from many different perspectives. Rooted in the philosophical traditions of Epistemology or Philosophy of Knowledge, the 20th century saw the emergence of specialized studies of science stemming from diverse disciplines. History, Sociology, Psychology and Political Science spawned subdisciplines eventually converging into the wider field of Science, Technology and Society Studies. Beyond the institutionalized domain of scientific and technological knowledge, the more recent area of knowledge studies or sciences of knowledge builds on the former to include not just other fields of academic research such as physics, cybernetics, management, and computer science but most distinctively other knowledge practices such as indigenous, ancient, pragmatic and non-institutionalized ones. Hence, in identifying the best stock of knowledge for facing the Anthropocene, we are counting on a wealth of frameworks to understanding knowledge as rich as they are diverse. They are not only stemming from different disciplines and specialized fields of research. They are also stemming from very different epistemologies, cultural traditions, political frameworks, social contexts and even historical and developmental stages. The first challenge to a unified theory of knowledge that best leverages the human predicament before the Anthropocene is to map out of the broader field of knowledge studies and to start working in a collaborative, transdisciplinary way. To socialize what is known to some and to collaborate on what is unknown to all.

REFERENCES

Adolf, M. T. (Ed.) (2018). *Nico Stehr: Pioneer in the theory of society and knowledge* (Vol. 16). Springer.
Antonacopoulou, E. P. (2007). Actionable knowledge. In Clegg, S., & Bailey, J. (Eds.), *International Encyclopaedia of Organization Studies* (pp. 14–17). Sage.
Arsenijević, O., Trivan, D., Podbregar, I., & Šprajc, P. (2017). Strategic aspect of knowledge management. *Organizacija, 50*(2), 163–176.
Audretsch, D. B., Bozeman, B., Combs, K. L., Feldman, M., Link, A. N., Siegel, D. S., ... & Wessner, C. (2002). The economics of science and technology. *The Journal of Technology Transfer, 27*(2), 155–203.
Barkhordari, S., Fattahi, M., & Azimi, N. A. (2019). The impact of knowledge-based economy on growth performance: Evidence from MENA countries. *Journal of the Knowledge Economy, 10*(3), 1168–1182.
Barth, F. (2002). An anthropology of knowledge. *Current Anthropology, 43*(1), 1–18.
Batra, S. (2012), Development perspectives of knowledge management. *Review of Knowledge Management, 2*(1), 17–23.
Bennet, A., & Bennet, D. (2014). Knowledge, theory and practice in knowledge management: Between associative patterning and context-rich action. *Journal of Entrepreneurship, Management and Innovation, 10*(4), 7.
Berger, P. & Luckmann, Y. (1966). *The Social Construction of Reality*. Anchor Books.
Bernal, J. D. (1971) *Science in History* (4 volumes). MIT Press.
Bindé, J. (2005). *Towards Knowledge Societies: UNESCO World Report*. UNESCO.
Burke, P. (2015). *What is the History of Knowledge?* John Wiley & Sons.
Carrillo, F. J. (2016). Drivers and processes: experiences of implementation of knowledge management in public administration in Mexico. In *Experiências internacionais de implementação da gestão do conhecimento no setor público*. IPEA.
Carrillo, F. J. (2014). What 'knowledge-based' stands for? A position paper. *International Journal of Knowledge-Based Development, 5*(4), 402–421.
Carrillo, F. J. (2006). From transitional to radical knowledge-based development. editorial. *Journal of Knowledge Management, 10*(5), 3–5.
Carrillo, F. J. (2001). Meta-KM: A program and a plea. *Knowledge and Innovation: Journal of the KMCI, 1*(2), 27–54.

Carrillo, F. J. (1998). Managing knowledge-based value systems. *Journal of Knowledge Management, 1*(4, June), 280–286.

Carrillo, F. J. (1983a). *El Comportamiento Científico*. Limusa-Wiley.

Carrillo, F. J. (1983b, September 16–18). *Psychology of science: A matter of choice* [Conference paper]. Joint EASST/STSA conference: Choice in Science and Technology. London, UK.

Carrillo, F. J. (1983c, July 26–30). *Empirical epistemology: A behavioural programme for science* [Conference paper]. First European Meeting of Experimental Analysis of Behavior. Liege, Belgium.

Carrillo, F. J., Edvardsson, B., Reynoso, J., & Maravillo, E. (2019). Alignment of resources, actors and contexts for value creation. *International Journal of Quality and Service Sciences, 11*(3), 1–31.

Carrillo, F. J., & Galvis-Lista, E. (2014). Procesos de gestión de conocimiento desde el enfoque de sistemas de valor basados en conocimiento. *Ideas CONCYTEG, 9*(107), 3–22.

Carrillo, J. (1997). Managing knowledge-based value systems. *Journal of Knowledge Management, 1*(4), 280–286.

Crick, M. R. (1982). Anthropology of knowledge. *Annual Review of Anthropology, 11*(1), 287–313.

Dang, D. & Umemoto, K. (2009) Modeling the development toward the knowledge economy: A national capability approach. *Journal of Knowledge Management, 13*(5), 359–372.

Drucker, P. F. (1994). *Post-capitalist Society*. Routledge.

Durkheim, E., & Swain, J. W. (2008). *The Elementary Forms of the Religious Life*. Courier Corporation.

Edvardsson, I. R., Yigitcanlar, T., & Pancholi, S. (2016). Knowledge city research and practice under the microscope: A review of empirical findings. *Knowledge Management Research & Practice, 14*(4), 537–564.

Fachinelli, A. C., Giacomello, C. P., Larentis, F., & D'Arrigo, F. (2017). Measuring the capital systems categories: The perspective of an integrated value system of social life as perceived by young citizens. *International Journal of Knowledge-Based Development, 8*(4), 334–345.

Faggian, A., Modrego, F., & McCann, P. (2019). Human capital and regional development. In Capello, R., & Nijkamp, P. (Eds.), *Handbook of Regional Growth and Development Theories* (pp. 149–171). Edward Elgar Publishing.

Fazey, I., Schäpke, N., Caniglia, G., Hodgson, A., Kendrick, I., Lyon, C., … & Verveen, S. (2020). Transforming knowledge systems for life on Earth: Visions of future systems and how to get there. *Energy Research & Social Science, 70*, 101724.

Feist, G. J., & Gorman, M. E. (Eds.) (2012). *Handbook of the Psychology of Science*. Springer.

Felt, U., Fouché, R., Miller, C. A., & Smith-Doerr, L. (Eds.) (2017). *The Handbook of Science and Technology Studies* (4th edition). MIT Press.

Fuller, S. (2019). *Philosophy of Science and its Discontents*. Routledge.

Garcia, B. C. (2008). Global KBD community developments: The MAKCi experience. *Journal of Knowledge Management, 12*(5), 91–106.

Gaviria-Marin, M., Merigo, J. M., & Popa, S. (2018). Twenty years of the Journal of Knowledge Management: A bibliometric analysis. *Journal of Knowledge Management, 22*(8) 1655–1687.

Grundmann, R., & Stehr, N. (2012). *The Power of Scientific Knowledge: From Research to Public Policy*. Cambridge University Press.

Heidenblad, D. L., Nilsson, A., & Östling, J. (Eds.) (2020). *Forms of Knowledge: Developing the History of Knowledge.* Nordic Academic Press.

Hermo, X. G., Giráldez, S. L., & Saura, L. F. (2014). A systematic review of the complex organization of human cognitive domains and their heritability. *Psicothema, 26*(1), 1–9.

Hodgson, G. (2012) *From Pleasure Machines to Moral Communities: An Evolutionary Economics Without Homo Economicus.* University of Chicago Press.

Holton, G. J. (1988). *Thematic Origins of Scientific Thought: Kepler to Einstein.* Harvard University Press.

Illich, I. (1981). *Shadow Work.* Marion Boyars Publishers.

Johnson, S. (1989). *National Assessment: The APU Science Approach.* Her Majesty's Stationery Office.

Kahneman, D. (2011). *Thinking, Fast and Slow.* Allen Lane.

Kandel, E. R., Schwartz, J. H., Jessell, T. M., Siegelbaum, S. A., & Hudspeth, A. J. (Eds.) (2012). *Principles of Neural Science.* McGraw-Hill.

Kirchhoff, C. J., Lemos, M., & Dessai, S. (2013). Actionable knowledge for environmental decision-making: Broadening the usability of climate science. *Annual Review of Environment and Resources, 38*, 393–414.

Klausen, S. H. (2020). 9 challenges for an anthropology of knowledge. In Mizumoto, M., & Ganeri, J. (Eds.), *Ethno-Epistemology: New Directions for Global Epistemology*, 201–215. Routledge.

Kuhlmann, S., & Rip, A. (2019). Next generation science policy and Grand Challenges. In Simon, D., Kuhlmann, S., Stamm, J., & Canzler, W. (Eds.), *Handbook on Science and Public Policy* (pp. 12–25). Edward Elgar Publishing.

Lajul, W. (2018). Reconstructing African fractured epistemologies for African development. *Synthesis Philosophica, 33*(1), 51–76.

Lambe, P. (Ed.) (2014). *Knowledge Management Special Issue: Connecting Theory and Practice.* Nowy Sacz School of Business-National-Louis University.

Lane, J. I., Fealing, K. H., Marburger III, J. H., & Shipp, S. S. (Eds.) (2011). *The Science of Science Policy: A Handbook.* Stanford University Press.

Lässig, S. (2016). The history of knowledge and the expansion of the historical research agenda. *Bulletin of the German Historical Institute, 59*(Fall), 29–58.

Laszlo, K. & Laszlo, A. (2002), Evolving knowledge for development: The role of knowledge management in a changing world. *Journal of Knowledge Management, 6*(4), 400–412.

Mach, K. J., Lemos, M. C., Meadow, A. M., Wyborn, C., Klenk, N., Arnott, J. C., ... & Stults, M. (2020). Actionable knowledge and the art of engagement. *Current Opinion in Environmental Sustainability, 42*, 30–37.

Machlup, F. (1962). *The Production and Distribution of Knowledge in the United States.* Princeton University Press.

Malabou, C. (2017). The brain of history, or, the mentality of the Anthropocene. *South Atlantic Quarterly, 116*(1), 39–53.

Martínez, A. (2010). Personal knowledge management by the knowledge citizen: The generation aspect of organizational and social knowledge-based development. In Metaxiotis, K., Carrillo, F. J., & Yigitcanlar, T. (Eds.), *Knowledge-Based Development for Cities and Societies: Integrated Multi-Level Approaches* (pp. 131–140). IGI Global.

McCarthy, E. D. ([1996] 2005). *Knowledge as Culture: The New Sociology of Knowledge.* Routledge.

Merton, R. K. (1973) *The Sociology of Science. Theoretical and Empirical Investigations*. University of Chicago Press.

Morales, C. & Carrillo, F. J. (2008a, July 18–20). *Psychology of science from an interbehavioral perspective*. II International Society for the Psychology of Science and Technology. Berlin; Germany.

Morales, C. & Carrillo, F. J. (2008b, August 20–23). *Shifting from data base to knowledge base: A proposal to enhance science studies and science development*. 4S-EASST Joint Conference. Acting with science, technology and medicine. Rotterdam, the Netherlands.

Neef, D., Siesfeld, T., Siesfeld, G. A., & Cefola, J. (1998). *The Economic Impact of Knowledge*. Butterworth-Heinemann.

Osbeck, L. M., Nersessian, N. J., Malone, K. R., & Newstetter, W. C. (2010). *Science as Psychology: Sense-making and Identity in Science Practice*. Cambridge University Press.

Parker, S. T., Langer, J., & Milbrath, C. (Eds.) (2014). *Biology and Knowledge Revisited: From Neurogenesis to Psychogenesis*. Psychology Press.

Powell, W. W., & Snellman, K. (2004). The knowledge economy. *Annu. Rev. Sociol.*, 30, 199–220.

Regnier, D., & Astuti, R. (2015). Introduction: Taking up the cognitive challenge. *Social Anthropology*, 23(2), 131–134.

Renn, J. (2020). *The Evolution of Knowledge: Rethinking Science for the Anthropocene*. Princeton University Press.

Romer, P. M. (1990). Endogenous technological change. *Journal of Political Economy*, 98(5, Part 2), S71–S102.

Rowley, J. (2003). Knowledge management – the new librarianship? From custodians of history to gatekeepers to the future. *Library Management*, 24(8/9), 433–440.

Sakaiya, T. (1991). *The Knowledge-Value Revolution, or, a History of the Future*. Kodansha America, Inc.

Scheler, M. (2012). *Problems of a Sociology of Knowledge*. Routledge.

Schmidt-Wellenburg, C., & Bernhard, S. (Eds.) (2020). *Charting Transnational Fields: Methodology for a Political Sociology of Knowledge*. Routledge.

Serenko, A. & Bontis N. (2017). Global ranking of knowledge management and intellectual capital academic journals: 2017 update. *Journal of Knowledge Management*, 21(3), 675–692.

Simon, D., Kuhlmann, S., Stamm, J., & Canzler, W. (Eds.) (2019). *Handbook on Science and Public Policy*. Edward Elgar Publishing.

Simon, H. A. (1982). *Models of Bounded Rationality*. MIT Press.

Skyrme, D. J., & Amidon, D. M. (1997). *Creating the Knowledge-Based Business. Key Lessons from an International Study of Best Practice*. Business Intelligence Limited.

Smith, P. & Max-Neef, M. (2011). *Economics Unmasked: From Power and Greed to Compassion and the Common Good*. Green Books.

Stefik, M. (2014). *Introduction to Knowledge Systems*. Elsevier.

Stephan, P. E. (1996). The economics of science. *Journal of Economic Literature*, 34(3), 1199–1235.

Stimson, D. (Ed.) (1962). *Sarton on the History of Science: Essays*. Harvard University Press.

Thaler, R. H., & Sunstein, C. (2008). *Nudge: Improving Decisions about Health, Wealth, and Happiness*. Yale University Press.

Van Wezemael, J. (2012). Afterword: Concluding: Directions for building prosperous knowledge cities. In Yigitcanlar, T., Metaxiotis, K., & Carillo, F. J. (Eds.), *Building Prosperous Knowledge Cities* (pp. 374–382). Edward Elgar Publishing.

von Mutius, B. (2005). Rethinking leadership in the knowledge society learning from others: How to integrate intellectual and social capital and establish a new balance of value and values. In Bounfour, A., & Edvinsson, L. (Eds.), *Intellectual Capital for Communities* (pp. 151–163). Butterworth-Heinemann.

Wang, X. (2009). *Knowledge-Based Urban Development in China*. [Doctoral dissertation] Newcastle University.

Weber, M. (1946). Science as a vocation. In Tauber, A. (Ed.), *Science and the Quest for Reality* (pp. 382–394). Palgrave Macmillan.

Yigitcanlar T. & Inkinen T. (2019). Theory and practice of knowledge cities and knowledge-based urban development. In Yigitcanlar, T., & Inkinen, T. (Eds.), *Geographies of Disruption* (pp. 109–133). Springer.

Zuboff, S. (2019). *The Age of Surveillance Capitalism: The Fight for a Human Future at the New Frontier of Power*. Public Affairs.

19. Knowledge as world capital: global knowledge
Alexander Ruser

INTRODUCTION

Knowledge, in particular scientific knowledge addresses global problems, provides the context (and sometimes the cause) for conflict, informs international politics and fuels global activism. However, this "globalization of knowledge" should not be confused with the rationalization of public discourse and politics. Nor should we expect that information and expertise lead to a "depoliticization" of politics or rationalization public debate. Similar things can be said about the impact of knowledge on national economies and the global economy: Despite the fact that Peter Drucker's early depiction of knowledge as a "vital economic resource" (Drucker 1968) has been adopted by key international actors such as the World Bank which identifies "knowledge [as] the main engine of economic growth" (Chen, Dahlam 2006:1) and predicts "new demands on citizens, who need more skills and knowledge to be able to function in their day-to-day lives" (World Bank 2003: xvii) the precise impact of knowledge on economies, societies and individuals remains somewhat unclear. Even worse, the very definitions of "knowledge," let alone "global knowledge" seem to be contested.

In this contribution we will ask in which sense global knowledge can be regarded as world capital. We will suggest to move beyond narrow definitions of knowledge as a component of economic or "human" capital. Instead of treating knowledge like a commodity or an individual asset, we suggest to define (global) knowledge as a capacity for action. Such a definition not only fits more comprehensive understanding of knowledge-based capital as "social capital" (Inkpen, Tsang 2005) but stresses its empowering dimension and focus on its ability to transform social realities. Investigating these enabling and transformative capacities of knowledge is particularly acute in the context of political debates about the "Anthropocene" (Machin 2019), an era in which "humans have become geologic agents" (Delanty, Mota 2017: 10). Knowledge gains were instrumental in bringing about new technological capabilities, the creation of unprecedented risks and unintended consequences that turned humans into geological agents. Moreover, the Anthropocene, the very prospect of human activity having an unprecedented magnitude, demands "Anthropocene knowledge" (Edwards 2017: 35), and thus knowledge that is inevitably "global."

We will begin our discussion with the question of what knowledge and global knowledge in particular actually is. Touching upon different concepts of knowledge allows us to estimate the impact of global knowledge as a commodity, capital or capacity and outline the prospects and promises as well as the impediments and challenges of "global knowledge spaces" (Stehr, Adolf 2017: 150–151).

To do so we will then discuss how (global) knowledge is distributed, before we turn to the implications of epistemic cultures for the spread and acquisition of knowledge before we turn to knowledge inequalities. We will finally engage with contestation and competition (fake news, politicization) in global knowledge societies and outline the significance and limitations of global knowledge in dealing with the challenges of the Anthropocene.

WHAT IS GLOBAL KNOWLEDGE?

Depicting knowledge as a vital resource makes it sound as if it could be described as something tangible, countable and inviting to conflate "knowledge" with related concepts such as "information" and "data" (Stehr, Ruser 2017: 484–5). The term "global knowledge" is even more ambiguous and defies any straightforward definition. One possible interpretation is that global knowledge refers to "what is globally known." Read this way the term describes a "global stock of knowledge" and repository of "true information." The desire to catalogue what is known seems to be deeply ingrained in the history of science and was the driving force behind the attempts of the European enlightenment to compile comprehensive dictionaries and collections of everything that was known. The most ambitious of these attempts, Diderot and D'Alembert's "Encyclopédie, ou dictionnaire raisonné des sciences, des arts et des métiers" (Shackleton 1970: 390) was a manifestation of this desire to record the global state of knowledge:

> The purpose of this Encyclopaedia is indeed to collect knowledge scattered over the face of the earth; to expose that general system to the men with whom we live, and to transmit it to the men who will come after us; so that the work of the past centuries is not useless works for the centuries which will succeed; that our nephews, becoming more educated, become at the same time more virtuous and happier, and that we do not die without having deserved well of the human race. (Darnton 1984)

As a project of the enlightenment era the Encyclopaedia was not a purely academic venture, undertaken to satisfy the needs of scholars and academics. Its purpose was to serve present and future generations, educate them and allow them to become "more virtuous and happier."

However, such a stocktaking of a vast amount of knowledge proves to be very difficult. Early attempts to compile an overview on what is known like Ephraim Chambers' 1728 "Cyclopedia" (Shackleton 1970: 389) took the form of an (alphabetic) listing (Wernick 2006: 32) of isolated information. Such an exhaustive list is of very limited use. In order to serve present and future generations, information needs to be ordered, systematized and related to disciplines, arts or crafts to allow the extraction of useable knowledge. The need to explore the structure of knowledge and to invent complex schemes of presenting bits of knowledge and their relations to other parts of the comprehensive system of human knowledge (ibid.) already point to some fundamental weaknesses of the equation of global knowledge with an internationally available *stock* of knowledge.

First, it is very challenging to develop a comprehensive and consistent system of knowledge. Second, even the most comprehensive and complete list of what is known, cannot guarantee that this knowledge is absorbed or accepted by all: Are the heliocentric system and the spheric shape of the Earth globally known? Is anthropogenic climate change a "fact"? Scientific research which proves the shape of the Earth is easily accessible. Moreover, an international community of climate scientists has produced strong evidence for human made climate change. And yet, the "Flat Earth Society" hosted a two-day international conference in Dallas in 2019, not to speak of the activities of well-organized climate change deniers (Powell 2011; Ruser 2018).

To understand what global knowledge *is* therefore requires not so much stocktaking – for every list could be contested for being incomplete or include disputed and therefore not "truly global" knowledge – but a reflection of what knowledge *does*. We therefore suggest a second depiction of "global knowledge" that derives from an understanding of knowledge as a "capacity for action" (Stehr 2016). In this perspective "knowledge" should not be conceived of "as something that is," but rather "as a model for reality, or as the ability to set something in motion" (Stehr, Ruser 2017: 475). If we ask Diderot and D'Alembert for the value of knowledge, we should conceive of knowledge as something that facilitates alternatives and potentially transforms reality (Stehr 2016: 22).

Conceptualizing knowledge as a capacity for action thus defies the conflation of knowledge and information. Information is something that can be quantified (e.g. as "data"), possessed or lost and that can hence be treated like a product (Stehr, Ruser 2017: 485). Knowledge in contrast refers to a process (ibid.) and requires active agents (Stehr 2016: 46). Global knowledge should therefore not be conflated with the sum of globally available information. It is thus questionable if the notion of global knowledge (in the singular) holds (Stehr, Adolf 2017: 142). Global knowledge should be conceived of as an assemblage of processes, models of reality and specific action capacities. Moreover, such an understanding begs the question whether global knowledge can be depicted as a consequence of the global dissemination of *scientific* knowledge (ibid.). The abovementioned "discovery" of global climate change, for instance, was mainly accomplished by Western scholars in the global north and then communicated to the rest of the world – despite the fact that for instance, farmers across the globe, had been well aware of changing rainfall patterns, temperatures or other unusual weather events without being able to relate their local observations to a global phenomenon.

Although the context of knowledge production is not necessarily affecting the content and validity of the scientific discoveries, it remains a question if scientific knowledge can simply be disseminated. The abovementioned frames may or may not accept the diagnosis of "global climate change" and they might use this knowledge to demand global climate justice or swell the ranks of climate change deniers.

It is therefore highly questionable, whether the globalization of scientific knowledge can foster the convergence towards one global knowledge world. We therefore suggest to distinguish between distinct global knowledge worlds (Stehr, Adolf 2017: 146) and ask for patterns of "horizontal" in addition to the "vertical" distribution of knowledge (ibid.: 14).

KNOWLEDGE DISTRIBUTION: GLOBAL KNOWLEDGE INEQUALITIES

"Knowledge is like light" states the World Bank (World Bank 1999 cited after: Stehr, Adolf 2017: 146) adding that "it can easily travel the world, enlightening the lives of people everywhere." This poetic depiction of knowledge should not distract from the many impediments that obstruct the travel and acquisition of knowledge. Karin Knorr-Cetina reminds us that "a knowledge society is not simply a society of more knowledge [but] also a society permeated with knowledge settings, the whole sets of arrangements, processes and principles that serve knowledge and unfold with its articulation" (Knorr-Cetina 2007: 361–362). Moreover, she states that "[if] knowledge is now a productive force, the production cultures and the larger knowledge-related cultures that sustain them become the primary object of cultural investigation" (ibid.). Accordingly, if we want to understand how knowledge is created and more importantly how it "travels the world" we need to understand these knowledge-related or epistemic cultures. Knorr-Cetina defines them as "sets of practices, arrangements and mechanisms bound together by necessity, affinity and historical coincidence which, in a given area of expertise, make up how we know what we know" (Knorr-Cetina 2007: 363). The concept of epistemic cultures fits our understanding of knowledge as practice, as a capacity for action. It provides thus an ideal starting point for investigating the vertical and horizontal integration of knowledge and subsequent knowledge-induced inequalities (Stehr [1994]2018: 186). Scientific expertise can be described as an expression of (epistemic) cultural division (Knorr-Cetina 2007: 364) that translates into social differences and hierarchies. However, even if we concede the penetration of "a substantial part of the world" by scientific-technological knowledge for instance (Stehr, Adolf 2017: 151) we cannot conclude that global knowledge brings about a convergence of epistemic cultures and thus a homogenization of patterns of vertical integration. The social role of experts, the significance of local or indigenous knowledge may still differ. Distinct epistemic cultures and the subsequent status of "folk-knowledges" (Knorr-Cetina 2007: 369) can, for example, explain why climate scientists are more vulnerable in the U.S. than in other Western countries (Ruser 2020).

Global knowledge should therefore not be confused with the equal distribution of horizontally and vertically integrated knowledge. Neither should we conclude from the existence of international knowledge elites (e.g. a global financial sector, international communities of scientists) or structures (such as increasingly international higher education systems) that such a uniform global world of knowledge is in the making.

STRUCTURES OF GLOBAL KNOWLEDGE WORLDS: COMPETITION AND CONTESTATION

Since a uniform global knowledge world is not on the horizon, investigators into the role of "global knowledge" have to take into consideration structural differences between distinct knowledge worlds. Even if we assume an increased horizontal integration of *some* social bases of knowledge production (Stehr, Adolf 2017: 151), for example the standardization of academic education – we cannot jump to the conclusion that knowledge is traveling lightly. "The distribution of research capacity world-wide is highly stratified," writes Simon Marginson (2010: 6974) for instance, thus pointing towards structural inequalities in global production of scientific research. This might serve as a reminder that scientific knowledge (e.g. patents) is increasingly being treated as a commodity rather than a public good. Defining (scientific) knowledge as a public good, however, would be a prerequisite for the non-competitive, "light-like" distribution of knowledge.

A standard definition of "public goods" names two central properties: "nonrivalry in consumption and nonexcludability" (Kaul et al. 1999: 3). Defining (global) knowledge as a "global public good" would be particularly challenging, since it would require to define the global public (ibid.: 10), that is the "reach" of knowledge and to describe its relation to other (tangible and intangible) global public goods such as peace, financial stability or an intact nature (see Kaul et al. 1999: 13).

Current infrastructures of knowledge production and dissemination, however, are mainly structured around for-profit principles and competition. Despite some initiatives that promote knowledge sharing (e.g. "open access"), scientific knowledge production is driven by rivalry and its products are often commodified by reserving intellectual property rights or patents. Likewise, global players in knowledge economies feel little desire to share knowledge developed in corporate research facilities and lose competitive advantages.

But it is not only economic actors who have an interest in erecting barriers to the free traveling of knowledge. Political actors have good reasons to control the availability of knowledge. Controversies about the use of user-data, for instance, have pitted national and supranational actors such as the EU against powerful tech-companies (Esteve 2017), while burdening national policy-makers with the maintenance of competitive advantages in global knowledge economies and the (alleged) "global competition for talent" (Florida 2007). But states not only have to defend citizens' rights and show concern for the needs of knowledge-based economies. The very authority of states might be at stake. Blocking the spread of knowledge can be crucial for states in times in which "[k]nowledge-based politics also becomes knowledge-based resistance to political action" (Ericson, Stehr 2000: 8). Denying people at large access to knowledge (be it scientific or non-scientific) can be vital to prevent some people from developing capacities (and an appetite) for political protest.

The discovery of human-made climate change, for instance, has not only influenced scientific debate and high-level negotiations but has fuelled local and global activism by believers in and deniers of climate change. The spread of knowledge hence had a very visible impact on the political landscape and created specific problems of governance and governing (Ruser 2018, 2021).

In particular states that seek to limit their population's capacity for action – such as authoritarian regimes – have strong incentives to limit the access to global knowledge and to prevent knowledge from becoming globally known. The very tendency of knowledge societies to "produce and enable many more, and much more knowledgeable, actors to enter the political system" (Stehr 2016: 252) might erode the capabilities of states to "monopolize and deny access to resources" (ibid.) and hence to exert control.

GLOBAL KNOWLEDGE AS CAPITAL?

The role of knowledge "as capital" on the global stage is therefore likely to be ambivalent. The growing importance of knowledge as a productive force and means to create and transform realities could benefit an emerging global class "whose power is a function of their privileged access to productive political economic, or cultural knowledge that can ... be commanded through their ability to mobilize relevant experts" (Stehr 2016: 205). From such a perspective, knowledge – like any other form of capital – tends to be concentrated in the hands of a small social strata. In consequence, knowledge inequalities might add to and deepen existing global social inequalities. However, knowledge has the potential to increase the capacity for action of marginalized and disadvantaged people. Distinct epistemic cultures as well as traveling knowledge can contribute to the demystification of expertise and (state) authority. International movements such as "Occupy Wallstreet," "Fridays for Future" and many other international social movements that protest political, environmental or economic injustices are enabled and driven by global knowledge. Moreover, attempts by authoritarian regimes to prevent knowledge from traveling to their respective countries indicates that knowledge indeed has the potential for becoming a "weapon of the weak" (Stehr 2016: 320).

However, we should not expect the increasing significance of knowledge – both for privileged elites and disadvantaged people or political activists – to translate into a rationalization public debate or political conflict. In particular the importance of scientific knowledge in identifying global challenges – such as anthropogenic climate change, global health risks and challenges to globally integrated economies – and in coping with their consequences can nurture the false hope of the emergence of "fact" or "evidence" based politics. As mentioned above, we can neither assume the existence of global epistemic cultures that would benefit the global "acceptance" of scientific knowledge claims, nor presume that scientific knowledge is universally "evident." The fragility of expertise (Stehr 2016: 316), concerns about a rise of "technocratic" decision making by scientists and the fact that scientific research itself is increasingly caught up in political conflict limit the prospect of evidence-based decision making. Moreover, potential gaps between scientific and common-sense knowledge (Stehr 2016: 303) pose serious challenges for any attempt to rationalize policymaking. Moreover, the deliberate fabrication (or widening) of such gaps, by discrediting scientific expertise, manufacturing "counter-evidence" or challenging the factual basis of any debate (by introducing "alternative facts" and decrying "fake news") the *knowledge base* of knowledge-based societies itself becomes a political battlefield (Ruser 2021).

KNOWLEDGE AND/FOR THE ANTHROPOCENE

The significance of knowledge in the Anthropocene is likely to increase since, "the dynamic coupling of human societies and the Earth system not only depended on the generation of knowledge, but enhanced, in turn, the significance of that generation" (Renn 2020: 365). But what role will global knowledge play?

Based on our inquiry, knowledge will not be playing one role, but many. This conclusion fits in with the debates about the Anthropocene and in particular the role of humans in shaping the future of a period in which they can have a lasting (and potentially catastrophic) impact on planet Earth. Most recently historian Jürgen Renn proposed the concept of the ergosphere to describe the new role of humans in the Earth system:

> The concept of the ergosphere is meant to capture both: the transformative power of human interventions beyond their intentions and the planetary limits under which these interventions unfold. Although I call it a sphere because of its global extent, the ergosphere concept should not invite us to overlook the immense heterogeneity of humanity, its striking inability to act collectively, the basic tensions and conflicts of interest tearing it apart, and its asymmetries of power (…). These asymmetries of power also concern the generation of knowledge and science on a global scale. (Renn 2020: 382)

This depiction fits in our conceptualization of global knowledge as an unevenly distributed, promoted and hindered capacity to act. In particular on the global stage asking for the future role of knowledge remains an open question. However, new challenges and risks, summarized in the declaration of the "Anthropocene" urge us to engage with the global roles of knowledge. Global challenges like climate change, pandemics, mass extinction, global social inequality or resource depletion demand the production, dissemination and application of knowledge. However, as for instance the existence of climate change deniers and anti-vaccination activists indicate, (scientific) knowledge will not eradicate political and normative disagreement. Moreover, the sheer magnitude of the consequences of human-made problems in the Anthropocene highlights the ethical dimension of knowledge. Knowledge cannot be reduced to an economic resource. Moreover, global knowledge shouldn't be confused with a global stock of information but rather be seen as the capacity to act, transform and shape realities across the globe and, perhaps, at the global level. Like all capital, knowledge as world capital needs to be "invested" wisely and the "returns" will be uncertain.

REFERENCES

Chen, Derek, H.C., Dahlman, Carl J. (2006) *The Knowledge Economy, The KAM Methodology and World Bank Operations*. The World Bank

Darnton, Robert (1984) *The Great Cat Massacre and Other Episodes in French Cultural History*. Basic Books, 191–213 (http://artflsrv02.uchicago.edu/cgibin/philologic/getobject.pl?c.4:1252.encyclopedie0513)

Delanty, Gerard, Mota, Aurea (2017) "Governing the Anthropocene: Agency, Governance and Knowledge" *European Journal of Social Theory* 20(1): 9–38

Drucker, Peter F. ([1968] 1972). *The Age of Discontinuity Guidelines to our Changing Society*. Harper & Row

Edwards, Paul N. (2017) "Knowledge Infrastructures for the Anthropocene" *The Anthropocene Review* 4(1): 34–43

Ericson, Richard, Stehr, Nico (2000) *Governing Modern Societies*. Toronto University Press

Esteve, Asunción (2017) "The Business of Personal Data: Google, Facebook, and Privacy Issues in the EU and the USA" *International Data Privacy Law* 7(1): 36–47

Florida, Richard (2007) *The Flight of the Creative Class: The New Global Competition for Talent*. Harper Business

Inkpen, Andrew C., Tsang, Eric W.K. (2005) "Social Capital, Networks, and Knowledge Transfer" *AMR*, 30, 146165, https://doi.org/10.5465/amr.2005.15281445

Kaul, Inge, Grunberg, Isabelle, Stern, Marc A. (1999) *Global Public Goods. International Cooperation in the 21st Century*. Oxford University Press

Knorr-Cetina, Karin (2007) "Culture in Global Knowledge Societies: Knowledge Cultures and Epistemic Cultures" *Interdisciplinary Science Reviews* 32(4): 361–375

Machin, Amanda (2019) "Democracy and Agonism in the Anthropocene: The Challenges of Knowledge, Time and Boundary" *Environmental Values* 28(3): 347–365

Marginson, Simon (2010) "Higher Education in the Global Knowledge Economy" *Procedia Social and Behavioral Science* 2: 6962–6980

Powell, James L. (2011) *The Inquisition of Climate Science*. Columbia University Press

Renn, Jürgen (2020) *The Evolution of Knowledge: Rethinking Science for the Anthropocene*. Princeton University Press

Ruser, Alexander (2018) *Climate Politics and the Impact of Think Tanks. Scientific Evidence in Germany and the US*. Palgrave

Ruser, Alexander (2020) "A Mission for MARS: The Success of Climate Change Skeptic Rhetoric in the US" *Res Rhetorica* 7(2): 48–63

Ruser, Alexander (2021) "Widening the Gap: US Think Tanks and the Manufactured Chasm between Scientific Expertise and Common Sense on Climate Change" in: Julien Landry (ed.), *Critical Perspectives on Think Tanks: Power, Politics, Knowledge*. Edward Elgar Publishing, 196–214

Shackleton, Robert (1970) "The 'Encyclopédie' as an International Phenomenon" *Proceedings of the American Philosophical Society* 114(5): 389–394

Stehr, Nico (2016) *Information, Power, and Democracy. Liberty is a Daughter of Knowledge*. Cambridge University Press

Stehr, Nico ([1994]2018) "The Culture and Structure of Social Inequality" in: Marian Adolf (ed.) *Nico Stehr: Pioneer in the Theory of Society and Knowledge*. Springer: 181–198

Stehr, Nico, Adolf, Marian (2017) *Knowledge*. 2nd Edition. Routledge

Stehr N., Ruser A. (2017) "Knowledge Society, Knowledge Economy and Knowledge Democracy" in: E. Carayannis, D. Campbell, M. Efthymiopoulos (eds), *Handbook of Cyber-Development, Cyber-Democracy, and Cyber-Defense*. Springer, 475–494 https://doi.org/10.1007/978-3-319-06091-0_16-1

Wernick, Andrew (2006) "Comte and the Encyclopedia" *Theory, Culture & Society* 23(4): 27–48

World Bank (2003) *Lifelong Learning in the Global Knowledge Economy: Challenges for Developing Countries*. The World Bank

20. Adaptive value of traditional knowledge
Michael Blakeney

IMPACTS OF CLIMATE CHANGE ON AGRICULTURE

A number of the reports of the International Panel on Climate Change (IPCC) and the Food and Agricultural Organization of the United Nations (FAO) have identified the negative impact of climate change on agricultural productivity through variations in water supply, in heat and in salinity and through increasing insect infestations, plant diseases and reduction of pollinators (IPCC, 2018, 2019; FAO, 2018; FAO et al., 2018). Crop yields report a reduction in wheat yields in India by 5.2 per cent from 1981 to 2009 (Gupta et al., 2017) and similar negative effects on wheat yields were noted in Australia (Innes et al., 2015). Studies of agriculture in the Hindu-Kush Himalayan region of India, Nepal, Pakistan, and China report disrupted agricultural yields from more frequent floods as well as prolonged droughts (Manzoor et al., 2013; Hussain et al., 2016). Similar results have been obtained in studies of agriculture in Africa (Ketiem, 2017) and South America (Saxena et al., 2016). As is discussed below, the drying of the land and general aridity attendant on climate change has also contributed to bushfires which in addition to their impact upon agriculture also have substantial economic and health impacts.

The foundation international convention on climate change was the United Nations Framework Convention on Climate Change (FCCC), which entered into force on 21 March 1994. The parties committed themselves to limiting greenhouse gas emissions. A complementary strategy, which was mentioned in the FCCC was adaptation to climate change through developing climate resilient agricultural practices (IPCC, 2019, 5–45). However, it was not until the Fifth IPCC Report in 2014 that adaptation strategies were considered in any detail (Noble et al., 2014). The IPCC Synthesis Report, 2014 noted that 'indigenous, local and traditional knowledge systems and practices, including indigenous peoples' holistic view of community and environment, are a major resource for adapting to climate change, but these have not been used consistently in existing adaptation efforts' (IPCC, 2014, para 4.4.2). It has been pointed out that incorporating indigenous knowledge into climate change policies can lead to the development of effective mitigation and adaptation strategies that are cost-effective, participatory, and sustainable (Robinson and Herbert, 2001).

This chapter looks at the contribution which traditional knowledge (TK) can make to climate change adaptation, commencing with a case study TK and bushfires in Australia, before surveying TK and adaptation to climate change in agriculture. The chapter then surveys proposals to protect TK.

TRADITIONAL KNOWLEDGE (TK) AND BUSHFIRES IN AUSTRALIA

Australia is an arid country which is susceptible to bushfires, but the bushfires of 2019–2020 were the most catastrophic since white settlement commenced in 1788. Between June 2019 and the end of January 2020 the fires had burnt an estimated 11 to 18.6 million hectares (110,000 to 186,000 square kilometres) (Baldwin and Ross, 2020). By way of comparison, the 2019 California wildfires burnt 800,000 hectares and the 2019 Amazon rainforest wildfires burnt 900,000 hectares of land (Mihai, 2020). By 7 January 2020, smoke from the fires had travelled approximately 12,000 kilometres across the South Pacific Ocean to Chile and Argentina (SBS, 2020) and UNEP (2020) reported 900 million tonnes of CO_2 emissions, estimated to be higher than the annual emissions of Germany and equivalent to Australia's typical annual emissions in just three months. Apart from the economic impact of the fires in terms of the destruction of property and commerce and their health impacts, the fires are estimated to have killed nearly three billion terrestrial vertebrates, driving a number of species to extinction (Legge et al., 2020).

In addition, the fires had a very significant impact on the Gondwana rainforests in eastern Australia which include species that date back to the time of the Gondwana supercontinent, before the modern continents split apart around 180 million years ago (UNESCO, 2020). The Gondwana rainforests have developed over millions of years in a fire-free state and were considered too damp to burn. More than half of these rainforests were affected by the 2019–20 bushfires and as they are not well adapted to recovery from fire, it is predicted that it will be unlikely for them to return to their previous ecological state (Hughes et al., 2020, 12).

The Australian Bureau of Meteorology reported that that 2019 was the hottest year on record across Australia with mean temperature 1.52°C above average and mean maximum temperature 2.09°C above average and it was also the driest year on record across Australia with rainfall 40 per cent below average (BoM, 2020). Bushfires are caused by a complex of factors, but the most recent research suggests that some of the drivers of these catastrophic fires show 'an imprint of anthropogenic climate change' (van Oldenborgh et al., 2020).

In addition to climate, a very significant driver of bushfires is fuel load. For over 50,000 years, Australia's indigenous community had used traditional burning as a system of land management. Fire was used to encourage native grasses to regenerate and produce new feed for the animals which the Aboriginal peoples hunted as well as to reduce scrub and fuel to prevent intense bushfires (Gammage, 2011). Traditional fire management involves low-intensity fires which preserve flora and fauna allowing animals, including beetles and ant colonies to escape and allowing young trees to survive and keeping seeds intact for regrowth. The fires are set in a mosaic pattern to reduce their intensity and they are self-extinguishing.

European settler colonisers dismissed the importance and viability of indigenous fire management practices, in part because of colonial assumptions about property ownership (Langton, 1998) and in part because of the departure after the Second World War of Aboriginal peoples from their ancestral lands for employment on cattle stations and in the towns. This meant that traditional fire management practices ceased. This had severe consequences in the summer and autumn months when lightning strikes ignited the abundant fuel (Robinson et al., 2016). The principal method adopted to deal with the bushfire danger was hazard-reduction burns to reduce fuel loads (Russell-Smith et al., 2007) however, these tended to be much hotter than the cool burns of Aboriginal peoples and they led to a build-up of Bracken Fern or Casuarina which fuelled very intense and destructive bushfires (Creative Spirits, 2020).

Following a series of devastating fires, attention has begun to be paid to the application of Aboriginal TK to fire management practices. In 2019, Forest Fire Management Victoria and Dja Dja Wurrung Clans Aboriginal Corporation jointly brought back cultural burning following the March 2018 fires across the Bega Valley in New South Wales which stopped where the Bega Local Aboriginal Land Council had done cultural burns the year before (Forest Fire Management Victoria, 2019).

The contemporary fire management plans developed with the Assistance of Aboriginal peoples acknowledge the spiritual significance of their TK. Thus, the Fire Management Plan of the Kuku-Thaypan people from Cape York Peninsula states that 'fire is a very sacred tool that was given to the people from their ancestors and carries with it laws governing its use and application' (Standley and Felderhof, 2011, 11). This holistic approach to fire knowledge emphasises the importance of linkages between diverse components including people, law, spiritual significance and knowledge of plants, animals and country, contrasting with 'the body of literature on Indigenous fire that preceded the entry of Indigenous voices and perspectives' (Robinson et al., 2016, 9). The simple adoption of approximations to Aboriginal TK is counselled against as this may ignore culturally embedded signals, which indicate the propitious time for burning such as the flowering of trees to which has to be added the kinship relationships that determine who can light fires: law, economies, social relations, ecology and diverse technologies, such as seasonal indicators (Hill et al., 2013).

Aboriginal fire management knowledge is being classified as traditional ecological knowledge (TEK), by scholars who are documenting TK for the purposes of legal protection. In the ecological context it is argued that maintaining natural resources is closely related to the diversity and sustainability of indigenous social-ecological-cultural systems (Berkes et al., 2000; Danielsen et al., 2005; Poe et al., 2014).

TRADITIONAL KNOWLEDGE AND CLIMATE CHANGE ADAPTATION

As one of the principal impacts of climate change is upon rain-fed agriculture, indigenous communities dependent on agricultural livelihoods are likely to be the most affected (Nyong et al., 2007; Chang'a et al., 2010; Sewando et al., 2016). Responding to climate stress, indigenous peoples have developed coping mechanisms through experience and experimentation which has been transmitted both orally and in practice from one generation to the next. Their knowledge and practices would seem to be an obvious source of information for dealing with the contemporary challenge of climate change, particularly because they have to survive in marginal agricultural conditions (Below et al., 2014; Maldonado et al., 2016; Etchart, 2017; ILO, 2017).

Given the limited access of rural communities to agricultural science information, a lack of resources and inadequate support services and limited skills to make use of the little information at their disposal (Pawluk et al., 1992; Nyong et al., 2007) farmers in those communities tend to rely upon indigenous knowledge to make agricultural decisions concerning climate risks (Mapfumo et al., 2016). A number of anthropological studies demonstrating the usefulness of local ecological knowledge adapted to agricultural practices in agricultural crisis situations. Gadgil et al. (1998) refer to the maintenance of buffer areas of Sahelian rangelands, which are protected from grazing except in the case of emergencies. Surveys of farmers in Burkina Faso and Nigeria showed how indigenous methods of weather forecasting complemented the planning of agriculture in those countries (Roncoli et al., 2001; Ajibade and Shokemi, 2003).

TRADITIONAL KNOWLEDGE (TK) AND SCIENTIFIC KNOWLEDGE

During the post-war decades, the knowledge and associated natural resource management practices of indigenous peoples were considered to be outmoded, inefficient and environmentally damaging. The successful adaptation to climate change in traditional communities requires an understanding of the processes of biophysical change and their interactions within socio-ecological systems. Thus, scientific data can be supplemented by indigenous knowledge of practical adaptive practices to promote resilience to environmental changes (Murphy et al., 2015; Pearce, 2018). For example, it was observed that in the Lushoto district, northern Tanzania, weather forecasts using indigenous knowledge were considered more reliable and specific to their location compared to scientific forecasts (Mahoo et al., 2015). To improve accuracy, the systematic documentation of that knowledge and the establishment of a framework for integrating it with the official weather forecasting authority was recommended and the need to establish an information dissemination network. To entrench weather forecasting within the local government agricultural development programmes was recommended (Mahoo et al., 2015).

On the other hand, it should be recognised that the largely biological indicators on which local and indigenous knowledge is traditionally based are also subject to increased climate variability, limiting the scope of this knowledge as a basis for policy formulation and decision making (Mapfumo et al., 2016).

DEFINITION OF TRADITIONAL KNOWLEDGE (TK)

The first reference to TK in an international instrument occurs in the 1992 Convention on Biological Diversity (CBD), which did not define traditional knowledge but in Art 8j required Contracting Parties subject to their national legislation, to 'respect, preserve and maintain knowledge, innovations and practices of indigenous and local communities' in relation to the conservation and sustainable use of biological diversity.

The *Intergovernmental Committee on Intellectual Property and Genetic Resources, Traditional Knowledge and Folklore* (IGC) of the World Intellectual Property Organization (WIPO) is currently settling the text of a treaty to protect TK. However, after 40 sessions of the IGC over a period of 20 years the draft treaty providing for its protection has not yet arrived at a settled definition of TK (WIPO, 2019). This, in part, reflects the competing interests of industrialised countries and developing countries (Blakeney, 2015/2020).

TRADITIONAL KNOWLEDGE (TK) AND SELF-DETERMINATION

A significant political obstacle to the protection of TK is that it has become associated with the right of peoples to self-determination. The 1993 report of Erica-Irene Daes, Special Rapporteur of the UN Sub-Commission on Prevention of Discrimination and Protection of Minorities and Chairperson of the Working Group on Indigenous Populations observed that 'the protection of cultural and intellectual property is connected fundamentally with the realisation of the territorial rights and self-determination of indigenous peoples' (Daes, 1993, para. 4). To Daes, the concept of 'indigenous' embraced the notion of a distinct and separate culture and way of life, based on long-held traditions and knowledge that are connected, fundamentally, to a specific territory.

Indigenous peoples themselves have conflated the protection of TK with self-determination. For example, the First International Conference on the Cultural and Intellectual Property Rights of Indigenous Peoples which was convened by the Nine Tribes of Mataatua in the Bay of Plenty Region of Aotearoa, New Zealand in June 1993, the resultant *Mataatua Declaration on the Cultural and Intellectual Property Rights of Indigenous Peoples* declared that 'Indigenous Peoples of the world have the right to self-determination and in exercising that right must be recognised as the exclusive owners of their cultural and intellectual property' (Mataatua, 1993).

Self-determination has remained a constant theme of calls by indigenous peoples to protect TK, thus on 19 June 2012, on the occasion of the Rio+20 Earth Summit the *Kari-Oca 2 Declaration* was issued by 500 representatives of indigenous peoples from around the world in the following terms:

> As peoples, we reaffirm our rights to self-determination and to own, control and manage our traditional lands and territories, waters and other resources. Our lands and territories are at the core of our existence – we are the land, and the land is us; we have a distinct spiritual and material relationship with our lands and territories, and they are inextricably linked to our survival and to the preservation and further development of our knowledge systems and cultures, conservation and sustainable use of biodiversity and ecosystem management. (Kari-Oca, 2012)

The calls for self-determination in the context of TK, probably explains why the WIPO negotiations for a TK treaty have been so protracted.

The political difficulties associated with the conflation of TK with self-determination can be seen in the fate of the United Nations Declaration on the Rights of Indigenous Peoples 2007 (UNDRIP) in Australia. Article 31 of this non-binding resolution recognises the rights of indigenous peoples to maintain, control, protect and develop their intellectual property over their cultural heritage, TK and traditional cultural expressions (UN, 2007). 147 UN Member States voted in favour of the resolution adopting UNDRIP, but four (Australia, New Zealand, Canada and the United States) voted against the Resolution. Senator Marise Payne, now Australia's Foreign Minister, offered a number of reasons for the Australian Government's opposition to UNDRIP, pointing out that 'as our laws here currently stand, we protect our Indigenous cultural heritage, traditional knowledge and traditional cultural expression to an extent that is consistent with both Australian and international intellectual property law, and we are not prepared to go as far as the provisions in the text of the draft declaration currently do on that matter' (Payne, 2007).

The UN Charter, in Article 1 refers to 'the principle of equal rights and self-determination of peoples' and the International Covenant on Civil and Political Rights and the International Covenant on Economic, Social and Cultural Rights refers to the 'right of all peoples to self-determination'. However, as General Assembly Resolution 1514 (XV) on the Granting of Independence to Colonial Countries and Peoples, subsequently provided, the rights of peoples are subordinated to the sovereignty of states. This statist interpretation of the rights of peoples is also a barrier to the recognition of various political and property rights, including intellectual property rights, of indigenous peoples.

TRADITIONAL KNOWLEDGE (TK) AND INTELLECTUAL PROPERTY

The contemporary debates concerning the conservation and protection of TK have been framed as an aspect of intellectual property law. The attempt by the IGC of WIPO to formulate a TK treaty assumes that all of the beliefs and practices of the world's indigenous peoples can be embraced by a single instrument applicable to all such peoples. The report on the regional fact-finding missions conducted by WIPO in 1988–89 illustrates some of the diversity of the belief systems of traditional and indigenous peoples (WIPO, 2001). Even within a single country, such as Australia, there are many hundreds of tribes and clans, each with their own beliefs and spiritual practices (Blakeney, 2013).

Daes suggests that 'it is clear that existing forms of legal protection of cultural and intellectual property, such as copyright and patent, are not only inadequate for the protection of indigenous peoples' heritage but inherently unsuitable' (Daes, 1993, para. 32) In her review of 'Intellectual Property and Other Intangibles' Dr Jessica Lai amplifies this observation to note that incompatibility of western notions of property and ownership with indigenous concepts of communal relationships (Lai, 2014, ch. 3). Notwithstanding this lacuna, the first clause of the Mataatua Declaration states that in the development of policies and practices, indigenous peoples should: 'define for themselves their own intellectual and cultural property.' The Australian Aboriginal peoples went a little further than this in November 1993 when the peoples of the Daintree Forest region issued the Julayinbul Statement on Indigenous Intellectual Property Rights which stated that:

> Aboriginal intellectual property, within Aboriginal Common Law, is an inherent, inalienable right which cannot be terminated, extinguished, or taken.... Any use of the intellectual property of Aboriginal Nations and Peoples may only be done in accordance with Aboriginal Common Law, and any unauthorised use is strictly prohibited. (Blakeney, 2015, 184)

The adoption by the Māori and Australian Aboriginal peoples of the terminology of western legal systems in debating the issue of TK has had the effect of framing the debate in an IP context, or implying IP-related solutions. In contrast, the 2012 *Kari-Oca 2 Declaration* stated that:

> We reject the assertion of intellectual property rights over the genetic resources and traditional knowledge of Indigenous peoples which results in the alienation and commodification of Sacred essential to our lives and cultures. (Kari-Oca, 2012)

Paterson and Karjala suggest that 'in many cases, legitimate concerns of indigenous people can be accommodated without going to the extreme of recognizing new intellectual property rights, either through modest reinterpretation of existing legal regimes concerning contract, privacy, and unfair competition law' (Paterson and Karjala, 2003, 635). Another area which might be more suitable for the protection of TK is the law pertaining to the conservation of cultural heritage. Yu (2008) listed as examples of intangible cultural heritage 'the practices, representations, expressions, knowledge, skills as well as the instruments, objects, artefacts and cultural spaces associated therewith that communities, groups and, in some cases, individuals recognise as part of their cultural heritage referred to in the 2003 UNESCO Convention for the Safeguarding of the Intangible Cultural Heritage' (Yu, 2008, 443). Professor Yu also referred to the 2001 Universal Declaration on Cultural Diversity which defined culture as embracing 'the set of distinctive spiritual, material, intellectual and emotional features of society or a social group, and that it encompasses, in addition to art and literature, lifestyles, ways of living together, value systems, traditions and beliefs' (Yu, 2008, 443–444). However, Yu concedes that 'advocates of strong protection for intangible cultural heritage often combine different objectives to craft their proposals and refers to 'cultural privacy', 'authenticity', 'recognition', 'benefit sharing', 'conservation', 'access' and 'resistance' (Yu, 2008, 483). Given these disparate objectives, some of which overlap and conflict with one another, Yu predicts that 'in the near future, achieving consensus is likely to remain a challenge' (Yu, 2008, 483).

CONCLUSION

A final problem with proposals for the protection of TK is the fact that it is very closely related to the protection of traditional cultural expressions (TCEs). At the fifth session of the IGC it was acknowledged that discussions had frequently stressed

> ...the holistic nature of traditional cultural and knowledge systems, and the need to recognize the complex interrelations between a community's social and cultural identity, and the specific components of its knowledge base, where traditional technical know-how, cultural expressions and traditional narrative forms, traditional ecological practices, and aspects of lifestyle and spiritual systems may all interact, so that attempts to isolate and separately define particular elements of knowledge or culture may create unease or concern. (WIPO, 2003, para. 36)

At the fifth session the IGC identified a tension which has permeated the entire course of its deliberations, that is, 'between an approach to defining TK/TCE subject matter that aimed at inclusiveness and recognition of the diverse local characteristics of traditional knowledge and cultures, and an approach that saw value in establishing a common set of terms and a general understanding of their signification at the international level' (WIPO, 2003, para. 36). It observed that the 'holistic quality of protection' was most apparent within the traditional context, where legal protection 'was often embedded in deeper cultural norms and practices, and integrated in the life of the community' and that it was generally only when TK or TCE subject matter was removed from that context, such as in relation to commercial or research interests, that community concerns and intellectual property policy issues arose along with the 'perceived need for distinct new forms of IP protection' (WIPO, 2003, para. 40).

The IGC explained that its draft text was concerned with the protection of TK 'against misappropriation and misuse beyond its traditional context and should not be interpreted as limiting or seeking externally to define the diverse and holistic conceptions of knowledge within the traditional context' (WIPO, 2007, 28). Coombe attributes the crisis of legitimacy in the intellectual property system to the failure of the western IP system to meet indigenous peoples' yearnings and aspirations for the preservation of their knowledge and cultural integrity (Coombe, 2001, 285).

REFERENCES

Ajibade, L.T. and Shokemi, O. (2003) 'Indigenous approaches to weather forecasting in Asa L.G.A. Kwara State, Nigeria', *Indilinga: African Journal of Indigenous Knowledge Systems* 2(1), 37–44.

Baldwin, C. and Ross, H. (2020) 'Beyond a tragic fire season: a window of opportunity to address climate change?', *Australasian Journal of Environmental Management* 27(1), 1–5.

Below, T.B., Schmid, J.C. and Sieber, S. (2014) 'Farmers' knowledge and perception of climatic risks and options for climate change adaptation: a case study from two Tanzanian villages', *Reg. Environ. Change* 15(7), 1169–1180.

Berkes, F., Colding, J. and Folke, C. (2000) 'Rediscovery of traditional ecological knowledge as adaptive management', *Ecological Applications* 10, 1251–1262.

Blakeney, M. (2013) 'Protecting the spiritual beliefs of indigenous peoples – Australian case studies', *Pacific Rim Law & Policy Journal* 22(2), 391–427.

Blakeney, M. (2015) 'Protecting the knowledge and cultural expressions of aboriginal peoples', *University of Western Australia Law Review* 39(2), 180–207. Also available at: guides.slsa.sa.gov.au/content.php?pid=278586&sid=4704011, accessed 18 August 2020.

Bureau of Meteorology (BoM) (2020) *State of the Climate, 2020*, Canberra, AGPS.

Chang'a, L.B., Yanda, P.Z. and Ngana, J. (2010) 'Indigenous knowledge in seasonal rainfall prediction in Tanzania: a case of the south-western highland of Tanzania', *Journal of Geography and Regional Planning* 3(4), 66–72.

Coombe, R.J. (2001) 'The recognition of indigenous peoples' and community traditional knowledge in international law', *St. Thomas L. Rev.* 14, 275–285.

Creative Spirits (2020) 'Cool burns: key to Aboriginal fire management', www.creativespirits.info/aboriginalculture/land/aboriginal-fire-management, accessed 15 August 2020.

Daes, E-I. (1993) 'Discrimination against indigenous peoples', E/CN.4/Sub.2/1993/28, 28 July.

Danielsen, F., Burgess, N.D. and Balmford, A. (2005) 'Monitoring matters: examining the potential of locally-based approaches', *Biodiversity and Conservation* 14, 2507–2542.

Etchart, L. (2017) 'The role of indigenous peoples in combating climate change', *Palgrave Communications* 3, 17085.

FAO (2018) 'The future of food and agriculture: trends and challenges', http://www.fao.org/3/a-i6583e.pdf, accessed 15 August 2020.

FAO, IFAD, UNICEF, WFP, and WHO (2018) *The State of Food Security and Nutrition in the World 2018, Building Climate Resilience for Food Security and Nutrition*, www.fao.org/publications, accessed 15 August 2020.

Forest Fire Management Victoria (2019) 'Celebrating the return of traditional burning', www.ffm.vic.gov.au/fuel-management-report-2016-17/what-we-achieved-statewide/traditional-owners-partnerships/celebrating-the-return-of-traditional-burning, accessed 15 August 2020.

Gadgil, M., Hemam, N.S. and Reddy, B.M. (1998) 'People, refugia and resilience', in Berkes, F. and Folke, C. (eds) *Linking Social and Ecological Systems: Management Practices and Social Mechanisms for Building Resilience*, 30–47. Cambridge University Press.

Gammage, W. (2011) *The Biggest Estate on Earth: How Aborigines made Australia*. Allen and Unwin.

Gupta, R., Somanathan, E. and Dey, S. (2017) 'Global warming and local air pollution have reduced wheat yields in India', *Climate Change 140*, 593–604.

Hill, R., Pert, P., Davies, J., Robinson, C., Walsh, F. and Falco-Mammone, F. (2013) 'Indigenous land management in Australia: extent, scope, diversity, barriers and success factors', Cairns, CSIRO Ecosystem Sciences.

Hughes, L. Steffen, W., Mullins, G., Dean, A., Weisbrot, E and Rice, M. (2020). *Summer of Crisis*, Canberra, Climate Council of Australia.

Hussain, A., Rasul, G., Mahapatra, B. and Tuladhar, S. (2016) 'Household food security in the face of climate change in the Hindu-Kush Himalayan region', *Food Security* 8(5), 921–937.

ILO (2017) *Indigenous Peoples and Climate Change: From Victims to Change Agents Through Decent Work*. ILO Geneva.

Innes, P.J. et al. (2015) 'Effects of high-temperature episodes on wheat yields in New South Wales, Australia', *Agricultural and Forest Meteorology 208*, 95–107.

IPCC (2014) AR5 Synthesis Report: Climate Change 2014, https://www.ipcc.ch/report/ar5/syr/, accessed 28 June 2021.

IPCC (2018) *Global IPCC Warming of 1.5°C. An IPCC Special Report on The Impacts of Global Warming of 1.5°C Above Pre-Industrial Levels and Related Global Greenhouse Gas Emission Pathways, in the Context of Strengthening the Global Response to the Threat of Climate Change, Sustainable Development, and Efforts to Eradicate Poverty*, report.ipcc.ch/sr15/pdf/sr15_spm_final.pdf, accessed 15 August 2020.

IPCC (2019) *IPCC Special Report on Climate Change, Desertification, Land Degradation, Sustainable Land Management, Food Security, and Greenhouse Gas Fluxes in Terrestrial Ecosystems*, https://www.ipcc.ch/report/srccl/, accessed 15 August 2020.

Kari-Oca (2012) indigenous4motherearthrioplus20.org/kari-oca-2-declaration/ accessed 17 October 2020.

Ketiem, P. (2017) 'Integration of climate change information into drylands crop production practices for enhanced food security: A case study of Lower Tana Basin in Kenya', *African J. Agric. Res. 12*(2), 1763–1771.

Lai, J.C. (2014) *Indigenous Cultural Heritage and Intellectual Property Rights: Learning from the New Zealand Experience*. Springer.

Langton, M. (1998) *Burning Questions: Emerging Environmental Issues for Indigenous Peoples in Northern Australia*. Darwin, CINCRM, NTU.

Legge, S. Woinarski, J., Garnett, S., Nimmo, D., Scheele, B., Lintermans, M., Mitchell, N., Whiterod, N. and Ferris, J. (2020) 'Rapid analysis of impacts of the 2019–20 fires on animal species, and prioritisation of species for management response – preliminary report' Report prepared for the Wildlife and Threatened Species Bushfire Recovery Expert Panel, 9 www.environment.gov.au/biodiversity/bushfire-recovery/research-and-resources, accessed 25 June 2020.

Mahoo, H., Mbunga, W., Yonah, I. and Recha, J. (2015) 'Integrating indigenous knowledge with scientific seasonal forecasts for climate risk management in Lushoto district in Tanzania', CCAFS Working Paper no. 103, CGIAR Research Program on Climate Change, Agriculture and Food Security (CCAFS).

Maldonado, J. Bennett, T. M. B. Chief, K., Cochran, P., Cozzetto, K., Gough, B., Redsteer, M.H., Lynn, K., Maynard, N. and Voggesser, G.(2016) 'Engagement with indigenous peoples and honoring traditional knowledge systems', *Clim. Change* *135*(1), 111–126.

Manzoor, M. Bibi, S., and Jabeen, R. (2013) 'Historical analysis of flood information and impacts assessment and associated response in Pakistan (1947–2011)', *Res. J. Environ. Earth Sci.*, *5*, 139–146.

Mapfumo, P., Mtambanengwe, F. and Chikowo, R. (2016) 'Building on indigenous knowledge to strengthen the capacity of smallholder farming communities to adapt to climate change and variability in southern Africa', *Climate and Development 8*(1), 72–82.

Mataatua (1993) *The Mataatua Declaration on Cultural and Intellectual Property Rights of Indigenous Peoples*, www.wipo.int/export/sites/www/tk/en/databases/creative_heritage/docs/mataatua.pdf, accessed 17 August 2020.

Mihai, A. (2020) 'Rain extinguishes Australian wildfires', www.zmescience.com/ecology/rain-australian-wildfire-10022020/, accessed 15 August 2020.

Murphy, B.P., Cochrane, M.A. and Russel-Smith, J. (2015) 'Prescribed burning protects endangered tropical heathlands of the Arnhem Plateau, northern Australia', *Journal of Applied Ecology* 52, 980–991.

Noble, I.R., Huq, S., Anokhin, Y.A., Carmin, J., Goudou, D., Lansigan, F.P., Osman-Elasha, B. and Villamizar, A. (2014) *Climate Change 2014: Impacts, Adaptation, and Vulnerability. Part A: Global and Sectoral Aspects. Contribution of Working Group II to the Fifth Assessment Report of the Intergovernmental Panel on Climate Change*. Cambridge University Press, 833–868.

Nyong, A., Adesina, F. and Elasha, B.O. (2007) 'The value of indigenous knowledge in climate change mitigation and adaptation strategies in the African Sahel', *Mitigation and Adaptation Strategies for Global Change 12*(5), 787–797.

Paterson, R.K. and Karjala, D.S. (2003) 'Looking beyond intellectual property in resolving protection of the intangible cultural heritage of indigenous peoples', *Cardozo J. Int'l & Comp. L. 11*, 633–670.

Pawluk, R.R., Sandor, J.A. and Tabor, J.A. (1992) 'The role of indigenous soil knowledge in agricultural development', *Journal of Soil and Water Conservation 47*(4), 298–302.

Payne, M. Senator (2007) Commonwealth, Parliamentary Debates: United Nations Declaration on the Rights of Indigenous Peoples, parlinfo.aph.gov.au/parlInfo/genpdf/chamber/hansards/2007-09-10/0075/hansard_frag.pdf fileType=application%2Fpdf, accessed 15 August 2020.

Poe, M.R., Norman, K.C. and Levin, P.S. (2014) 'Cultural dimensions of socioecological systems: key connections and guiding principles for conservation in coastal environments', *Conservation Letters 7*, 166–175.

Pearce, T.C.L. (2018) 'Incorporating indigenous knowledge in research' in McLeman, R. and Gemenne, F. (eds.) *Routledge Handbook of Environmental Migration and Displacement*, 125–134. Taylor & Francis Group.

Robinson, C.J. Barber, M., Hill, R., Gerrard, E., and James, G. (2016) *Protocols for Indigenous Fire Management Partnerships*. CSIRO.

Robinson, J.B. and Herbert, D. (2001) 'Integrating climate change and sustainable development', *International Journal of Global Environmental Issues*, *1*, 130–149.

Roncoli, C., Ingram, K. and Kirshen, P. (2001) 'The costs and risks of coping with drought: livelihood impacts and farmers' responses in Burkina Faso', *Climate Res.* *19*, 119–132.

Russell-Smith, J., Yates, C.P., Whitehead, P.J., Smith, R., Craig, R., Allan, G.E., Thackway, R., Frakes, I., Cridland, S., Meyer, M.C.P., and Gill, A.M. (2007)

'Bushfires "down under": patterns and implications of contemporary Australian landscape burning', *Int. J. Wildland Fire 16*, 361–377.

Saxena, A. K. Fuentes, X.C., Herbas, R.G. and Humphries, D.L. (2016) 'Indigenous food systems and climate change: impacts of climatic shifts on the production and processing of native and traditional crops in the Bolivian Andes', *Frontiers in Public Health, 4*, Art. 20.

SBS (2020) 'Australia bushfire smoke travels 12,000 kms to Chile', www.sbs.com.au/news/dateline/australia-bushfire-smoke-travels-12-000-kms-to-chile, accessed 15 August 2020.

Sewando, P.T., Mutabazi, K.D. and Mdoe, N.Y.S. (2016) 'Vulnerability of agro-pastoral farmers to climate risks in northern and central Tanzania', *Development Studies Research 3*(1), 11–24,

Standley, P. and Felderhof, L. (2011) *Kuku-Thaypan (Awu Laya) Mo Fire Management Research Project*, site.emrprojectsummaries.org/2012/01/26/three-action-research-projects-itraditional-knowledge-revival-pathways-fire-program-iikuku-thaypan-fire-management-research-project-and-iiithe-importance-of-campfires-to-effective-conservation-2/, accessed 15 August 2020.

UN (2007) United Nations Declaration on the Rights of Indigenous Peoples, GA Res 61/295, UN GAOR, 61st sess, 107th plen mtg, Supp No 49, UN Doc A/RES/61/295 (13 September 2007).

UNEP (2020) 'Ten impacts of the Australian bushfires', www.unenvironment.org/news-and-stories/story/ten-impacts-australian-bushfires, accessed 15 August 2020.

UNESCO (2020) 'Gondwana rainforests of Australia', whc.unesco.org/en/list/368/, accessed 15 August 2020.

van Oldenborgh G.J., Krikken, F., Lewis, S., Leach, N. J., Lehner, F., Saunders, K. R., van Weele, M., Haustein, K., Li, S., Wallom, D., Sparrow, S., Arrighi, J., Singh, R. K., van Aalst, M. K., Philip, S. Y., Vautard, R., and Otto, F. E. L. (2020) 'Attribution of the Australian bushfire risk to anthropogenic climate change', *Nat. Hazards Earth Syst. Sci. Discuss.*, doi.org/10.5194/nhess-2020-69, in review, accessed 15 August 2020.

WIPO (2001) *Intellectual Property Needs and Expectations of Traditional Knowledge Holders. Report on Fact-finding Missions on Intellectual Property and Traditional Knowledge (1998–1999).* WIPO Geneva.

WIPO (2003) 'Overview of activities and outcomes of the Intergovernmental Committee', WIPO doc., WIPO/GRTKF/IC/5/12, 3 April.

WIPO (2007) 'The Protection of Traditional Knowledge: table of written comments on revised objectives and principles', WIPO doc. WIPO/GRTKF/IC/11/5(b), 18 May.

WIPO (2019) 'The protection of traditional knowledge: draft articles' WIPO doc., WIPO/GRTKF/IC/40/4, 9 April.

Yu, P. (2008) 'Cultural relics, intellectual property, and intangible heritage', *Temp. L. Rev. 81*, 433–492.

PART VI

Imagination in the Anthropocene

21. Designing post-human futures
Raphaële Bidault-Waddington

THE ANTHROPOCENE, A POST-HUMANIST FUTURE CHALLENGE

The concept of Anthropocene epitomizes a period of history where the actions of mankind on Earth have reached the point of modifying its ecosystems, atmosphere and life conditions, creating an era similar to a geological era, no matter where its precise starting point is. The Anthropocene is a critical anthropological paradigm marking that core planetary evolution turn, asking mankind to be aware of and responsible for its impact, and calling for a civilizational transformation toward resilient uses and ways of living. The Anthropocene is both a critical wake-up call and an impulsion to redesign the life conditions on earth. Humanity needs to emancipate from anthropocentric and predatory economies, policies and lifestyles, most of which are rooted in the western humanist, progressive, liberal and modernist value system. These conceptual frames gave full license to men's freedom to innovate in all matters and have shaped most of globalization standards. For that reason, the future is bound to be "post-humanist" and we paradoxically need to invent it, which means redefining the way we project ourselves toward the future, create and lead a sustainable and "future-proof" change. The earthquake of the COVID-19 pandemic and the radical adaptation of the entire humanity has shown how systemic change is possible and has started prototyping what could be the "new normal," somehow paving the way for a now urgent environmental transition. It is time to act and eventually enforce futures, not just predict and anticipate them as an outside performance we would watch. Whether we like it or not, we are inevitably included in the picture.

In order to explore and design Anthropocene futures and to open alternative perspectives to the collapse narratives, we cannot keep the same strategic tools and foresight methods, most of them being also rooted in western modernity and cognitive frames. Imagination is a resource that needs to be mobilized to open the future horizon and engage in it. The Anthropocene forces us to rethink and reinvent the human condition at large, from the everyday gestures to shared lifestyles, from all social organizations to the atmospheric sustainability and space expansion. The paradigm shift touches on the cosmological order, as it touches on humanity's visions of the world, its place in the universe, and its relation to knowledge, truth, existence, space and time, whether past, present or future. Simultaneously, science, technology and the digital transition at large also challenge the limits and definition of human nature and societies, calling for an anthropological redefinition and leading to innovative post-human philosophies. Bioengineering, synthetic biology, neuroscience, AI and deep fakes bring profound ontological issues and uncertainties, making human-beings as alien as other natural and artificial species, on a planet where the natural co-living contract seems broken. "We have become viruses for the planet" (Descola, 2020) says the iconic anthropologist during the COVID-19 crisis. Digital humanities research shows the impact of technologies on every aspect of life from work to intimate life and health, from economy to urban life, from education to community building and democracy. The information overload and the *data deluge* (Anderson, 2008) have entailed the post-truth paradigm and created a multiverse effect where reality becomes far more ambiguous and versatile. Resulting from a mix of these multi-dimensional, multi-scalar and entangled transitions, the future will be both post-humanist and post-human. It however remains an absolutely complex challenge to address and design these holistic futures and bigger pictures, requiring cognitive and methodological innovations. Although various disciplines are anticipating, formulating or programming futures, the future remains an art and a fiction that cannot be demonstrated. The future is an evolving frontier to explore with imagination, knowledge and experimentation. For this purpose, this chapter will introduce a still imperfect cognitive guidance tool, called Toward Alien Cosmologies, to be read as a methodological prototype to design possible post-human futures.

DESIGN OF A POST-HUMAN FORESIGHT METHODOLOGY

In her essay "Toward a post-humanist methodology, a statement" (Ferrando, 2012), philosopher Francesca Ferrando calls for an emancipation from western essentialism, anthropocentrism, biocentrism and rationale to address the post-human and post-colonial paradigm. She writes: "Post-humanism has to acknowledge the whole human experience in order to be receptive to the non-human and be open to unknown possibilities" (op. cit., p. 17). And "According to cosmology, the universe is expanding at an accelerating rate" (ibid., p. 11). Also:

> A post-humanist methodology should not be sustained by exclusive traditions of thought. It should be dynamic and shifting, engaging in pluralistic epistemological accounts, not in order to comply with external requirements of political correctness,

but to pursue less partial and more extensive perspectives, in tune with a post-human future which will radically challenge human comprehension. (ibid., p 17)

The future guidance cognitive tool presented in this chapter, is meant to support the exploration of these plural and somehow floating and versatile perspectives. But Toward Alien Cosmologies is also informed by more pragmatic, strategic foresight and design-oriented perspectives to remain connected to earthly realities and support resilient action for the Anthropocene.

From the field of foresight, we want to draw upon its methodologies and forward-looking cognitive resources, such as systemic and morphological analysis, horizon scanning and building, or scenario-planning, each allowing to address a certain level of complexity, build future perspectives or show future potentials. But on the other hand, it is important to keep in mind that the discipline itself was created after the Second World War, mostly to guide both corporate investments and public policies (such as industrial programs, energy and mobility infrastructures planning, etc.), which was certainly not neutral in regard to the rise of the Anthropocene and human impact on the planet. Although some new voices and more hybrid approaches currently arise – as we will show with design, anthropology, or science-fiction, and attempt to do in this chapter – the foresight discipline and community remains today very western, masculine and geared toward strategic purpose.

On the contrary, weak-signals research, that is, the detection of emerging signs (information, scientific discovery technology, project, representation, concept, etc.) considered as indicators of the future, meant to become change drivers and form trends, is an under-valued foresight practice that we will favor. Although quite intuitive and subjective, its diversity, openness, accessibility and collaborative dimensions make it especially inclusive, empowering and efficient, if done thoroughly and in-depth. The collection of weak signals leads to the design of trends as an evolutive map.

Following Ferrando, the foresight approach developed in this chapter offers a multi-scalar and multi-perspectives canvas (from the microbiome to the cosmos), where a rich ecosystem of trends can be highlighted as a kind of borderless climate rather than an instrumental and framed system. The resulting live future panoramas will show micro and mega trends emerging from weak signals detected with a 360°, post-colonial, and non-hierarchical approach. For instance, a designer's creative experiment can be associated with an important academic publication or with a film to highlight and articulate a trend.

From the field of design, an acknowledgement of the pragmatic power of design to translate creative ideas into new uses and innovations and somehow "make the future" is necessary. Design is an engine to prototype and install resilient uses and ultimately change lifestyles. Its use- and human-centric pragmatism shows however some limits in regard to the Anthropocene and post-human paradigm. Design requires more pluralism, broader lifestyle and eco-systemic perspectives to better include impact concerns in its process. It also needs a deeper ontological, anthropological and cultural questioning beyond its pragmatic endeavor. In that regard, art productions, which reveal cultural imaginaries and values, will be complementary weak signals. Combining pragmatism, imagination and criticality enables finding relevant weak signals at the new frontiers of the design discipline addressing these issues such as in *critical design, speculative design, fiction design, instance design*, and the emerging and future-oriented *design anthropology* (Smith et al., 2016; Pink & Salazar, 2017). This latest promising discipline is also highly relevant to bring forth new understanding of human existential relation to the future, a topic we will address again in the conclusion. These emerging disciplines help get a perspective on how the future is being used whether by foresight and design professionals, or by businesses, institutions, politicians, cultural and spiritual leaders. The contribution of anthropology to the Anthropocene foresight and design research should, however, include a critical awareness of its conceptual and historical fundamentals, which are rooted in western colonial times with the depiction of "other" and past cultures. Now the other, the alien, is us all in the future.

A common challenge and new frontier to both design (Bonnet et al., 2019) and foresight, is the need to redesign holistic worldviews, value systems and meta-narratives (cosmologies emancipated from classic western anthropological frames) in and of the Anthropocene era. Both disciplines are currently renewing their interest in science fiction (Zaidi, 2019), which has more traditionally been creating rich and complex future worlds, and now serves as a future world-building approach. Although they result in vibrant narratives giving an emotionally rich sense of possible futures, world-building methodologies tend, however, to create closed worlds as in video games, novels or films and still require methodological innovation. Science-Fiction and world-building do not show and discuss their underlying systems. Referring to SF authors Gibson and Delaney who coined that concept, Zaidi recalls that SF eventually allows for a reflection on implicit *cultural superstructures.* She explains: "by engaging with alternative future worlds, science fiction readers develop an understanding of the modular, foundational components of a culture. They build their capacity to engage with alternative systems and ways of living" (Zaidi, 2019, p. 21).

Following both Ferrando and Zaidi, the post-human Anthropocene paradigm requires to design open and dynamic future multiverse and cosmologies, allowing for a diversity of co-existing worlds, lifestyles, value-systems and meta-narratives, existing both in the tangible and informational reality. As a critical topology and canvas, Toward Alien Cosmology is meant to help design, question, realize or clarify, both world-building and cultural superstructures. It is a dual and imperfect future world-building and -unbuilding cognitive tool that doesn't claim any sense of truth.

LIID Future Lab Experiences and Conceptual Tools

Taking into account these insights from post-human, foresight, design, anthropology and science-fiction studies, I will now introduce the latest prototype (Bidault-Waddington, 2020a, 2020b, 2020c) of the conceptual matrix used at LIID Future Lab to investigate and design post-human futures, as open and holistic cognitive spaces to support both imagination and action. *Toward Alien Cosmologies (V6)* works as a fluid methodological canvas to guide the design of creative future projections at multiple scales and horizons, still giving it a knowledge-intensive solid support and a plural and multi-purpose orientation. In that regard, the conceptual model capitalizes on twenty years of future research and eco-systemic change analysis carried by LIID at the crossroads of art, economy, urban design, academia, digital humanities and foresight. Since its inception in 2000, LIID has designed and experienced *future lab* methodologies to explore the future, which, among other formats (texts, images, conferences, workshops, art installations), include the design of heuristic tools and diagrams, such as a creative conceptual canvas to, for instance, envision the future of large-scale metropolis (Bidault-Waddington, 2012). A diagram synthesizes a cognitive architecture, which helps to mentally spatialize, cross and interrelate a wide scope of critical topics or possible facets – a *polygonal semiospace* in LIID's vocabulary (Bidault-Waddington & Menetrey, 2016) – still remaining open and borderless, with no specific beginning or end (vs narratives or scenarios). Often used in foresight methods, conceptual diagrams are efficient foresight guidance tools as they facilitate shared mental representations, collaborative research, inclusive pedagogy, future empowerment and the design of inspirational projects and scenarios. *Toward Alien Cosmologies (V6)* also takes into account how the mind needs to make simplifications to organize reasoning around complex issues, to produce and ground a sense of understanding, if not a certitude. It is one of the challenges to address the Anthropocene cosmological paradigm. As a reflexive tool, the canvas also helps become aware and clarify tacit cultural superstructures and value-systems. And on a more pragmatic level, the conceptual matrix also serves as a classifier of information, research material and weak signals, to then draw and map trends, as we will now experience (see Figure 21.1).

268   *Knowledge for the Anthropocene*

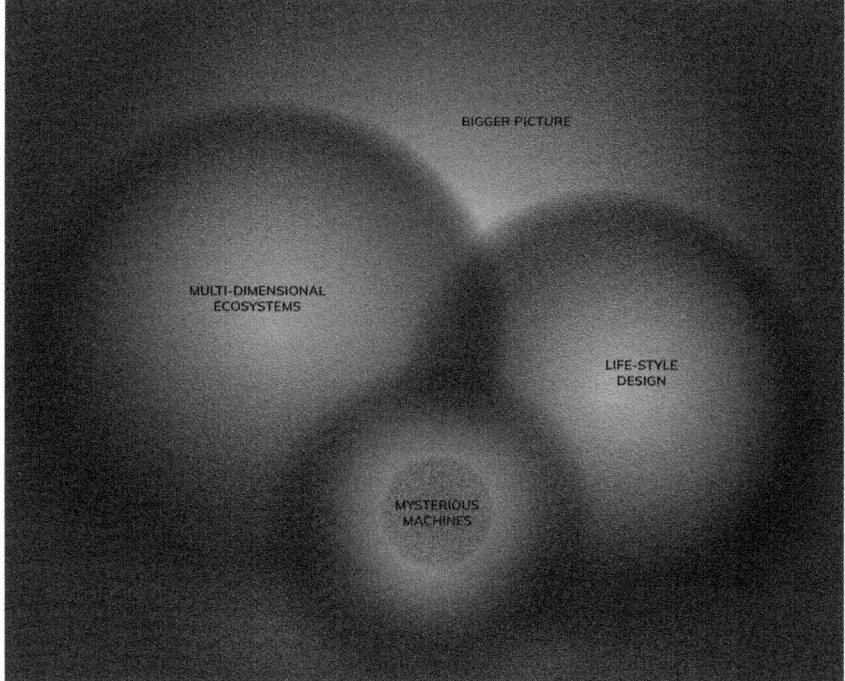

Notes: Original diagram is a large-scale panorama (80 x 100 cm) in colors, including a myriad of keywords distributed in each of the four spheres, corresponding to the critical issues and future trends raised in the chapter.
Source: Simplified black and white version, © Raphaële Bidault-Waddington, 2020.

Figure 21.1   Toward Alien Cosmologies, critical future topology (V6)

TOWARD ALIEN COSMOLOGIES, A CRITICAL FUTURE TOPOLOGY (V6)

The Toward Alien Cosmologies conceptual matrix is designed as a critical future topology visually organized as four floating spheres used to map weak signals at four different scales, but with no specific order or hierarchy. Each sphere is border-less like a cloud, interweaves and collides with the others, but still carries specific critical topics, vocabulary and perspectives. With this blurry conceptual canvas, weak signals, information, and ideas of all kinds can be mapped in a large and open panorama, the vagueness leaving space for infinite appropriation and adaptation. From these weak signals, it becomes possible to highlight future trends, a full diversity of macro and micro trends, within and across spheres as virtual constellations to be drawn. Trends and weak signals give insights of future potentials, still leaving space for speculation and imagination. To give an example, below are presented for each sphere, key critical topics, a selection of weak signals and pivotal trends that should be continuously enriched and updated.

Then each user can design all-inclusive future cosmologies as large-scale, dynamic and open cognitive panorama where to draw a diversity of inspirational scenarios giving them texture and somehow staging them. With its four spheres, the canvas obliges users to expand from their own mindset and bubble and consider all dimensions and perspectives rather than staying in limited frames (whether the one of design, foresight or any domain).

With its title and distinctive semantic, Toward Alien Cosmologies is also meant to be thought provoking from the first instant to stimulate, liberate and empower. To feed that impulse, the four spheres are called "Mysterious machine"; "Life-Style Design"; "Multi-dimensional ecosystems"; and "Bigger Picture." As Ferrando suggests, "A post-humanist methodology has to be adaptable and sensitive; it has to indulge in its own semiotics, hermeneutics, pragmatics, meta-linguistics, in order to be aware of the possible consequences which they might enact on a political, social, cultural, ecological level" (Ferrando, 2012, p. 11). The blurriness should paradoxically create a new form of clarity where everyone can somehow "see" at different levels and in different ways but still keeping a shared canvas, allowing to feel, give and make sense, but also imagine, speculate and project toward the future. We are now going to take a journey into each of the spheres to apprehend their critical topology with a selection of weak signals. As the reader of this chapter will experience it, and that should be the demonstration made live, it will give a new clarity and more sense to the complexity of the future of the Anthropocene, and to the critical topics already expressed in this chapter.

Mysterious Machines, Being and Ontologies: Bodily and Cognitive Diversity

In this first cloud, we want to take full awareness and explore the current ontological uncertainty and insecurity human beings have in front of themselves and in front of new material and cognitive entities such as AIs, all reaching the status of *aliens*, as foreign and not fully knowable or controllable beings. Whether organic, viral, informational or technological, new families of fluid species are emerging, transforming and taking part in the extended post-human Anthropocene society. Are we viruses? as Descola suggests or Are we humans? are asking curators Beatriz Colomina and Mark Wigley at the Istanbul Design Biennale in 2017 (Colomina & Wigley, 2019), showing how humanity has historically been shaped by design. Or else are we becoming *humax*, that is, humans maximizing themselves to better sustain in the future (whether augmented or not)? as design researcher Tony Fry explores in his referential book *Becoming Human by Design* (Fry, 2012). Even gender distinction seems to become unclear with transgender realities showing how ambiguous the genetic expression and mutable body functions can be.

Along with the augmented and bio-engineered bionic body, which tends to add capabilities to the human body (e.g., sportsman Victor Pistorius), the Cyborg foundation created by artists Neil Harbisson and Moon Ribas, shows how humans can hack their own body to add sensorial aptitudes. Whether intentionally or not, and via any form of cultural influence, it is now clear that human sensorial, emotional and creative mechanisms can be manipulated, as part of the design of subjective "bubbles," broader attention ecology (Citton, 2014) and transduction process (Simondon, 1989). Continuing on the cognitive side, even though neurosciences now allow men to intervene on the brain and the mind, these remain mostly terra incognita. Following prominent philosopher Catherine Malabou, even the human freewill, consciousness and creative autonomy are questioned, as the brain seems to have its own automatism (Malabou, 2017), a thesis also defended by Yval-Noah Harari in his influential book *Homo Deus, a Brief History of the Future* (Harari, 2016). Beyond the human live experience, similar ontological questions touch the animal and vegetal realms, raising broad ethical and legal issues (topics to be furthered in the three other spheres). Same issues arise from AIs, these "mysterious machines" as *MIT Technology Review* called them on its May 2017 cover page, or new generation of synthetic biology beings such as self-healing xeno-bots designed by the University of Vermont (Brown, 2020), bots or deep-fakes, these quasi-personas we interact with online. The hypothesis of the emergence of a Singularity (Kurzweil) is also on this critical *topos*. So is the status of viruses such as COVID-19 that humanity seems to have silently given the right to stop its planetary activity. Who or what instances are entitled to tackle or lead the Anthropocene?

To redefine the diversity of modes of existence on earth and beyond, we suggest, from the perspective of this first sphere, to revisit this extended family of being and agents under the fundamental prism of the bodily and cognitive diversity to help us create a sense of ontological order. The implicit anthropological core-trend in this sphere is the convergence between the organic, the cognitive and technological species. Another one is the declining possibility of a clear "discontinuity" (cognitive and/or bodily separation) among this diversity, only degrees and potentials for differentiation, versatility, ubiquity, autonomy, agency and resilience.

Lifestyle Design and Cultural Reset: Relational, Collaborative and Spatial Diversity

In this second sphere of future critical topics, we want to zoom in on the design of uses, and how they transform and shape the lifestyles of tomorrow: food, agriculture, production, work, business, innovation, education, art, culture, sport, leisure, care, health, sexuality, love, religion, dating, hospitality, mobility, urbanity, and so on are the domains of uses that we want to analyze without separating them in silos as we see how again, the historical anthropological categories are evolving and hybridizing. What is work when a passive leisure practice of watching a film for free generates value in a firm? What are equitable co-working conditions between humans and AIs? is another question already and seriously raised by the workers syndicate TCO in Sweden.

What new form of relation, collaboration and sentiment do we develop with caring smart co-bots as with animals? How is it that people fall in love with sexbots? Can a supposedly "useless" art experiment be a resilient production for the Anthropocene? What is the *permaeconomics* (Bidault-Waddington, 2018) of *third places* where expenses simultaneously become productive in another repertoire of use? Many shifts among these repertoires are in progress, such as, to give a very concrete example, the current massive repositioning of hotel rooms into day-work spaces. In this regard, design, anthropology and innovation are pivotal disciplines to capture the weak signals and pragmatic prototypes of possible futures. Instead of a simple focus on uses, this sphere offers the possibility of experiencing how cultural lifestyles are transforming with and for the Anthropocene, and of (mentally) staging them in digital and physical space as live pictures including characters.

The transformation of social uses goes along a transformation of related value(s) and utility, a concept at the heart of the whole societal organization and economy. How relevant is it to travel when the associated emotion is not pleasure or efficiency anymore but a sense of pollution? Where and what are the new forms of value creation around resilient uses and impacts? What are the new loops of local uses and circular economies? The Anthropocene includes new value(s) systems and symbolic economies that will guide the transformation of capitalism in the future, as we see for instance with the Tech For Good innovation movement. At the heart of these social uses' transformation, we want to capture via weak signals how innovative cultural uses can be absorbed, such as via educational processes and co-learning mechanisms, whether happening in schools, online, in cultural places, incubators, community centers, or any peer-to-peer interface. The learning dynamics are going to be leading the shift toward resilient lifestyles.

Behind this pragmatic focus on societal uses, utility and resilience, we should also apprehend in this sphere, the explicit and implicit hierarchies and subordination mechanisms operating in concrete uses and lifestyles. AI research, post-colonial studies and feminism are useful filters to debunk power structures encrypted in inherited anthropological frames and categories (e.g., the economy of educational, domestic and care work) to restore equity in the future.

Multi-dimensional Ecosystems: Mapping Planetary Impact Chains

As said in the introduction, resilient innovation for the post-human Anthropocene cannot be approached solely with a focus on experiential social uses and lifestyles. It should include systemic and eco-systemic perspectives to acknowledge chains of positive and negative impacts at a planetary level. Since these chains are borderless and so is the realm of information, there is no point in choosing intermediary scales as frames of analysis. This does not prevent from programming systemic change at a narrower scale such as the one of cities or bioregions, but it has to be with full concern for extended planetary impact chains. So, in this third critical sphere, the tool should help to capture the large scales and systemic transformations (agriculture and natural resources, mobility, energy, economy, academia, cultural soft-power, public bodies, legal frames), and, as for the sphere of uses, keep them entangled, as hybridization is also a key movement of transformation. Global resilience will come from a smart adaptation of the material, social and informational ecosystems, and, as the Green New Deal (Rifkins, 2019) prescribes it, infrastructures will be pivotal to implement an Anthropocene sustainable ecology. A core future trend is (again) in the hybridization of infrastructures and especially energy, mobility and data infrastructures.

Then the question is: what level of man-made geo-engineering and space occupancy is acceptable and truly resilient? Emerging digital infrastructures such as the Gaia X recently launched by the EU, or the OneWeb satellite infrastructure carried by SpaceX, show how the data-ecosystems are at the center of the global equation with attempts to recreate autonomous continents to protect sovereignty. They even take a lead on energy and transportation infrastructures, as they tend to capture users' interfaces. The polemic around TikTok and Huawei is at the heart of this new global equation.

The post-truth paradigm, democracy and science crisis are also part of this eco-systemic vulnerability. Another eco-systemic future trend is around the legal and accounting frames' transformations for the Anthropocene, which will impact political, economic and academic systems, and large parts of human activity on Earth. The World Economic Forum is now calling for a Great Reset of global capitalism and a deep metamorphosis is on its way, with new value, purpose and truth regimes to be designed for an Anthropocene oriented *global equity* (both as equitable and as shared capital). How can we design and operate the material and immaterial global commons? How to mobilize global collective intelligence to redefine our shared life conditions on the planet? How to design relevant instances to represent all stakeholders (humans, non-humans, organizations) properly? Or can we imagine a fair post-human algorithmic democracy? These matters touch on our core societal, existential and cosmological value-systems and the fourth sphere will allow us to open a more in-depth critical space for this purpose. But in this sphere, we want to remain close to change and adaptation capacities, and foresight research will still be very relevant here. Finally, another strong future trend to be highlighted in this eco-systemic and entangled sphere is around *de-mobility*, a concept becoming a solid reality and an accelerated transformation mechanism with the COVID-19 crisis. How to pragmatically reshape our global ecosystem around that pivotal idea of a dramatic reduction of mobility? What are our urbanities, community cultures and territorial integration when most activities go online? What has become the Global Village, when business, leisure and educational tourism stop or go virtual?

Bigger Picture: Meta-narratives, Cultural Superstructures and Art of the Future

In this last sphere that will also serve as a conclusion toward alien cosmologies, we want to take another more holistic perspective and explore the richness of anthropological and future meta-narratives that help human beings to give meaning to the life experience on Earth. Behind any rationale lie belief and value systems but also stories and emotional drives (such as fear and hope) that shape mankind's relation to existence and to the world (cosmologies). If historically religions have been the holistic frames to give meaning to life or if western science has tried to rationalize it, many other meta-narratives are emerging, such as the Anthropocene itself. This last sphere allows analyzing the mechanisms and cultural superstructures behind the new meta-narratives such as Ray Kurzweil's Singularity messianic scenario, or James Lovelock's Novocene (Lovelock, 2019), which furthers his Gaia theory and stages algorithmic beings. Then, why do creative theories, such as Afro futurism seem to address so well the hopes of new generations? What could be new *cosmovisions* and *pluriverse* inspired by native American holistic cosmologies (Escobar, 2017), or speculative *extro-science fiction* (Meillassoux, 2015), that is, structured but borderless meta-narratives that include what science hasn't rationalized in its frame? With a focus on possible bigger pictures, this sphere is meant to feed the more pragmatic rationale of the two previous spheres, expand them to non-earthly and more paradoxical or invisible realities, whether in their imaginary, philosophical or quantic dimensions. Ultimately, this last sphere rebuilds a bridge with the first ontological sphere.

As Ferrando recommends (2012, p. 11), we want to indulge in creativity and hermeneutic potentials, to oblige us to a proper critical thinking and as a safety-gate to keep the Anthropocene and the future open, diverse and impossible to fully master. The future belongs to everyone and the critical topology here presented is meant to design multiple future perspectives and empower everyone to resist to the way the future (and the Anthropocene) is also becoming an instrument of power. The ecology and economy of fear, is a key emotion driving the future, currently challenging the ecology of free desire (libidinal economy), which has prevailed in liberal times. Fear is a biological process deeply nested in the human body, mind and life condition. In 1929, Heidegger marked a turn toward modern existential philosophy by introducing the limit and fear of death in metaphysics heritage. The Anthropocene turn is marked by the fear of the planet's destruction and more generally of the future, as pivotal life reconditioning forces. The Anthropocene is shifting our minds toward the future and obliges us to be future-conscious, as a new form of consciousness not centered on the bubble of the ego, but on multiple spheres.

With this in mind, Toward Alien Cosmologies stands as a cognitive and pedagogical tool to support humanity's Futures Literacy (Miller, 2018), an emerging bottom-up foresight approach promoted by UNESCO, as much as a contribution to the resetting of the anthropology discipline toward the future. Whatever foresight methodology or predictive technologies we might have, the future is not written and will remain a fiction. But addressing and cultivating a creative relation to the future at all scales, is what can bring meaning and help overcome fear. It provides a paradoxical sense of clarity and a form of emancipatory inner trust rather than certitude, something we generally call hope, ultimately comforting mankind in its most singular, intelligent and imaginative capacity.

REFERENCES

Anderson, C. (2008). The End of Theory: The Data Deluge Makes the Scientific Method Obsolete, *Wired Magazine*, June 2008.

Bidault-Waddington, R. (2020a). Toward Alien Cosmologies, AI and the Human Frontier (V3), *Susch Muzeum Magazine #1*.

Bidault-Waddington, R. (2020b). Toward Alien Cosmologies (V4), Prototyping a Future Conceptual Canvas of the AI Era [conference presentation], "Prototyping Futures" Winter Symposium, Study Circle "Cybioses: Shaping Human-Technology Future," Nordic Summer University, Diffract, Berlin, March 12–14, 2020.

Bidault-Waddington, R. (2020c). Vers des Cosmologies Alien, Topologie Prospective d'une Refondation Anthropologique (V5), *FuturHebdo Anthologies Prospectives #4*.

Bidault-Waddington, R. (2018). *The Rising City, the New Frontier of Social and Urban Innovation*, future trend report (in French), Peclers Paris.

Bidault-Waddington, R. and Menetrey S. (2016). *Semiospace, a Spaced-Out Artistic Experiment*, Clinamen publishing, Geneva.

Bidault-Waddington, R. (2012). Paris Galaxy Inc., a Conceptual Model and Holistic Strategy Toward Envisioning Urban Development, *Parsons Journal for Information Mapping*, 4, no. 1.

Bonnet, E., Landivar, D., Monnin, A. and Allard, L. (2019). Le Design, une cosmologie sans monde face à l'Anthropocène, *Science du Design*, 10 (2): 97–104.

Brown, J.E. (2020). Team Builds the First Living Robots, University of Vermont, uvm.edu, retrieved on December 14, 2020. https://www.uvm.edu/uvmnews/news/team-builds-first-living-robots

Citton, Y. (2014). *Pour une Ecologie de l'Attention*, Le Seuil, 2014.

Colomina, B. and Wigley, M. (2019). *Are We Human? Notes on an Archaeology of Design*, Lars Mueller Publishing.

Descola, P. (2020). Nous Sommes Devenus des Virus pour la Planète, interview in *Le Monde*, May 20.

Escobar, A. (2017). *Design for the Pluriverse, Radical Interdependent, Autonomy and the Making of Worlds*, Duke University Press.

Ferrando, F. (2012). Toward a Post-Humanist Methodology, a Statement, *Frame, Journal for Literary Studies*, Issue "Narrating Posthumanism" 25 no. 1 (May 2012): 9–18, Utrecht University.

Fry, T. (2012). *Becoming Human by Design*, Berg Publishers.

Harari, Y-N. (2016). *Homo Deus, a Brief History of the Future*, Harvill Secker Publishing.

Lovelock, J. (2019). *Novacene, the Coming Age of Hyperintelligence*, Allen Lane Publishing.

Malabou, C. (2017). *Métamorphose de l'Intelligence, que faire de leur cerveau bleu ?* Presses Universitaires de France.

Meillassoux, Q. (2015). *Science Fiction and Extro-Science Fiction*, Univocal Publishing.

MIT Technology Review, Issue May 2017.

Miller, R. (2018). *Transforming the Future: Anticipation in the 21st Century*, UNESCO and Routledge.

Pink, S. and Salazar, J-F. (2017). Anthropologies and Futures: Setting the Agenda, in Salazar, J-F., Pink, S., Irving, A. and Sjöberg, J. (eds), *Anthropologies and Futures, Researching Emerging and Uncertain Worlds*, Bloomsbury Academic Press (pp. 3–22).

Rifkins, J. (2019). *The Green New Deal: Why the Fossil Fuel Civilization Will Collapse by 2028, and the Bold Economic Plan to Save Life on Earth*, St Martin's Press.

Simondon, G. (1989–2007). *L'individuation psychique et collective*, Editions Aubier.

Smith, R.C., Vangkilde, K.T., Kjærsgaard, M.G., Otto, T., Halse, J. and Binder, T. (eds) (2016). *Design Anthropological Futures*, Bloomsbury Academic Press.

Zaidi, L. (2019). Worldbuilding in Science Fiction, Foresight, and Design, *Journal of Futures Studies*, June 2019, 23(4): 15–26.

22. Integral ecology: reconnecting nature, culture, and knowledge

Sam Mickey

There are many ways of knowing the myriad habitats and inhabitants of Earth. Consider water lilies. When ecologists tell you what they know about water lilies, they might mention taxonomic categories and empirical facts, describing water lilies as *Nymphaeaceae*, a family of flowering plant that thrives in the freshwater ecosystems of temperate and tropical climates. When artists tell you what they know about water lilies, they might mention images and styles, like the water lilies depicted in several paintings by the French impressionist Claude Monet, which can be contrasted with the water lilies that appear in *Marsh with Water Lilies*, the drawing by the Dutch post-impressionist painter Vincent Van Gogh. If you ask climate scientists to tell you what they know about water lilies, they might mention that, as temperatures rise around the globe due to anthropogenic climate change, some relatively cooler climates will become warmer and thus more hospitable to water lilies, which means that an increasing presence of water lilies is a sign of a heating climate. If you ask a philosopher like Luce Irigaray about water lilies, she might mention the historical, symbolic, and ontological meanings of water lilies, whose family name (*Nymphaeaceae*) derives from "nymphs"—female nature deities found in ancient Greek mythology (Marder 2014, p. 216). Who is right? Ecologists and climate scientists, or artists and philosophers? Who decides whose knowledge is the best or the truest?

One of the challenges facing humans in the Anthropocene is to navigate the multiplicity of perspectives whereby people think, feel, and act in relationship to the more-than-human world. As the ecological impacts of humans have reached planetary proportions, disrupting Earth's systems of life, land, air, and water, the challenge of integrating humans with one another and with the planet has never been more urgent. One of the stumbling blocks on the road to integration is the fragmentation of knowledge. Ways of knowing are often partitioned into disciplinary silos, with interdisciplinary and transdisciplinary perspectives relatively marginalized in academia, research, and public discourse. More specifically, following the Scientific Revolution and Industrial Revolution in the modern historical period, knowledge tends to be bifurcated, split into two opposing domains, nature and culture, which correspond respectively to the split between facts and values. Human subjects (culture) are seen as fundamentally separated from material objects (nature). In that dualistic schema, epitomized in René Descartes' famous separation of thinking (res cogitans) from the extended matter (res extensa), nature is a realm of mechanical substance with no intrinsic value. In modern ways of knowing, scientists have knowledge of the facts of nature, while philosophy, literature, and the arts appear to express cultural values. The experimental methods of the sciences are often considered superior to the speculative, imaginative work of arts and humanities, but even if one were to consider arts and humanities superior, there is still a problematic split, which is to say, a failure to integrate. If the Anthropocene involves a complex entanglement of human knowledge and culture with nature, then the question remains as to whether and how it is possible to integrate the varieties of ecological knowledge and thus render obsolete the modern bifurcation separating humans from their material conditions. This chapter describes this question in terms of integral ecology.

THE QUEST FOR AN INTEGRAL ECOLOGY

"The quest today is increasingly for an *integral ecology*," as the theologians Leonardo Boff and Virgil Elizondo put it (Boff and Elizondo 1995, p. ix). When they made that remark in 1995, Boff and Elizondo were among the first to use the phrase "integral ecology," but the quest that they describe is not entirely new. It is an increasingly urgent quest, but the quest itself has been part of the entire history of ecology. That history extends to the earliest attempts of human beings to understand and respond to the relations and patterns between their environmental conditions.

Some forms of ecological inquiry are more integral than others. In his account of the history of ecology from antiquity through the twentieth century, the environmental historian Donald Worster discerns two distinct approaches to understanding and responding to the interactions between organisms and environments: first, the "*arcadian*" approach, which is oriented toward "*peaceful coexistence*" with organisms and environments, and second, an "anti-arcadian tradition," which fosters an "*imperial*" view that separates humans from nature, focusing on objectifying Earth's resources and exploiting them for human ends (1994, pp. 2, 29). Integral approaches to ecology resonate more with the former. The tension between integration and separation continues through the development of the formal science of ecology in the nineteenth century.

For Worster, Charles Darwin is the "single most important figure in the history of ecology over the past two or three centuries" (p. 114). Responding to Darwin's evolutionary theory, the German biologist Ernst Haeckel coined the term *oecologie* in 1866, conceiving of this new field as an extension of biological science. The principle of natural selection suggests that species evolve in relationship to environmental conditions. While biology studies organisms, ecology focuses on the relationships between organisms and environments. Ecology can thus be understood as a more complex and comprehensive way to study life, situating life within the "household" (*oikos*) of nature. Haeckel is also a cautionary figure insofar as he did not question some racist assumptions about an evolutionary progression of racial superiority. Nonetheless, by focusing on organism–environment interrelations, Haeckel's ecology developed more thorough explanations of the conditions of existence for living beings, including explanations that would lead later theorists to criticize pseudoscientific ideas of race. Theories and methods of ecology became increasingly thorough in subsequent generations, and the tension between fragmentation and integration continued.

The twentieth century saw the emergence of a "new ecology," which included biophysical and socioeconomic sciences to provide "an energy-economic model of the environment," whereby ecologists like Charles Elton and Arthur Tansley used thermodynamics and economic models of efficiency, production, and consumption to describe the flow of energy through an ecological "community" (Elton) or "ecosystem" (Tansley) (Worster 1994, p. 311). This approach to ecology was further refined with the inclusion of systems theory and chaos theory into ecology during the 1970s and 1980s. Applied to ecology, those theories showed the important role of disorder and natural disturbances in ecological relationships, such that the energy flows of ecosystems need to be understood not as harmonious or static systems but as changing, unpredictable, unruly, and complex. However, this new ecology was fragmentary, reducing ecosystems to their constituent parts. In contrast, in 1977, Eugene Odum (2000, p. 198) proposed an updated version of the new ecology, which would be an "integrative discipline," where "integrative" reflects a commitment to complexity and an opposition to oversimplification. Affirming holism and opposing reductionism, "the new ecology links the natural and the social sciences," both in theory and in practice, seeking "to raise thinking and action" to a holistic encounter with ecosystems (p. 199). Odum follows the energy-economic model of ecology in working toward the "integration of economic and environmental values," but he also goes further, including not only economics but also politics and legal issues within the holistic discipline of integrative ecology (p. 201). However, Odum's ecology still contains aspects of the reductionism it proclaims to avoid.

Odum envisions an interdisciplinary and engaged ecology, but his vision fails to address imaginative, speculative, or spiritual perspectives on ecology. It fails to include arts and humanities. Neglecting those perspectives means neglecting a vast array of experiences, ideas, symbols, artistic expressions, and ways of being in the world. Those perspectives have gradually become more well-represented in environmental theory and practice since the 1970s, specifically with the emergence of philosophical, religious, and literary discourses on the natural world (e.g., environmental ethics, ecofeminism, ecocriticism, and the field of religion and ecology). Such discourses are collectively known today as the environmental humanities. The promise of integral ecology is to include and coordinate evermore perspectives, even conflicting and contradictory perspectives.

A deeply inclusive sense of integral ecology was proposed by Boff and Elizondo (1995, pp. ix–x) in a jointly written introduction to an issue of the theology journal *Concilium*, where they say the following:

> The quest today is increasingly for an *integral ecology* that can articulate all these aspects with a view to founding a new alliance between societies and nature, which will result in the conservation of the patrimony of the earth, socio-cosmic wellbeing, and the maintenance of conditions that will allow evolution to continue on the course it has now been following for some fifteen thousand million years.
>
> For an integral ecology, society and culture also belong to the ecological complex. Ecology is, then, the relationship that all bodies, animate and inanimate, natural and cultural, establish and maintain among themselves and with their surroundings. In this holistic perspective, economic, political, social, military, educational, urban, agricultural and other questions are all subject to ecological consideration. The basic question in ecology is this: to what extent do this or that science, technology, institutional or personal activity, ideology or religion help either to support or to fracture the dynamic equilibrium that exists in in the overall system?

THREE ECOLOGIES

Boff has since elaborated on more details of his idea of integral ecology, specifically by situating integral ecology in relationship to three ecological registers: environmental, social, and mental (Boff 2016). The environmental approach engages ecological issues through biophysical sciences and the development of technologies. The social approach includes humans and society within ecological issues, addressing problems of social justice and cultivating sustainable social institutions (education, healthcare, economic development, etc.). The mental approach focuses on subjectivity, showing how ecological problems call not only for a healthier and more sustainable society and environment, but also for healthier modes of thought, perception, emotion, desire, and imagination. Like ecopsychology and deep ecology, what Boff calls mental ecology speaks to the soul, whereas the previous two ecologies speak to science and society.

Those three ecologies (environmental, social, and mental) represent the multiple perspectives that have emerged from the biophysical sciences, social sciences, and humanities, respectively. The integral approach brings together those multiple ecologies to present a new vision of planetary coexistence (Mickey, Kelly, and Robbert 2017). It is a vision of humans and Earth emerging from the evolutionary becoming of the universe, which is to say, processes of cosmogenesis, which include three aspects: (a) complexity and differentiation, referring to the objective or exterior facets of things; (b) self-organization and consciousness, meaning the subjective depth or interior facets of things; and (c) reconnection and relation, which involve the ways things come together not merely as a collection of different objects but as communing agents, communicating subjects.

Boff's idea of three aspects of cosmogenesis draws upon the integral vision developed by Berry (1999, p. 162), who similarly articulates "three basic principles: differentiation, subjectivity, and communion." Those principles are presented as the "cosmogenetic principle" in Berry's work with the cosmologist Brian Swimme, *The Universe Story* (Swimme and Berry 1992, pp. 66–78). In terms of the cosmogenetic principle, all evolutionary processes in the universe involve an objective side, that is, an exterior that differentiates things from one another, as well as a subjective side. The subjective side includes the activities of self-organization that constitute the interiority or agency of things, and the relational interactions whereby all subjects in the universe exist in community or communion. Differentiation can be seen in the diversity of life and the uniqueness of every single being. The subjective dimension of things can be understood in terms of scientific conceptions of self-organization (autopoiesis) as well as in terms of religious traditions that articulate the ensouled or numinous element of reality. An experience of this numinous quality is crucial for Berry's integral vision. "What is important is the attainment of a conscious realization of the spiritual nature of human development. Only then can a truly integral human experience be achieved" (Berry 2009, p. 15).

For Berry, the integral ecologist guides our awakening to the profound complexity and numinous mystery of the Earth community. Along these lines, Berry proposes "an ecological spirituality with an integral ecologist as spiritual guide" (p. 135):

> The integral ecologist can now be considered a normative guide for our times. The integral ecologist would understand the numinous aspect of a universe emergent from the beginning. [...] The integral ecologist is the spokesperson for the planet in both its numinous and its physical meaning, just as the prophet was the spokesperson for the deity, the yogi for the interior spirit, and saint for the Christian faith. In the integral ecologist, our scientific understanding of the universe becomes a wisdom tradition. (p. 136)

Bringing together wisdom, experience, and know-how, the integral ecologist communicates narratives whereby we humans might "accept that we exist as an integral member of this larger community of existence" (p. 138). Moreover, the spirituality of this endeavor is by no means otherworldly. "Religion takes its origin here in the deep mystery of what we see, hear, touch, taste, and savor" (p. 147).

While drawing on Berry's work, Boff also draws on another lineage of radical thought, that of the twentieth-century French philosopher and psychoanalyst Félix Guattari, who is a well-known philosopher and political activist in his own right while also being well-known for co-authoring several books with the philosopher Gilles Deleuze. Guattari reconfigures Marxist thought and practice for an era of globalization, multiculturalism, communication and information technologies, and environmental degradation.[1] With Guattari, integral ecology is the antidote to the psychospiritual, economic, and environmental catastrophes of global capitalism.

Guattari proposes a "generalized ecology" or "ecosophy" that would seek to reinvent human practices in their relationship to material conditions, social relationships, and subjectivity (Guattari 2000, pp. 28–37, 52). Just like Boff's environmental, social, and mental ecologies parallel Berry's cosmogenetic principle, they also parallel Guattari's three ecologies. Boff is explicit on this point. Integrating the three ecologies requires what Guattari calls "transversal tools"—experimental practices whereby individuals and communities can cross boundaries to achieve communication between multiple levels or registers of meaning (p. 69).

Guattari also develops his concept of ecosophy in his final book, *Chaosmosis*, which poses a fundamental question to guide ecosophy:

> how do we change mentalities, how do we reinvent social practices that would give back to humanity—if it ever had it—a sense of responsibility for its own survival, but equally for the future of all life on the planet, for animal and vegetable species, likewise for incorporeal species such as music, the arts, cinema, the relation with time, love and compassion for others, the feeling of fusion at the heart of the cosmos? (Guattari 1995, pp. 119–120)

Guattari's mental ecology not only includes ideas and cognition, but the full spectrum of cognitive and affective processes whereby subjectivity articulates itself and participates in embodied engagements with the world and with "the 'mysteries' of life and death" (Guattari 2000, p. 35). Guattari proposes that mental ecology focus on "the promotion of innovatory practices" and "alternative experiences," which respect the unique singularity of subjects and create appropriate relations between subjects and society (p. 59).

Social ecology addresses the collective processes of subjectivity. Addressing events such as "sudden mass consciousness-raising," transformative social struggles, technology, media, and labor, social ecology promotes creative subjectivity that overcomes exploitative and oppressive powers (p. 62). Between mental and social ecology, the question of ecosophy becomes one of "the whole future of fundamental research and artistic production," a question of "how to encourage the organization of individual and collective ventures" that care for the singularity of subjectivity (p. 65). Guattari's environmental ecology attends to the complexities and uncertainties of environmental processes, affirming that "anything is possible—the worst disasters or the most flexible evolutions" (p. 66). Drawing on systems sciences, Guattari attends to the complexity and openness of self-organizing (autopoietic) systems as affective assemblages, which have interrelated parts and enable different ways of acting and being acted upon. He thus avoids any reduction of human or nonhuman beings to mere objects.

The scope of the three ecologies embraces all assemblages and the complex relations in and between them, spanning the human and the nonhuman, across all scales of existence. This supports the creation of new ethical and political practices that integrate humans and nonhumans into resilient forms of solidarity. Integrating "the tangled paths of the tri-ecological vision," Guattari's ecosophy aims for creative transformations in both the collective unity and singular differences between individuals (including human and nonhuman individuals), such that ecosophy aims for all individuals to "become both more united and increasingly different" (pp. 67–69). In other words, ecosophy is about building alliances across differences, including difference that humans have with one another as well as the differences between humans and the rest of the universe. Ecosophy puts the cosmos back in cosmopolitanism.

During the current period of ecological and social change, immensely complex challenges are facing life on Earth, including pollution, deforestation, water scarcity, climate change, and mass extinction. Such challenges cannot be sufficiently addressed through a single perspective alone. What is needed is dialogue and integration among diverse perspectives. Planetary problems call for globally coordinated responses. Integral ecology is an invitation to deepen one's ways of knowing the natural world, including multiple perspectives on the environmental, social, and subjective complexities of human–Earth relations in the Anthropocene. Integral ecology welcomes humans to undertake collective experiments in intellectual, imaginative, emotional, and physical intimacy with the universe. Without such experiments, knowledge in the Anthropocene is doomed to fragmentation, which renders collective cooperation unachievable. With such experiments, knowledge in the Anthropocene can attain an integrative coherence that affords the building of collective alliances across the different perspectives that make up the Earth community.

NOTE

1. For a detailed account of the ecological implications of the writings of Deleuze and Guattari in comparison with Berry and Boff, see Mickey (2015).

REFERENCES

Berry, Thomas. (1999). *The great work: Our way into the future.* New York, NY: Bell Tower.

Berry, Thomas. (2009). *The sacred universe: Earth, spirituality, and religion in the twenty-first century.* M. E. Tucker (Ed.). New York, NY: Columbia University Press.

Boff, Leonardo. (2016). "Interview: Leonardo Boff, a founder of liberation theology." *Kosmos Journal*, September 6. Accessed March 2, 2020. https://www.kosmosjournal.org/news/interview-leonardo-boff-a-founder-of-liberation- theology

Boff, Leonardo, & Elizondo, Virgil. (1995). Ecology and poverty: Cry of the earth, cry of the poor. *Concilium: International Journal of Theology* 5: ix–xii.

Guattari, Félix. (1995). *Chaosmosis: An ethico-aesthetic paradigm.* P. Bains & J. Pefanis (trans.). Bloomington: Indiana University Press.

Guattari, Félix. (2000). *The three ecologies.* I. Pindar & P. Sutton (trans.). London: Athlone Press.

Marder, Michael. (2014). *The philosopher's plant: An intellectual herbarium.* New York: Columbia University Press.

Mickey, Sam. (2015). *Whole earth thinking and planetary coexistence: Ecological wisdom at the intersection of religion, ecology, and philosophy.* New York: Routledge.

Mickey, Sam, Kelly, Sean, & Robbert, Adam. (Eds.) (2017). *Integral ecologies: Nature, culture, and knowledge in the planetary era.* Albany, NY: SUNY Press.

Odum, Eugene. (2000). The emergence of ecology as a new integrative discipline. In D. Keller & F. Golley (Eds.), *The philosophy of ecology: From science to synthesis* (pp. 194–203). Athens, GA: University of Georgia Press.

Swimme, Brian, & Berry, Thomas. (1992). *The universe story: From the primordial flaring forth to the ecozoic era—A celebration of the unfolding of the cosmos.* San Francisco, CA: HarperCollins.

Worster, Donald. (1994). *Nature's economy: A history of ecological ideas.* 2nd ed. New York: Cambridge University Press.

23. Visuality conditions under the Anthropocene[1]

Irmgard Emmelhainz

The Anthropocene is the era in which man's impact on the earth has become the single force driving change on the planet, thus giving shape to nature, shifting seas, changing the climate, and causing the disappearance of innumerable species, including placing humanity on the brink of extinction. The Anthropocene thus announces the collapse of the future through "slow fragmentation towards primitivism, perpetual crisis and planetary ecological collapse" (Srnieck & Williams, 2013, para. 23). Instead of being conceived as speculative images of our future economic and political system, the Anthropocene has been reduced to an apocalyptic fantasy of human finitude, world finitude, and the manageable problem of climate change. In the last decade, films about the end of the world have been characterized by an apocalyptic and doomsday narrative that may end with moral redemption—from *The Day After Tomorrow* (2004), and *2012* (2009), to Lars von Trier's *Melancholia* (2011) and *World War Z* (2013). In parallel, we have seen in the mass media a narrative presenting climate change as a fixable catastrophe, just like any other (such as the 2008 financial crisis, or the 2010 BP gulf oil spill). Neither our condition of finitude nor the world after the human has been imagined, and the massive environmental impact from the industrial era onward, with its long-term geomorphic implications, has become unintelligible.

The Anthropocene has meant not a new image of the world, but rather a radical change in the conditions of visuality and the subsequent transformation of the world into images. These developments have had epistemological as well as phenomenological consequences: while images now participate in forming worlds, they have become forms of thought constituting a new kind of knowledge—one that is grounded in visual communication, and thereby dependent on perception, demanding the development of the optical mind (Brakhage, 1978, p. 120). The radical changes in the conditions of visuality under the Anthropocene have brought a new subject position, announced by the reformulating trajectories between impressionism and cubism, and those between cubism and experimental film. While cubism culminated with the antihumanist rupture of the picture plane and converted the visual object, along with surrealism, into "manifestation," "event," "symptom," and "hallucination," experimental film introduced a mechanical, posthuman eye conveying solipsistic images at the sensorimotor level of perception. The consequence of these developments is that images, as opposed to being subject to our "beliefs," or being objects of contemplation and beauty, came to be perceived as "the extant." This involves a passage from representation to presentation, that is, instead of showing a perpetual present in a parallel temporality in order to make the absent partially present, the image has become sheer presence, immediacy: the here and now in real time. Made up of particles of time, wrested out of sensation and turned into cognition, the image deals more with concepts and saying than with intuition and showing.

With its break from the Renaissance point of view, cubism decomposed anthropomorphism. Based on linear perspective, Renaissance perspective had normalized a viewing position as a centered, one-eyed static entity within a mathematical, homogenous space. Creating the illusion of a view to the outside world, Renaissance perspective made the pictorial plane analogous to a window. Images constructed with traditional perspective bestow identities and subjects given a priori, configured by the point of view provided by the picture plane. Cubism, in contrast, turned space, time, and the subject upside down, redefining spatial experience by rupturing the picture plane (Didi-Huberman, 2007, p. 5). If classical representation conveys a continuous space, cubism invented discontinuous space by subverting the relations between subject and object, making identity and difference relative, questioning classical metaphysics. The cubist image renewed the image of the world by dissociating gaze, subject, and space, but without estranging them from each other, bringing about a new, antihumanist subject position (Didi-Huberman, 2007, p. 6). Moreover, with cubism, temporality—duration—and a multiplicity of points of view became embedded in the picture plane.

With North American experimental (or structural) film in the 1960s and 1970s, notably influenced by Andy Warhol's filmic work, duration became a key component of aesthetic experience, grounded in an exploration of the filmic apparatus and seeking to make it analogous to human consciousness. By creating cinematic equivalents or metaphors of consciousness, experimental film brought about a prosthetic vision giving way to solipsistic visual experiences.[2] A futuristic technoscape, Michael Snow's experimental film *La Région centrale* (1971) is exemplary in this regard. In the film (as in all of his work), Snow explores the genetic properties of the filmic apparatus, using it to intensify and diminish aspects of normal vision. *La Région centrale* shows images from the wilderness collected by a machine specifically designed to shoot the film (De La). The machine was able to move in all directions, turn around 360 degrees, and zoom in and out, reaching places no human eye could perceive before. The resulting footage was independent of any human decisions and framing vision: a three-and-a-half-hour topological exploration of the wilderness, a "gigantic landscape" (Snow, 1994, p. 56).

Because De La extracts gravity from the situation as well as human (pre-constructed or given) referential points of view, *La Région* hypostasizes the cubist relativization of identity and difference and its rejection of a priori space. Furthermore, the film puts forth an experience of matter within, decentering the subject, which is constituted by the experience of the work itself. To paraphrase Rosalind Krauss on minimalism (Krauss, 1999), the film subverts the notion of a stable structure that could mirror the viewer's own self—a self that is completely constituted prior to experience. That is to say, the film formulates a notion of self that exists only at the moment of externality of that particular experience. By presenting every possible position of the framing-camera in relationship to itself, *La Région* releases the subject from its human coordinates, creating a "space without reference points where the ground and the sky, the horizontal and the vertical inter-exchange" (Deleuze, 1986, p. 84). The references to human coordinates are the screen's rectangular frame and the breaks made by the intermittent appearance of a big glowing yellow "X" against a black screen. Every time the X comes up, it fixes the screen and transfers the movement in a different way or direction; thus, the Xs are the point of view *embodying* the *apprehension of the passage from chaos to form*. In viewing the film, the present is experienced as immediacy, a pure phenomenological consciousness without the contamination of historical or a priori meaning; the world is thus experienced as self-sufficient, pure presence, foregrounding an awareness of the presence of the viewer's own perceptual processes. As Snow stated:

> My films are experiences: real experiences ... The structure is obviously important, and one describes it because it's more easily describable than other aspects, but the shape, with all the other elements, adds up to something which can't be said verbally and that's why the work is, why it exists. (Snow, 1994, p. 44)

In general, experimental film sought to posit alternatives to the mimetical inscription of lived experience into simulacral images (signs) by artistic neovanguardist practices that came to be embedded within the logic of spectacle—not in order to dislodge subjectivity (early modernism) or to constitute subjects by mapping out signs (postmodernism), but by exploring through film the conditions of cognition and perception. And while there is something in the image delivered by *La Région* that shares something with the condition of thought, it yields a solipsistic subject at the genetic level of perception; beyond auditory or optical perceptions, it delivers motor-sensory perceptions (Deleuze, 1986, p. 85). Therefore, the machine delivers a posthuman, prosthetic enhancement of vision, inaugurating three important developments in the history of perception.

First, the machine introduces the incipient normalization of perception as augmented reality and the solipsistic visualization of data. Second, as Donna Haraway posited, the prosthetic enhancement of vision brings about the notion of limitlessness and an "unregulated gluttony" that desires to see everything from nowhere, spreading the assumption that anything can and is seen (Haraway, 1988, p. 582). Third, with *La Région*, machinic vision becomes an epistemological product of a centered human point of view (with the Xs) without stable reference points, foregrounding the conditions of contemporary visuality. While cubism embodies the antihumanist scission of the subject and the possibility of the construction of many psychical planes, *La Région* embodies the displacement of the human agent from the subjective center of operations (Didi-Huberman, 2007, p. 9). Both epitomize modernity's fragmentation by mechanization, its alienating character, its inability to give back an image or to serve as a reflective mirror—it can never do this because the antihumanist image is *indifferent to me* (McMahon, 1997, p. 4). And yet, this was always going to be the fate of an image and of art based on contemplation. These works also attest to the fact that the foundational experience of modernity is to refuse, in advance, the "given" as a ground for thought (McMahon, 1997, p. 8).

THE TRANSFORMATION OF EVERYTHING INTO DATA-IMAGES

As previously explained, the Anthropocene era implies not a new image of the world, but the transformation of the world into images. Humanity's alteration of the biophysical systems of earth is parallel to the rapid modifications of the receptive fields of the human visual cortex announced by cubism and experimental film. This alteration is also accompanied by an unprecedented explosion in the circulation of visibilities, which are actually making the outcome of these alterations opaque (Nixon, 2011, p. 12). For instance, the exhaustive visualization and documentation of wildlife is actually rendering its ongoing extinction invisible. Aside from having become shields against reality, images are not only substitutes for first-hand experience, but have also become certifiers of reality, and, as Susan Sontag points out, they have extraordinary powers to determine our demands on reality (Sontag, 1977, p. 80). In discussing the democratization of tourism in the 1970s, Sontag further described tourists' dependence on photographic cameras for making real their experiences abroad:

> Taking photographs ... is a way of certifying experience, [but] also a way of refusing it—by limiting experience to a search for the photogenic, by converting experience into an image, a souvenir ... The very activity of taking pictures is soothing,

and assuages general feelings of disorientation that are likely to be exacerbated by travel. (Sontag, op. cit., p. 177)

Almost forty years later, posing for, taking, sharing, liking, forwarding, and looking at images are actions that are not only integral to tourism; they actually *give shape* to contemporary experience. Arguably, representation has ceased to exist in plain view and manifests itself as experience, event, or the appropriation and sharing of a mediatic space. Instead of representation, we have media objects (i.e., a twitterbot) that purport to provide vague participatory representational events that ground our cultural and social experience. Thus, as Stephen Shaviro points out, in the contemporary world, the opposition between reality-based and image-based modes of presentation breaks down, and the most intense and vivid reality is precisely the reality of images (Shaviro, 2010, p. 12).

In other words, images have in themselves become opaque cognitive and empirical experiences. Each episode of the recent British science-fiction television series *Black Mirror* explores the implications of this precise phenomena—of images becoming not only an intrinsic part of our empirical experience but also our cognitive experience at large. The "black mirror," then, is nothing other than the LCD screen through which we give shape to reality.

One of the show's early episodes, "The Entire History of You" (2011), imagines a world in which almost everyone has a "grain" implanted behind their ear. This grain has the capacity to transform human eyes into cameras that record reality and projectors that can reproduce it, thereby amalgamating lived experience, memory, and image. In a later episode, "Be Right Back" (2013), a woman is able to revive her dead partner with a program that rebuilds him—first his writing habits, then speech patterns, and eventually his very self via a cloned, synthetic body—solely from the proliferation of information he uploaded on the internet when he was alive. The deathless and bodiless information, images, and signs—the inert map of a life—becomes embodied by an avatar that exists in actual, not virtual, reality, and that has the (albeit limited) capacity to exist and interact directly with humans. In the episode, the fabrication of subjectivity from data—which implies the automatization of subjectivity—foreshadows the relationship between determinist automatisms and cognitive activity, which, according to Franco Berardi, is the core goal of the Google Empire: to capture user attention and to translate our cognitive acts into automatic sequences. The consequence is the replacement of cognition by a chain of automated connections, seeking to automatize the subjectivities of users (Berardi, 2014, paras. 21–23.

Aside from the fact that images and data are taking the place of or giving form to experience, automating our will and thought, they are also transforming things into signs by welding together image and discourse, bringing about a tautological form of vision. With the widespread use of photography and digital imaging, all signs begin to lead to other signs, prompted by the desire to see and to know, to document and to archive information. Thus, the fantasy that everything is or can be made visible coexists with the increasing automation of cognition, which, following Franco Berardi, is the basic condition of semiocapital (the valorization and accumulation of signs as economic assets) (Berardi, op. cit., par. 3).

In the pilot episode of *Black Mirror*, "The National Anthem" (2011), an alleged terrorist group kidnaps a nationally beloved British princess in the early morning hours. In order to free her, the anonymous group demands that the prime minister have sex on live television with a pig at four o'clock that same afternoon. The video in which the princess announces the "price" of the ransom goes viral and the whole nation pressures the prime minister to fulfill the kidnappers' demands. At the end of the episode, postcoitus, it is revealed that the kidnapping was a singular artist's gesture, intended in its successful implementation to point critically to the obscenely inflated role the media has in shaping public opinion and official policy. The artist's action, in other words, illuminates the highly visceral shift in power brought on almost instantaneously by the ransom video's circulation in the infosphere. Insofar as the episode unfolds montages of the whole nation glued to televisions in the pub, workplace, and waiting room at four o'clock, the artist highlights how connective interfaces actually govern, as they have the direct capacity to manipulate and coordinate behavior on almost every level (Berardi, 2012, p. 15).

Under the conditions of semiocapitalism, images and signs acquire value and/or power by means of being seen, largely through "likes" and retweets. The fact that sign-value has supplanted exchange-value means moreover that we no longer consume material things, but rather swallow cognitive signs embedded in and around them. Aside from consuming "experiences" or "moods," we buy immaterial commodities (in the name of lifestyle and branding) and consume signs for "equality," "happiness," "wellness," and "fulfillment." In Don Delillo's *White Noise* (2002), Jack Gladney describes a trip to the supermarket and makes clear how the signs found in the brands and labels of products that he and his wife buy have the power to relieve them of the mysteries and anxieties brought about by everyday life:

> It seemed to me that Babette and I, in the mass and variety of our purchases, in the sheer plenitude those crowded bags suggested, the weight and size and number, the familiar package designs and vivid lettering, the giant sizes, the family bargain packs with Day-Glo sale stickers, in the sense of replenishment we felt, the sense of well-being, the security and contentment these products brought to some snug home in our souls—it seemed we had achieved a fullness of being that is not known to people who need less, expect less, who plan their lives around lonely walks in the evening. (Delillo, 2002, p. 20)

What becomes evident in this paragraph is Baudrillard's assertion that objects are no longer commodities whose message and meaning we can appropriate and decipher, but rather, tests that interrogate us. For him, commodities are a referendum, the verification of a code, circularity as well as sameness and homogeneity: here the commodities bring a well-being that reflects the well-being of the consumer and his or her lifestyle (Baudrillard, 1976). Furthermore, the acceleration in proliferation of cognitive signs since the time of Delillo's novel is another of the features of communicative capitalism's subjugation, submitting the mind to an ever-increasing pace of perceptual stimuli, and in so doing generating not only panic and anxiety, but also destroying all possible forms of autonomous subjectivation. Under communicative capitalism, images transformed into signs embody the current concatenation of knowledge and machines—that is, the technological organization of capitalism to produce value. With the enabling of the visualization of data by machines, images have become scientific, managerial, and military instruments of knowledge, and thus of capital and power (Bratton, 2013, para. 16). In this context, *seeing* means the accelerating perception of the fields of everyday experiences, or rather, the field of trivial visual analogies of experience: a kind of groundless, *accelerated tautological vision* derived from passive observation. This is for Berardi another of communicative capitalism's forms of governance, as this kind of vision generates technolinguistic automatisms by carrying information without meaning, automating thought and the will (Berardi, 2012, p. 41).

IMAGES AS COGNITION AND THUS FORMS OF POWER

Images circulating in the infosphere are also charged with affect, exposing the viewer to sensations that go beyond everyday perception. Hollywood cinema, for instance, delivers pure sensation and intensities that have no meaning. In Alfonso Cuarón's *Gravity* (2013), the main characters try to survive in outer space by solving practical and technical problems. The movie repudiates a point of view and a ground for vision in favor of immersion, transforming images into physical sensations mobilized by the visual and auditory (especially in its 3-D version), and thus into affect. The becoming-affect of images derives from communicative capitalism's ruthless conversion of sensation and aesthetic experiences into cognition: its transformation of these experiences into information, sensations, and intensities without meaning is precisely what enables them to be exploited as forms of work and sold as new experiences and exciting lifestyle choices (Shaviro, 2010, para. 14). One of the problems that arises is that affect cannot be linked to a larger network of identity and meaning. *Gravity* also presents itself as a symptom of the normalization of a groundless seeing brought about by modernity's decentering of the subject parallel to our exposure to aerial images (for example, Google Earth). The hegemonic sight convention of visuality is an empowered, unstable, free-falling, and floating bird's eye view that mirrors the present moment's ubiquitous condition of groundlessness (Steyerl, 2011, para. 6).

According to many thinkers, this groundlessness characterizes the Anthropocene. The current fragmentation and transience of sociopolitical movements attests to the fact that we are first of all lacking ground on which to found politics, our social lives, and our relationship to the environment. Second, as Claire Colebrook put it, with the Anthropocene we are facing human extinction, as well as causing other extinctions, thereby annihilating that which makes us human. We are thus all thrown into a situation of urgent interconnectedness, in which a complex multiplicity of diverging forces and timelines that exceed any manageable point of view converge (Colebrook & Wolfe, 2013). In this context, criticality is both in trouble and spinning on its head. Many questions arise: How do we redefine the ground of deterritorialized subjectivation beyond the subsumption of subjectivity by the modes of governance of accelerated tautological forms of vision and communicative capitalism? How can we transform our relationship to the indeterminacy of deterritorialization and the multiplicity of diverging points of view in order to provide a heightened sense of place, giving way to the possibility of collective autonomous subjectivation and thus a new sense of politics and of the image?

In an era of ubiquitous synthetic and digital images dissociated from human vision and directly tied to power and capital, when images and aesthetic experience have been turned into cognition and thus into empty sensations or tautological truths about reality, the image of the Anthropocene is yet to come. The Anthropocene is "the age of man" that announces its own extinction. In other words, the Anthropocene thesis posits "man" as the end of its own destiny. Therefore, while the Anthropocene narrative keeps "man" at its very center, it marks the death of the posthuman and of antihumanism, because there can be no redeeming critical antihumanist or posthuman figure in which either metaphysics or technological and scientific advances would find a way to reconcile human life with ecology. In short, *images* of the Anthropocene are missing. Thus, it is necessary to transcend our incapacity to imagine an alternative or something better. We can first do this: draw a distinction between *images* and imagery, or pictures. Although it is related to the optic nerve, the picture *does not make an image* (Daney & McMahon, 1999). In order to make images, it is necessary to make *vision* assassinate perception; it is necessary to *ground vision*, and then *perform* (as in artistic activity) and *think vision* (as in critical activity) (Didi-Huberman, 2007, p. 17).

IMAGES TO COME

Following Jean-Luc Godard, who operates in his work between the registers of the real, the imaginary, and art, only cinema is capable of delivering images as opposed to imagery, conveying not a subject but the supposition of the subject and thus the verb (substance) (Didi-Huberman, 2005, p. 8). Alterity is absolutely necessary for the image, as the *image* is an *intensification of presence*—this is why it is able to hold out against all experiences of vision (Daney & McMahon, 1999, para. 2). In this light, Godard's cinematic project can be interpreted as a conception of the image as a promise of flesh. For Godard, the image is incertitude, it is "trying to see" and the possibility of "giving voices back to their bodies." For the filmmaker, *images do not show*; rather, images are a matter of belief and a *desire to see* (which is different from the desire *to know* or *to possess*).

An essay-film Godard made with Anne-Marie Miéville, *The Old Place* (2001), addresses the Anthropocenic concerns of life after the extinction of man, the current groundlessness of vision, and the lack of *images* of the world and of humanity. While we see images from outer space, Miéville and Godard discuss "CLIO," the archaeological bird of the future, a microsatellite sent into space in 2001. The satellite is supposed to come back to earth in five thousand years to inform its future inhabitants about the past. Aside from carrying traditional human forms of knowledge, the bird will deliver messages written by the current inhabitants of the globe addressed to its future inhabitants. Miéville and Godard ponder whether humanist messages such as "Love each other," or "Eliminate discrimination against women," will be included in the satellite (they doubt it). Later on, they conclude:

> We are all lost in the immensity of the universe and in the depth of our own spirit. There is no way back home, there is no home. The human species has blown up and dispersed in the stars. We can neither deal with the past nor with the present, and the future takes us more and more away from the concept of home. We are not free, as we like to think, but lost.

Here Godard and Miéville paint the termination of a world, its exhaustion and estrangement from its conditions of possibility. As they underscore the lack of a home for the spirit, they highlight the loss of a sense of origin and destination, implying that the active principle of the world has ceased to function (Fox, 2009, p. 7). The last line is spoken while we see the image of a mother polar bear staring at her dead cub, followed by an image of Alberto Giacometti's sculpture *L'homme qui marche* (Walking Man, 1961): life persists irrationally, not given form by imagination, ceasing to cohere into a higher truth (Fox, op. cit., p. 70).

In *The Old Place*, Godard and Miéville explore the image of humanity throughout the Western history of art, underscoring the fact that for two thousand Eurocentric, Christian years, the image was sacred. We also see images of violence, torture, and death juxtaposed with beautiful sculpted and painted figures and faces created throughout all the ages of humanity: people by turns smiling, screaming, or crying.

For Godard and Miéville, the image is also something related to the origin that reveals itself as the new but that had been there all along: an originary landscape always present and inextricable from history. Marking the passage to the current regime of communicative capitalism, where images are permeated by discourse and tautological truths about reality, they state: "The image today is not what we see, but what the caption states." This is the definition of publicity, which they further link to the transformation of art into market and marketing represented by both Andy Warhol, and by the fact that "The last Citroën will be named Picasso," which has as a consequence that "The spaces of publicity now occupy the spaces of hope." And yet, in spite of the ubiquity of communicative capitalism, for them there is something that resists, something that remains in art and in the image. Meanwhile, we see a blank canvas held by four mechanical legs moving furiously.[3] This evokes the resisting image to come; this resisting image is a question of (sensible, un-automated) purity and, in post-Christian secular sense, of the sacred and redemption, of an ambivalent relationship between image and text, foreign to knowledge and intrinsically tied to belief. At the end of *The Old Place*, the filmmakers posit the Malay legend of A Bao A Qu as the paradigm of the image of these times in which "we are lost without a home," as they state. "The text of A Bao A Qu is the illustration of this film." The legend is rewritten by Jorge Luis Borges in his *Book of Imaginary Beings*:

> To see the most lovely landscape in the world, a traveler must climb the Tower of Victory in Chitor. A winding staircase gives access to the circular terrace on top, but only those who do not believe in the legend dare climb the tower. On the stairway there has lived since the beginning of time a being sensitive to the many shades of the human soul known as A Bao A Qu. It sleeps until the approach of a traveler and some secret life within it begins to glow and its translucent body begins to stir. As the traveler climbs the stairs, the being regains consciousness and follows at the traveler's heels, becoming more intense in bluish color and coming closer to perfection. But it achieves its ultimate form only at the topmost step, and only when the traveler is one who has already attained Nirvana, whose acts cast no shadows. Otherwise, the being hesitates at the final step and suffers at its inability to achieve perfection. As the traveler climbs back down, it tumbles back to the first step and collapses weary and shapeless, awaiting the approach of the next traveler. It is only possible to see it properly when it has climbed half the steps, as it takes a clear shape when its body stretches out in order to help itself climb up. Those who have seen it, say that it can look with all of its body and that at the touch, it reminds one of

a peach's skin. In the course of the centuries, A Bao A Qu has reached the terrace only once. (Borges, 1967, p. 2)

In their film, Godard and Miéville explore the imprint of the quest of what it means to be human throughout the history of images. Humanity transpires as a mark that is perpetually reinscribed in a form of an address. A Bao A Qu is an inhuman thing activated by the passage of humans wishing to see the most beautiful landscape in the world. The act of vision is a unique event, and what delivers the vision of the landscape and of the creature are the purity and desire of the viewer. A Bao A Qu is an image of alterity; it stares back with all of its body. A Bao A Qu is an antidote to the lack of imagination in our times: an inhuman vision that undermines the narrative that holds the human as the central figure of its ultimate form of vision and destruction.

In the voiceover of his most recent film, *Adieu au langage* (Farewell to Language, 2014), Godard quotes Rilke: "It is not the animal, which is blind, but man. Blinded by consciousness, man is incapable of seeing the world." With a strident palette and saturated sound, the film evokes abstract, fauvist, cubist, and impressionist painting, and is Godard's most radically experimental film (as in the genre, because all his work is experimental and radical) to date. Rilke's quote, together with an aphorism he attributes to Monet, frame Godard's quest in this film: "It is not about seeing what we see, because we do not see anything, but [it is about] painting what we cannot see." In parallel, Godard revives the romantic poet's wish to "describe" immediate reality, to hit the viewer with electroshocks that make a real visible and audible world emerge from language. In the film, as a way to enable a new form of communication beyond tautological digital communication (Godard points out that with texting there is neither the chance to interpret a code nor room for ambiguity) and to reestablish harmony between the couple in the movie who can no longer communicate face to face, Roxy Miéville's dog appears. Roxy becomes the metaphor for the possibility of an "other" post-anthropocentric language "between" humans. In the movie, the dog asks, "What is man? What is a city? What is war?" Rocky's comings and goings between the couple bear the possibility of giving back freedom to the face-to-face encounter. Godard compare's Roxy's "other" language to the lost language of the poor, the excluded, animals, plants, the handicapped—those who are out of the frame. In sum, the movie is a giant mirror that reflects a grammar of thought that no longer resides in enunciation (and this is the farewell to language): marking the absence of a relationship between the characters by using Roxy—the third person, the post-anthropocentric "other"—as a vehicle of communication.

In both *The Old Place* and *Adieu au langage*, Godard addresses spectacular modernity's (semiocapitalism's) crisis of visuality, which causes a lack of imagination, or even blindness. He also posits alternatives: an inhuman vision beyond a humanist-centered view, a post-anthropocentric "other." In contrast to posthumanism, the filmic camera and technology are not what enable vision in these films. Rather, vision is enabled by a mythical being (A Bao A Qu) and by Roxy the dog, which, at the end of *Adieu*, barks in unison with the cry of a newborn baby, announcing the new to come.

NOTES

1. Originally published in Heather Davis and Etienne Turpin (Eds.) *Art in the Anthropocene: Encounters Among Aesthetics, Politics, Environments and Epistemologies*. Open Humanities Press. This version is based on the online paper *Conditions of Visuality Under the Anthropocene and Images of the Anthropocene to Come*, published at *E-Flux* (Journal #63 – March 2015): https://www.e-flux.com/journal/63/60882/conditions-of-visuality-under-the-anthropocene-and-images-of-the-anthropocene-to-come/. Adapted and reproduced with permission from the publisher.
2. Anthology Film Archives is a theater in New York where in the late 1960s and early 1970s filmmakers and artists (Snow amongst them) would gather to watch films. At the time, the theater had wing-like chairs that isolated the viewer sensorially in order to "equate" her field of vision to the screen, thereby delivering a solipsistic experience.
3. I have been unable to locate the author, title, date and location of this evocative mechanical sculpture.

REFERENCES

Baudrillard, J. (1976). Toward a Critique of the Political Economy of the Sign. trans. Carl R. Lovitt and Denise Klopsch. *SubStance*, 5(15), 111–116.

Berardi, F. (2014, December). The Neuroplastic Dilemma: Consciousness and Evolution. *E-Flux Journal 60*. Retrieved from: https://www.e-flux.com/journal/60/61034/the-neuroplastic-dilemma-consciousness-and-evolution/

Berardi, F. (2012). *The Uprising: On Poetry and Finance*. Semiotexte.

Borges, Jorge-Luis (1967). *The Book of Imaginary Beings*. Penguin Group (2006).

Brakhage, S. (1978). From 'Metaphors on Vision'. In P. Adams Sitney (ed.), *The Avant-Garde Film: A Reader of Theory and Criticism*. Anthology Film Archives, article 17.

Bratton, B. (2013). Some Trace Effects of the Post-Anthropocene: On Accelerationist Geopolitical Aesthetics. *E-Flux Journal 46*. Retrieved from: https://www.e-flux.com/journal/46/60076/some-trace-effects-of-the-post-anthropocene-on-accelerationist-geopolitical-aesthetics/

Colebrook, C. and Wolfe, C. [Conference] (January 13, 2013). Is the Anthropocene … A Doomsday Device? In Haus der Kulturen der Welt (Ed.), *The Anthropocene Project: An Opening*, p. 35. Retrieved from: https://www.hkw.de/en/programm/projekte/veranstaltung/p_83894.php

Daney, S., & McMahon, M. (1999). Before and After the Image. *Discourse*, 21(1), 181–190.

Deleuze, G. (1986). *Cinema I: The Movement-Image*. University of Minnesota Press.

Delillo, D. (2002). *White Noise*. Picador.

Didi-Huberman, G. (2007). Picture = Rupture: Visual Experience, Form and Symptom According to Carl Einstein. *Papers on Surrealism*, 7.

Didi-Huberman, G. (2005). The Supposition of the Aura: The Now, the Then, and Modernity. In Andrew Benjamin (Ed.), *Walter Benjamin and History*. Continuum, pp. 3–18.

Fox, D. (2009). *Cold World: The Aesthetics of Dejection and the Politics of Militant Dysphoria*. Zero Books.

Haraway, D. (1988). Situated Knowledges: The Science Question in Feminism and the Privilege of Partial Perspective. *Feminist Studies*, *14*(3), 575–599.

Krauss, R. E. (1999). *Perpetual Inventory*. MIT Press.

McMahon, M. (1997). Beauty: Machinic Repetition in the Age of Art. *Canadian Review of Comparative Literature/Revue Canadienne de Littérature Comparée*, 453–459.

Nixon R. (2011). *Slow Violence and the Environmentalism of the Poor*. Harvard University Press.

Shaviro, S. (2010). Post-Cinematic Affect: On Grace Jones, Boarding Gate and Southland Tales. *Film Philosophy*, *14*(1), 1–102.

Snow, M. (1994). *The Michael Snow Project: The Collected Writings of Michael Snow*. Wilfrid Laurier University Press.

Sontag, S. (1977). *On Photography*. Farrar, Straus and Giroux.

Srnieck, N. and Williams, A. (May 14, 2013). #ACCELERATE MANIFESTO for an Accelerationist Politics. Retrieved from: https://criticallegalthinking.com/2013/05/14/accelerate-manifesto-for-an-accelerationist-politics/

Steyerl, H. (2011, April). In Free Fall: A Thought Experiment on Vertical Perspective. *E-flux journal 24*. Retrieved from: https://www.e-flux.com/journal/24/67860/in-free-fall-a-thought-experiment-on-vertical-perspective/

24. The aesthesis of plastic capitalism
Amanda Boetzkes

INTRODUCTION

As global warming becomes an accepted scientific reality accompanied by a seemingly endless set of ecological challenges, a field of contemporary art focused on waste has come to haunt the global cultural imagination. The emergence of *waste art* is symptomatic of a condition in which the global oil economy has occupied the planet through its production of indestructible materials. The aesthetics of contemporary art offer a materialist perspective of the impact of resource extraction on the planet. Capitalism's plastic waste has become a geological agent that conditions land, water, climate, and the future. As plastic collapses the boundaries between interiority and exteriority, and the dichotomy between human and nature, through art we witness the penetration of the economy's waste into earthly systems. I consider this process the *aesthesis* of plastic, a viruluent topological movement across the planet in patterns of expression and reaction.

I suggest that contemporary art's visions of plastic are indicative of the economy's reliance on a perpetual stockpiling of excess energy, and paradoxically, a *prohibition* to waste. Art visualizes the systemic ingestion of the oil economy's waste and the cultural logic by which the economy is driven by a contradictory impetus to both accumulate and burn oil. The enduring and pliant material of plastic has come to represent an economic regime that prevents energy wasting, even in the form of decomposition, the cycle on which biological regeneration relies.

PLASTIC CAPITALISM AS RESTRICTED ENERGY ECONOMY

Heap, a sculpture by L.A.-based artist Jim Shaw, is composed entirely of McDonaldland toys installed on an armature of metal rods and Styrofoam and assembled into a vaguely human form (Figure 24.1). The seemingly animate, gluey mass is both threatening and yet oddly sympathetic as it extends two closed fists outward to offer the viewer a single dead rose. On the one hand,

Source: Private collection. Image courtesy of the artist and Metro Pictures, New York, NY.

Figure 24.1 Jim Shaw, Heap, 2009. McDonaldland toys, Styrofoam, metal rods, 64 × 24 × 77 IN. (162.6 × 61 × 195.6 CM)

the hoard of toys is visually attractive: bright, shiny, lacquered. Figurines are still discernable amid a clamor of exaggerated facial expressions, pairs of eyes, protruding limbs, accessories, hairbrushes, vehicles, and an airplane propeller. Yet, the work fuses the childhood toys—with all their pleasures, dramas and affects—into a single plastic amalgam. Carefree memories have been absorbed by a monstrous material condition. The romantic gesture of offering a flower has been inverted. As though it possesses a plastic version of the Midas touch, the figure has deadened nature.

Shaw's sculpture is one of many contemporary artworks that signal the permeation of plastic waste into all domains of cultural life. More strongly, it speaks to the persistence of waste as an ecological dilemma preying at the heart of a global economy that has become increasingly dependent on oil since the latter half of the 20th century. The dilemma at stake, I would argue, is a contradictory logic by which energy is rendered for economic exchange through laborious and destructive methods of fossil fuel extraction, biochemical refinement, and warfare, while at the same time, it is destined to be accumulated and preserved as a form of profit, never to be expended and returned into general circulation. The logic of the global oil economy discloses its destructive effects precisely by re-ingesting its own environmental waste, a systemic process that destines the planet to totalizing forms of annihilation such as those articulated by the marker horizons of the Anthropocene: the extinction of species and the sedimentation of toxic carbon wastes.

In order to approach the global oil economy from the perspective of its forms of waste, I invoke George Bataille's distinction between general and restricted economies from his influential mid-20th-century essay, *The Accursed Share* (Bataille, 1991). Bataille speculated that all societies are inherently driven towards acts of "glorious expenditure" in order to burn off their surplus energy. He cites the ecstatic rituals of sacrifice in the Aztec civilization, or the potlatch of the Northwest Coast Native tribes as case studies of sacrificial rituals that unleash excess energy, or what he calls "heterogeneous" energy: a wholly chaotic force that cannot be directed or harnessed. For the duration of the ritual, the release of energy acts like a burst of air after opening a release valve, precipitating an orgiastic destabilization of social hierarchies, and even a radical undoing of the subject. In this sense, the discharge of heterogeneous energy is the condition for a radical ethics: it is the force and imperative by which the subject is violently given over to a relinquishment of the limits of the self (Stoekl, 2007). Acts of copious gift-giving and sacrifice break prohibitions and open these societies to the uncontrollable transfer of energies of the earth and sun, for example the sun's endless expenditure of light and heat, which is absorbed by plants, which are eaten by herbivores, which are eaten by carnivores, which rot and then are eaten by insects which die and return to the soil, and so on, in a violent and endless movement from one organism to the next.

By comparison to the "general economy" of solar societies whose rituals account for the need to periodically burn surplus energy, Bataille argues that the economic system of mid-20th-century bourgeois capitalism is entirely restricted. While it still produces surplus energy, the rule of profit prohibits all forms of energy expenditure. Though there is still a profound need to "burn off" excess, instead energy is re-routed back into the economy. Surplus energy is therefore suppressed, but inevitably discharges in unexpected and highly destructive forms of expenditure. The unprecedented pressures of this "restricted economy," had, he believed resulted in the unexpected explosion of energies, which took the form of two world wars, and the nuclear bomb.

Bataille was trying to account for the correspondences between the rise of capitalism and the extraordinary scale of the world wars, but there are more recent examples as well. We might think of the Deepwater Horizon Oil Spill in the Gulf of Mexico in 2010. Here, the highly competitive race for oil resulted in a massive offshore drilling project, which went horribly wrong when a major leak in the sea floor pipeline caused an explosion that killed eleven people, and then a three-month long gush of hundreds of millions of gallons of crude oil into the Gulf. This was but one of many oil disasters from pipelines and ocean liners. In the case of the Kuwait oil fires in 1991, when Iraqi troops ignited Kuwaiti oil lakes upon their forced retreat causing a ten-month long burn of 6 million barrels a day. The purposeful squandering of oil as an act of violence was geared towards sacrificing profitability. Of particular interest, then, is not simply the catastrophic environmental consequences of the fossil fuel economy, but the status of oil as a privileged form of energy that governs the market and the global political field at large.

It is precisely the oil economy that Allen Stoekl, the English translator of Bataille's writings, addresses in his case for a move towards postsustainability (Stoekl, 2007). He notes that oil is both a scarce resource and yet it is the life force of the world economy. From a Bataillean perspective, the perilous situation in which a society squanders a scarce resource, consistently waging war over it and thus driving towards its own destruction, is the trademark of a restricted economy. More pertinently, Stoekl points out the schematic continuity between economic restriction and environmental sustainability. Sustainability has emerged from an economic system in which energy is preserved in such a way that, ideally, its consumption never outstrips production, and correspondingly, wealth never plummets but grows indefinitely in micro-cycles of market highs and lows. It therefore both mirrors the law of profit (stockpiling energy) and promises to counter what are perceived to be the overindulgences of the global economy, which manifest as excremental pollution such as greenhouse gas emissions. In the face of the recent economic crash, the scarcity of oil and the consequent development of the exploitative industry of the tar sands, all compounded with the expectation of other resource shortages (including clean water, arable land, and rainforest among others), sustainability denounces all excess energy consumption: don't drive, turn off your lights, recycle, turn down the heat and put on a sweater, use less water and so forth. Evidently, the predicament is now a bad duality. On the one hand, environmentalism has come to focus on sustainable practices, with the goal of regulating the output of carbon emissions. On the other, the global market is wholly dependent on the consumption of the same non-renewable resource that creates those emissions, and it has waged an insatiable desire for that resource.

It is important to draw a distinction between consumption as we generally think of it, as simply the act of using up energy—burning gas when driving a car, going shopping, watching a movie, for example—and expenditure, which Bataille associates with an absolute release of heterogeneous energy that transgresses all limits and meaning. Expenditure in this expansive sense simply cannot be reclaimed into a system or recovered as profit in an economy. The visually spectacular plastic waste in art such as Jim Shaw's sculpture *Heap* airs the economic restriction that prohibits waste but paradoxically effects expenditure through the wasting of planetary ecologies. The Anthropocene is therefore the outcome of our failure to expend energy in ways that are ethical and ecologically sound. It is a testament to a paradigm that requires the continual reabsorption of excess energy rather than expenditure of it, which now takes the form of the planetary ingestion of plastic waste.

PLASTIC TOPOLOGY AS THE REINGESTION OF ENERGY

The philosopher Michael Marder argues that the schema at work in capitalism's current energy imaginary can be characterized by the desire to liquidate potential energy from the static, inward dwelling energy of all beings and things (Marder, 2017, p. 56). In its quest to harvest energy and keep it in perpetual expansion through exchange, the oil economy flattens ontological differences and displaces all boundaries. Under the spell of these energy dreams, the conceptual and spatial framing of the very concept of nature has collapsed. Further, the economy has expanded through a topological procession through planetary systems.

It is under this paradigm that plastic becomes both a signifier and an agent of the oil economy. Though the connection between plastic and the oil-based economy is perhaps not self-evident, their histories are intertwined. Plastic, which is derived from petroleum, was invented during the rise of commercial oil drilling in the mid-19th century. It featured in world fairs and international exhibitions alongside the internal combustion engine in the early 20th century. As oil became the primary source of global fuel in the 1950s, the development of new substances such as polystyrene (Styrofoam), polyurethane, polyvinyl and silicone made plastic the iconic material of postwar consumer culture, particularly in North America.

In its original production, plastic was designed to be *disposable*. Yet it has now so permeated planetary ecosystems—whether in atmospheric pollution, microplastics found accumulating in Arctic sea ice, or by inducing genetic defects in animals—that plastic's disposability must be understood as the very impossibility of waste. Plastic cannot be disposed of and is instead an all-encompassing material condition and measure of the Anthropocene. This is why cultural theorist Gay Hawkins argues that plastic must be understood through its patterns of emergent causation (Hawkins, 2013). Rather than analyzing how external structures situate plastic in time and space, instead she assesses how the chemical refinement of plastic co-evolved with planetary systems as well as social relations. Plastic became an increasingly informed material, whose chemical programming was thickened and enriched as it was designed to fulfill the demand to be cheap, available, sanitary, light, durable, and expendable (Hawkins, 2013, p. 55). It is this informational richness that destined it to persist but become increasingly toxic as it acts upon and reacts with oceans, rivers, open-air landfills, in animal bodies, in the air.

Plastic has thus become a vital synthetic agent, co-evolving with the earth's material substrate. Its force as a geocultural agent is therefore fully revealed in the way it exposes the unthought of the economy: the reingestion of energy in plastic's material recursion within the very economy that produced it. In other words, plastic waste has evolved in and through the energy economy's paradoxical vectors of disposability and eternal return as pollution. Plastic waste embodies the economic value of disposability so thoroughly that it began to express the economy itself as a self-consuming system. It not only stands for a new achievement of the capitalist drive that spurred its development, but it has also retroactively and quite uncontrollably, informed that drive as wasting operation. In other words, plastic expresses the self-eradication of the economy in and through the planetary ecologies it has produced.

THE AESTHESIS OF PLASTIC CAPITALISM

To describe plastic waste as a medium is also to suggest that it communicates and expresses. Inasmuch as plastic is an economic agent, it emerges in and through its interpenetration in social relations and subjectivities. It is in this sense that plastic can be thought as an economic *expression*; it was chemically refined in accordance with a market demand for disposability. But this very economic expression effected an ecological trajectory in much the manner of the genetic expression of a DNA code. As plastic has been wasted in waterways, atmosphere and dump sites, it effects chemical reactions that intervene on living beings, drawing them into a common topology.

It is through the economy's transformation of ecological systems into topologies of plastic waste that we see the full fruition of its restrictive logic. What becomes apparent in art is that capitalism expresses itself aesthetically, both in the sense of expression as we commonly think of it (a kind of spatial movement from the inner to the outer, from the economic system to its consumers, and outward to the planet at large as a communication of that very system), but also in and through the way plastic materiality appears in living ecologies. Thus, we can consider the aesthesis of plastic waste as an operation of economic expression and biological reaction as capitalism becomes a planetary condition. For example, plastic waste is mistaken for organic matter and is consumed by animals and microorganisms. Its deceptive appearance enables its passage from an economic system into living ecosystems. But this movement of plastic is also the dark antithesis of expression, pure reaction with no content, and no defined spatio-temporal destination, only informational coordinates and force. As in the case of a chemical reaction, the catalytic movement is not so much intentional and directed as it is encompassing, and substantively irreversible. Whereas an expression might be subject to interpretation—"what does this mean?"—a reaction is held to the basic mechanics of pure stimulus-response. The economy's expression in plastic's chemical reactions is the means by which it materializes in air, water, land and living beings.

Source: Photograph © Chris Jordan.

Figure 24.2   Chris Jordan, Midway: message from the gyre (albatross), 2011

We might consider Chris Jordan's film and photograph series, *Midway-Message from the Gyre* (2011–2017) which documents the effects of plastic accumulation in the North Pacific Ocean (Figure 24.2). Jordan has been documenting albatrosses on the island of Midway through film and photography for over ten years. The problem of plastic pollution in the ocean has come to roost in the albatross ecosystem, as the birds are hopelessly attracted to the colorful objects when they are fishing in the water, ingest them and then die from the obstruction. Jordan's photographs reveal plastic as an uncanny presence that remains still vibrant in the albatross's exposed digestive tract. Listed as an Endangered Species under the U.S. Fish and Wildlife Service, the albatross is one of millions of species that are threatened by anthropogenic change. His film *Albatross* (2017) tracks the trajectory of plastic as it enters the albatross's lifecycle and activates intense struggle, pain and death.

When one looks at Jordan's images of the albatross, it must be understood that the visual scenario is a product of a system in which economy, ecology and sensation are intimately entangled. Such a reflection might easily induce a depressive paralysis in the place of responsibility.[1] Or this paralysis can easily submit to the economic pressures of global capitalism: when confronted with catastrophes there is a well-established reflex to think transactionally, to calculate the cost and imagine an exchangeable equivalent (Davis & Turpin, 2015, p. 5). The reflex discloses that we are already habituated to the slow catastrophe of capitalism, an untenable system that nevertheless becomes the default mechanism as the economy reveals its reingestion of waste. But we might also take art's visualization of the economy's plastic topology as an opportunity to reflect on the intersection of economy and ecology.

Plastic's aesthesis is so wholly destructive that it cannot be contained by either knowledge of economy or science. It marks a complete lapse in meaningful interpretation of living beings and environments. Plastic marks the displacement of life with human economic expression. It persists after the animal body dies and will persist even after the species is extinct. It therefore behooves us to address the force of this displacement. As Jean-Luc Nancy suggests, "We are being exposed to a catastrophe of meaning...Let us remain exposed and let us think about what is happening [*ce qui nous arrive*] to us: Let us think that it is we who are arriving or are leaving" (Nancy, 2015, p. 8).

NOTE

1. For a longer discussion on ecology and depressive paralysis see Boetzkes, A. *Plastic Capitalism: Contemporary Art and the Drive to Waste* (Cambridge, MA: The MIT Press, 2019).

REFERENCES

Bataille G. (1991). *The Accursed Share: An Essay on General Economy. Volume 1 Consumption*, trans. Robert Hurley. Zone Books.

Davis H. and Turpin E. (2015). "Art & Death: Lives Between the Fifth Assessment and the Sixth Extinction," *Art in the Anthropocene: Encounters Among Aesthetics, Politics, Environments and Epistemologies*, pp. 3–30. Open Humanities Press.

Hawkins G. (2013). "Made to be Wasted: PET and Topologies of Disposability". *Accumulation: The Material Politics of Plastic*, Jennifer Gabrys, Gay Hawkins and Mike Michael (eds.), pp. 49–67. Routledge.

Marder M. (2017). *Energy Dreams of Actuality*. Columbia University Press.

Nancy J. L. (2015). *After Fukushima: The Equivalence of Catastrophes*, trans. Charlotte Mandell. Fordham University Press. Cited in Davis and Turpin, *Art in the Anthropocene*, p. 5.

Stoekl A. (2007). *Bataille's Peak: Energy, Religion, and Postsustainability*. University of Minnesota Press.

PART VII

Co-creating futures

25. Democracy in the Anthropocene
David W. Orr

I

In the Spring of 2020, Richard Schiffman quoted Peter de Menocal from the Lamont-Doherty Earth Observatory: "The tragedy and inconvenience we've seen from this pandemic pale in comparison to what's in store from climate change. There is a much bigger crisis knocking on our door" (Schiffman, 2020). Since the plague of Justinian in 541, there have been an estimated eighty global epidemics, all eventually thwarted by geography, distance, and time. The pandemic of 1918 that killed between 50 and 100 million worldwide including 675,000 Americans, however, was a preview of worse to come. The COVID-19 outbreak will not be the last nor likely the worst of those ahead. Even so, governments everywhere have been slow to recognize and respond to the threat despite repeated and authoritative warnings (Barry, 2004; Cartwright, 1972; Crosby, 1972; Garrett, 1994; McNeill, 1976; Rosen, 2007; Spinney, 2017; Zinsser, 1934). Unread, ignored, or forgotten, however, the history of previous epidemics can give neither warning nor instruction.

The same is true, even more emphatically, to the study of Earth systems, resilience, sustainability, and climate. Authoritative reports over many decades show the vulnerabilities of the modern world run like geologic fault lines below the surface of the daily news and exist at the periphery of our public consciousness (Meadows et al., 2013; Steffen et al., 2006). History, however, is an active seismic zone in which something is always about to slip, break out, erupt, ignite, erode, or collapse when that last straw breaks the proverbial camel's back. Nature, it is said, sets booby traps for unwary species (Sinsheimer, 1978). When the trap springs shut, however, disaster seldom travels alone. Even the four horsemen of the apocalypse (pestilence, famine, war, and death) are latecomers arriving like vultures to pick the bones clean only after others with names like climate chaos, genocide, racism, species extinction, deforestation, ocean acidification, toxic pollution, economic crises, technology run-amuck, terrorism, human fallibility, stupidity, greed, imperial overreach, and ecological overshoot—a veritable cavalry of woe—have done their work.

Our situation, at best, is precarious which makes our collective indifference to it curiously unnerving. Some dangers like soil erosion move by stealth toward critical thresholds. Others strike without warning. Some risks are statistically probable, what Yale sociologist Charles Perrow calls "normal accidents" (1984). The deployment of nuclear power plants, for example, has a statistical probability of catastrophic failure as a function of variables like the number of plants, quality of engineering, years in operation, level of management skill, operator error, human malice, and the devils lurking in the details like metal fatigue, embrittlement, broken pipes, and equipment failures. Events with low or unknown probability but high consequences, aka "black swans," are not so easily predictable. Hindsight, however, often reveals astonishing carelessness. The Fukushima disaster, for one, resulted from a decision to cluster four poorly designed nuclear power plants on the edge of the Pacific Ocean in an active earthquake zone with a history of tsunamis. What could possibly go wrong? These and many similar risks, nonetheless, are essential parts of the American economy, military policy, and the supporting technological apparatus resting on a rickety scaffolding assembled from rusty ideas built on a shaky foundation of technological hubris. What could possibly go wrong?[1] (Ellsberg, 2017; Perry, 2015).

The case for foresight, precaution, and design intelligence begins with the humility to know that trouble is always close at hand. When it strikes, havoc, suffering, and loss are one side of the ledger. The other side, however, can be heightened awareness and stronger commitment to avoid what can be avoided with foresight and planning, to fix what's broken, and improve the resilience of the system often by changing the system itself.

In this regard, the COVID-19 pandemic is rather like a sudden bolt of lightning on a dark night that for a brief moment illuminates the fractured landscape of our politics, economy, and public institutions. It is also a preview of what will happen in a world spiraling down into climate chaos. There is every reason of prudence and self-interest to see it as an opportunity to rethink, redesign, and build a more resilient, durable, fair, generous, and secure nation and world. There may not be another such crisis that is small enough to permit recovery, but large enough to get our full attention. What then can we learn from our current crisis? Among the possible answers these stand out:

1. Alert and competent governments could have headed off the worst of the crisis and plausibly saved hundreds of thousands of lives.
2. In the case of the United States, the multiple failures of the government to anticipate the pandemic and take early steps to head off its worst effects is a consequence of a long-standing deep antipathy to competent, active government and indifference to the public interest that is baked into the philosophies of Ayn Rand, "neo-liberal" economics, white supremacists,

hate spouting media commentators, and gun-toting libertarians. Birds of a feather.
3. At the global level, the forty-year war against democratic governments waged in the cause of free-market economics has taken a large toll on government's capacity to act effectively in any emergency or solve significant public problems. Ideas do have consequences and the upshot of bad ideas that excuse racism, violence, ecological exploitation, injustice, misogyny, inequality, and dereliction of duty is that people suffer and die unnecessarily when public capabilities such as foresight, prevention, emergency services, and compassion wither in the arid heat of a reckless philosophy that is conservative in name only.
4. Global supply chains and just-in-time production systems that eviscerate local and regional skills also devastate local economies. Otherwise, they work until they don't, and when they don't there is Hell to pay. They are, incidentally, very efficient channels by which to distribute lethal viruses worldwide.
5. In pandemics and most other emergencies, the poor, the elderly, and children are much more vulnerable than the well to do.
6. Corporations are very efficient creators of wealth, but unregulated have neither the wherewithal nor the incentive *on their own* to consistently act in the public interest. They have, instead, every reason to appear as if doing so which partly explains their large advertising expenditures.
7. It is no great revelation to point out that societal traumas like pandemics and war bring out the best and worst in human nature, but it bears repeating, nevertheless. Lurking in the shadows, always, are the grifters, profiteers, cheats, con-artists, snake-oil salesmen, and inept and corrupt leaders, who regard human suffering as an opportunity to cash in.

Most important, would-be dictators and authoritarians use disasters to expand their power, corrupt the law and the courts, and defile democratic institutions (Beauchamp, 2020; Klein, 2007).

In both the COVID-19 pandemic and climate change, what appears to be inexplicable, random, and unpredictable, is only the remorseless working out of the rules of a subsystem (a badly designed economy operating without moral instruction) that conflict with those of the four-billion-year-old ecosphere of which it is a part. Uncorrected, it will prove to be a lethal miscalibration. Consequences, in other words, have causes rooted in the ideas, however mistaken, hair-brained, or noble, for which we live and die. For better or worse, those same ideas form the paradigms and models for economies, global supply chains, the banking industry, political systems, public institutions, consumer-oriented societies, and systems of technologies with effects that no one truly understands. Globalization, for example, designed for maximum economic growth and minimal public oversight has performed miracles on one hand while creating pandemics, massive inequality, deforestation, pollution, acidification, widespread obesity *and* starvation, oceanic dead zones, and climate chaos on the other. The latter are no more accidental than the former. Given the rules of the system, time, and circumstance, both are predictable outcomes.

Though it will leave deep emotional wounds, political scars, and a damaged economy, we will outlast the current pandemic. In time, vaccines will protect survivors and the world will recover economically, not as before but prosperous enough. Disruptive as the cumulative effects of COVID-19 will be, however, they will be orders of magnitude less than those of a destabilizing climate. Even in the best scenarios, the history yet to be written will feature "cascading systemic failures" caused by record heat waves, massive fires, super storms, rising seas, loss of coastal cities, species extinctions, and eventually famines, wars, political upheaval, social breakdown, and economic collapse (Diamond, 2005; Steffen et al., 2018; Tainter, 1988). One thing will cause another and yet another rippling outward in an avalanche of upheaval. We will struggle to find a secure foothold, solid ground, and solidarity with others similarly caught in the maelstrom. Because of the longevity of heat-trapping gases in the atmosphere, much of this is already unavoidable over the next century and beyond—but not all, not yet.

II

As the temperature of the Earth continues to rise, beyond some unknown threshold our various systems of governance will suffer the equivalent of cardiac arrest. Rising heat and its uncountable collateral effects will overwhelm the systems by which societies preserve public order, maintain viable economies, and supply the public with food, energy, clean water, and security. Many functions of government and corporations will deteriorate and then cease altogether. Designed for different challenges, Western democracies and legal systems are unsuited to those posed by complex systems of biology, ecology, epidemiology, atmosphere, and the infinite ways they affect humans. The present pandemic and rapidly de-stabilizing climate ahead, however, will cause us to reckon with a more recalcitrant and complex reality that punishes self-deception, overreach, and governmental incompetence.

Rising seas will force millions of people to relocate to higher ground. Many coastal cities will become uninhabitable. Super storms will wreak havoc with infrastructure including the electric grid and municipal water systems. Lethal heat waves will decimate urban populations and wither crops. Tropical diseases will move north into populated areas. When, not if, disasters occur simultaneously, pandemics, nuclear meltdowns, category 5 storms, blackouts, terrorist attacks, and the like will overwhelm emergency response agencies everywhere. Hundreds of millions of people will flee the heat, drought, floods, and storms. Beyond some threshold, the fabric of societies will come undone (Diamond, 2005; Tainter, 1988; Wright, 2005).

In other words, we are in not a crisis but in a crisis of crises and not a single emergency but a "long emergency" that will extend far into the future. No patchwork reforms will be adequate to the scale and duration of the problems. No change in consumer behavior or market response alone will suffice unless it is part of larger changes in the structures of politics, law, governance, education, and economics as well as the hardscape of industrial societies: ports, utility grids, feedlots, water systems, refineries, agribusiness, railways, roads, shopping malls, and housing. We must learn to think and act as parts of an interconnected whole, what Donella Meadows called "thinking in systems."

The COVID-19 pandemic revealed the deep flaws in the political economy of "developed" nations. It showed the necessity for effective, competent, transparent, and agile governments capable of defending the health of people and nature. It revealed deep cracks in the corporate-dominated economy hidden beneath a veneer of unsustainable and grossly unequal prosperity. It revealed the importance of science and an uncompromising regard for truth. Patched up governments will prove no more adequate to remedy systemic problems than a patched-up Boeing 737 could fly safely as the 737 MAX. The looming climate disaster calls for a thorough redesign of political systems and economies to harmonize with the way the Earth works as a non-linear physical system with leads, lags, stocks, flows, and complex feedback loops. Neither time nor our political habits are on our side.

We are in a race and running a lap behind. The climate emergency will require a more imaginative, larger, faster, and more coordinated worldwide response to move beyond fossil fuels to renewable energy while rebuilding sustainable and just economies. These will require, among other things, competent and inspiring leadership. Broken and corrupted governments must be reconstituted and repurposed around a more unified agenda. But governments alone cannot succeed without delegating much of the work of transition throughout society and harnessing the creativity and energies of their populations. But to what end and how?

Answers, I think, are not as daunting as it may first appear. In the past half-century, a revolution in the integrated design of agriculture, energy, architecture, waste cycling, cities, economics, and transportation has been quietly building momentum, mostly without the help of governments. Instead, it has come from widely dispersed ingenuity, creativity, and public spirit. The result is the art and science of integral design emphasizing systems solutions to systemic problems of climate change, health, farming, land management, energy systems, building, and urban design.[2] In every case, the point is to solve multiple problems without causing new ones. Ecological design competence, already flourishing at local and regional levels, can reduce the need for government oversight and regulation, reduce or eliminate problems, reduce or eliminate pollution, develop cost-saving synergies between otherwise separate systems, and improve resilience by making localized solutions that are fast, cheap, repairable, replicable, redundant, and build local prosperity. Green urban design, for example, improves property values, enhances quality of life, tempers climate extremes, grows food, lowers pollution and greenhouse gases, provides parks and greenways, creates walkable communities and bikeways that eliminate cars thereby improving health, lowers blood pressure, lowers obesity levels, and gets kids outdoors (Beatley, 2020; Fairchild, 2020; Lovins & Lovins, 1982, Ch. 13). Whether the design of cities, or manufacturing, or agriculture, the benchmark is human health in its fullest sense. All of this will appear to future generations as merely obvious responses to existential crises. But why has it been so hard to do?

III

I do not know whether we are in the Anthropocene or the late stages of the Holocene or some other. But I do know that humankind is in a mess of its own making and that our words, concepts, philosophies, theories, and professions have not helped us to anticipate and forestall the crisis. Perhaps, as historian Ronald Wright writes, our problems result not from failure, but because we succeeded all too well. We have caught ourselves in a "progress trap." Bewitched by our own cleverness: "our practical faith in progress has ramified and hardened into an ideology .... But progress has an internal logic that can lead beyond reason to catastrophe. A seductive trail of successes may end in a trap" (Wright, 2005, pp. 4–5).[3]

Progress traps result from our lack of foresight and our inability to know when enough is enough. Cities can be great, sprawling megalopolises not so much. Legitimate concern for security brings forth weapons but nuclear weapons are a form of insanity. We have a gargantuan economy but have turned Earth into what Pope Frances calls "a pile of filth." Worldwide, militaries and military thinking predominate but we are less secure than ever. We control nature in ways unimaginable even a few decades ago, but we are devastated by a mere virus. We are wealthy beyond measure but are not necessarily happier or better for it. We dominate nature but the Earth is stormier and more capricious than ever. We confused the amount of stuff we consumed (a word that once meant "disease") and economic growth with human betterment. The ironies stack up like cordwood.

Economic growth is perhaps the ultimate progress trap. The economy is a subsystem of the Earth and the idea that it can grow indefinitely is, well, madness.[4] The fact is that we are already in overshoot, taking more resources, land, and biological productivity than Earth can regenerate. The challenge, then, is not to grow the economy until it collapses but to grow the wisdom and heart necessary to share what we have fairly within and between the nations of Earth and between current and future generations. The problem, in other words, is primarily moral and political, not purely economic.

Freeing ourselves begins with seeing the trap for what it is. Our indicators of economic growth, security, and technological progress are the stuff of epitaphs. We confused the volume of stuff with wealth, velocity with progress, power with national goodness, weapons with safety, and inequality as a measure of social worth. In truth, we were offloading a great deal of our ecological and moral debt onto others somewhere else or at some later time.[5]

The present pandemic and a destabilizing climate, in Wright's view are predictable results of too much progress, too little foresight, too little heart. The hard truth is that we will not technologize or grow our way to safe harbor. We will have to do it the hard way by redesigning households, neighborhoods, towns, cities and regions for sunshine-powered, locally grounded resilience and this requires a political revolution.

For many reasons, authoritarian governments cannot harness or sustain the creativity, flexibility, ingenuity, imagination, or public support necessary to avert global catastrophe. The far better course is a further expansion of democracy; to develop, equip, and support an ecologically and civically competent citizenry. To that end, it is time to build national and international forums for public deliberation, not siloes for ideologues. It is time to build and support competent public agencies doing the public business. It is long past time to eliminate all outside money for all political campaigns and banish purchased politicians, authoritarians, and ideologues once-and-for-all. It is time to guarantee the right to vote to everyone and mean it.

Democracy, however, comes at a price that balances privileges with the responsibilities of engaged citizenship. In political theorist Sheldon Wolin's words:

> The survival and flourishing of democracy depends ... upon the peoples changing themselves, sloughing off their political passivity and, instead, acquiring some of the characteristics of a demos. That means creating themselves, coming-into-being by virtue of their own actions ... democratization of politics remains merely formal without the democratization of the self. (Wolin, 2017, p. 289)

Growing evidence shows that all people, in the right settings, can deliberate thoughtfully, rationally, and civilly.[6] Deliberative democracy that brings people together across the political spectrum "promotes considered judgment and counteracts populism." In other words, we are not fated to futile and endless political gridlock. We've been riding in a horse and buggy democracy for too long; it's time to upgrade by deploying our creative talents to building new and robust institutions for democratic deliberation.

A final thought. The original democratic vision was not about getting rich but having the freedom to live life to the fullest in a democratic community. It was about justice, equality, and freedom balanced by the responsibilities that go with membership in a civil, civic community. That dream survives and urgently needs to be enlarged and energized. In other words, our only real choice is "coming to our senses" in Vaclav Havel's words "as humanity, as people, as conscious beings with spirit, mind, and a sense of responsibility" (Havel, 1992, p. 116).

NOTES

1. Well, accidental or intentional nuclear war for one.
2. See for example, Stuart Walker and Jacques Giard, *The Handbook of Design for Sustainability*. London: Bloomsbury, 2013.
3. Wright continues: "but we've reached a stage where we must bring the experiment under rational control ... if we fail—if we blow up or degrade the biosphere so it can no longer sustain us—nature will merely shrug and conclude that letting apes run the laboratory was fun for a while but in the end a bad idea" p. 31.
4. Kenneth Boulding, an esteemed economist and a former president of the American Economic Association, once said that "Anyone who believes that exponential growth can go on forever in a finite world is either a madman or an economist."
5. "The rule and root of all economy," John Ruskin wrote, "that what one person has, another cannot have; and that every atom of substance, of whatever kind, used or consumed, is so much human life spent ... the world is not to be cheated of a grain, not so much as a breath of its air can be drawn surreptitiously. For every piece of wise work done, so much life is granted; for every piece of foolish work, nothing; for every piece of wicked work, so much death." John Ruskin, *Unto This Last*. London: Everyman's Library, 1862/1968, pp. 192, 202.
6. John Dryzek et al., "The Crisis of Democracy and the Science of Deliberation," *Science*, March 15, 2019, pp 1144–1146. (Vol, 363 Issue 6432); Ilya Somin's perspective to the contrary in *Democracy and Political Ignorance*, 2nd edition. Stanford: Stanford University Press, 2016, pp. 58–62.

REFERENCES

Barry, J. (2004). *The Great Influenza*. Penguin.
Beauchamp, Z. (2020). "Hungary's 'Coronavirus Coup Explained'" *Vox*, April 16. https://www.vox.com/policy-and-politics/2020/4/15/21193960/coronavirus-covid-19-hungary-orban-trump-populism
Beatley T. (2020). Biophilia and Direct Democracy, in D. W. Orr, A. Gumbel, B. Kitwana, and W. Becker (eds) *Democracy Unchained*, 375–382. The New Press.
Cartwright, F. (1972). *Disease and History*. Signet.
Crosby Jr., A. (1972). *The Columbian Exchange*. Greenwood Press.
Diamond, J. (2005). *Collapse*. Viking.

Ellsberg, D. (2017). *The Doomsday Machine*. Bloomsbury.
Fairchild D. (2020). Powering Democracy Through Clean Energy, in D. W. Orr, A. Gumbel, B. Kitwana, and W. Becker (eds), *Democracy Unchained*, 263–272. The New Press.
Garrett, L. (1994). *The Coming Plague*. Farrar, Straus and Giroux.
Havel, V. (1992). *Summer Meditations*. Knopf.
Klein, N. (2007). *The Shock Doctrine*. Metropolitan Books.
Lovins, A. B. and Lovins, L. H. (1982). *Brittle Power*. Brickhouse Publishing.
McNeill, W. (1976). *Plagues and People*. Doubleday.
Meadows, D. H., Randers, J., & Meadows, D. L. (2013). *The Limits to Growth* (1972) (pp. 101–116). Yale University Press.
Perry, W. (2015). *My Journey at the Nuclear Brink*. Stanford University Press.
Perrow, C. (1984). *Normal Accidents*. Basic Books.
Rosen, W. (2007). *Justinian's Flea*. Penguin.
Schiffman, R. (2020, April 24) "The Lab that Discovered that Global Warming has Good News and Bad News," *The New York Times*. https://www.nytimes.com/2020/04/24/nyregion/lamont-doherty-earth-observatory-global-warming.html
Sinsheimer, R. (1978). The Presumptions of Science, *Daedalus 107*(2), 23–35.
Spinney, L. (2017). *Pale Rider*. Public Affairs.
Steffen, W., Rockström, J., Richardson, K., Lenton, T. M., Folke, C., Liverman, D., Summerhayes, C. P. et al. (2018). Trajectories of the Earth System in the Anthropocene. *Proceedings of the National Academy of Sciences 115*(33), 8252–8259.
Steffen, W., Sanderson, R. A., Tyson, P. D., Jäger, J., Matson, P. A., Moore III, B., … & Wasson, R. J. (2006). *Global change and the earth system: a planet under pressure*. Springer Science & Business Media.
Tainter, J. (1988). *The Collapse of Complex Societies*. Cambridge University Press.
Wolin, S. S. (2017). *Democracy Incorporated: Managed Democracy and the Specter of Inverted Totalitarianism* – New Edition. Princeton University Press.
Wright R. (2005). *A Short History of Progress*. Carroll & Graf.
Zinsser, H. (1934). *Rats, Lice and History*. Little Brown.

26. Envisioning scenarios for the Anthropocene
David Arthur Sampson

INTRODUCTION

What are Scenarios?

I suspect that, for most individuals, the noun "scenario" invokes a specific meaning depending on your perspective (and profession) but, in its simplest form, a scenario can be considered a postulated sequence in the development of events (Oxford dictionary https://www.lexico.com/definition/scenario). From an actionable perspective a scenario, then, may be a designed sequence or series of actions taken to explore the temporal trajectory of a system (towards a final outcome) to changes in the driver variables that impact system processes or function. Or, it could be a series of actions postulated to be needed to produce a temporal sequence of development in order to achieve a desired outcome, future, or set of futures. In the first case, the process derives from either inherent uncertainty in the response of the system to known factors that impact it, or from a lack of knowledge of the important variables that "should" be evaluated. In the latter, the process derives from known actions that "could" be taken over time to achieve the future or futures desired. Hopkins and Zapata (2007, pp. 9, 10) simply suggest that a scenario represents a possible future and that it "emphasizes a process of change, not just a point in the future." In the context of the Anthropocene scenarios would, then, be an assessment of one form or another of the impact that humans have had on a system or systems process, form, or function.

Scenarios can take on many forms, but they always exhibit a focus on the future. This focus typically revolves around storylines that are created or envisioned to explore potential, future, outcomes from a series of desired or expected changes in the structure or function of a system or systems. Essentially, a story of functional behavior (Alexander and Maiden 2004). These storylines are conceptual representations of the trajectory of policies, or practices needed (or modeled) to achieve a desired, expected, or yet unknown future. In the creation of a scenario these storylines may arise from "visioning" exercises that empower experts to brainstorm plausible (but sometimes unlikely) pathways needed to meet the goals of the exercise (e.g., Iwaniec et al. 2014; Iwaniec et al. 2020). Scenario storylines may also arise from simple, postulated changes in the important driver variables of a system to explore outcomes from variation in known, but uncertain, variables that impact a system (e.g., Gober et al. 2016; Sampson et al. 2016; Sampson et al. 2020). Alternatively, the storylines may be simple, practical constructs from a planning perspective created to meet current and future infrastructure needs of a community as influenced by, for example, expected population growth. Notwithstanding, the storyline remains foundational to any forecasting exercise.

Why we use Scenarios

From a planning perspective, scenarios are used to help city managers "plan" for future growth. Hopkins and Zapata (2007) suggest that urban (and regional) scenario planning exists on a continuum between normative and exploratory scenario planning. Holway et al. (2012, p. 10) give voice to the Hopkins and Zapata (2007) continuum construct, stating that "normative scenario planning is used to articulate the values of a community or region by eliciting people's opinions about different possible visions of the future." In this type of planning the scenarios are frequently applied to major policy decisions that could have a substantial impact on the future form of the community, such as funding and locating transportation infrastructure or changing land use regulations (Holway et al. 2012). Exploratory scenario planning, then, is used to anticipate the impact that different future conditions may have on values, policies, or goals that have been established or are being considered (Weber 2006). The desired end result of such a process is a set of robust or contingent strategies that policy makers can use to achieve agreed upon goals under a wide variety of possible but uncertain futures; there is a need "… to move toward developing and implementing planning tools and processes that foster anticipation and adaptation" (Holway et al. 2012, p. 2). When scenarios are used to anticipate the future, they become proactive instruments for planning rather than reactive responses to perceptions and, so, offer a chance to mold the future desired.

In the exploratory realm practitioners generate forecasts of expected changes in known variables that impact a system or process of interest. This could be population growth and, thus, water demand. Or, perhaps expected population growth and probable changes in zoning and the transportation infrastructure needed to accommodate growth. Quay (2010) mentions that the early scenario planning literature recognized the potential limitations of forecasting and even the need to address uncertainty in scenarios that characterize the future (e.g., Berger 1964; Kahn 1965; Kahn and Wiener 1967; and later

Ringland 1998, 2002, 2006; Schwartz 1991; and van der Helm 2007). Starting around the turn of the century Lempert and coauthors suggested that when there is high uncertainty, a broad range of possible futures must be considered (Lempert et al. 2003; Lempert and Schlesinger 2000); he and others proposed, and used, advanced scenario planning methods that considered hundreds of scenarios (Ahmed et al. 2003; Bankes 1993; Bankes et al. 2003; Burke and Ewan 1999). Of course, decision makers do not want to consider hundreds of options when planning for the future; a reduced set of plausible scenarios are more likely to influence policy and decision making. This may come from aggregated averages (Quay 2010) but risk assessment, sensitivity analysis of factors or decisions driving the scenarios, identification of unacceptable and worst-case scenarios, and assessment of common and different impacts among the scenarios are often considered (Quay 2010).

Scenario planning, in the context of urban planning for growth, may also include physical models used as a tool to bring a reality check to the planning process. The Lincoln Institute, as noted in Holway et al. (2012) has used physical models, in this case Lego™ bricks in workshop settings, to help planners visualize the impact that specific planning scenarios would have on the built environment. Large maps of the planning area under consideration for development are placed across several tables; participants use the bricks to site new development of various product types and their density. This process provides a visual representation of what the urban landscape could look like, making the conceptual ideas about growth and density very real. In the process three concepts are considered to be critical to the scenario planning and tool-building exercise: collaboration, capacity building, and creation of an open environment for engagement (Holway et al. 2012). This process helps to ensure that the design criteria and the policies and goals needed to achieve the expected or envisioned future create the greatest chance of success.

From an urban sustainability perspective, scenarios are a tool to develop plausible and coherent visions about a sustainable future (Iwaniec et al. 2020). In this framing the tool could be a specific methodological approach to design a future (or futures) based on collective anticipatory knowledge through collaboration of experts and stakeholders. In this case academics and practitioners (and often those likely impacted) gather, often over a period of years, in a workshop setting to create the structure and elements of processes and events through time needed to reach a designed set of futures. These visions bring together a variety of perspectives with the central goal to develop consensus among the participants using strategies that include collaboration, capacity building, and an honest open environment to draft co-developed pathways to an envisioned future (Iwaniec and Wiek 2014). Typically, a back casting approach, small groups are formed and given a specific focus with the instructions to brainstorm potential pathways needed for meeting a desired future given the subgroup goals. They derive policies and a timeline through iterative sessions, and, over time, they prioritize their policy choices using design sketches—scenario vignettes—and quantitative metrics to outline the pathways needed to achieve a desired future (e.g., Iwaniec et al. 2020). The result is a concrete set of policy decisions and the planning and governance needed in an attempt to meet these goals. This approach may incorporate adaptation strategies and co-developed systems modeling to highlight probable scenario outcomes (e.g., Iwaniec et al. 2014; Sampson et al. 2020); one example may be found by pointing your browser to https://sustainability.asu.edu/future-scenarios/.

Another approach from an urban sustainability perspective uses numerical simulation models to "create" a suite of potential futures using current knowledge about the system processes of interest and their impactful variables. In this instance the modelers have either created a suitable model for the system or have parameterized a proprietary model deemed suitable for the research questions at hand. Following parameterization and validation the researchers create the storylines that they wish to pursue to ask their questions. Often, the approach would be to use an application programmer's interface (API) that allows a modeler to run hundreds or thousands of scenarios by incrementally changing the driving variables, one at a time, to explore interactions and their impact on the system (e.g., Sampson et al. 2016; Sampson et al. 2020). This approach creates a "cone" of uncertainty in the temporal trajectories and the response of the system to changes in the inputs. Certainly, even when a researcher knows the variables that are important for a system there remains much uncertainty in the outcomes that changes may present. Typically, incremental changes to a reference scenario are used to define the ensuing ensemble of scenarios; the reference scenario maintains the most probable trajectory (at the time) but may not ultimately be the likely trajectory with hindsight or greater scrutiny (i.e., Burgess et al. 2020; Hausfather and Peters 2020).

Often, a researcher (or research team) may have a fundamental understanding of the driving variables of a system but, rather than examining incremental influences of variables on a system they want to study specific influences, perhaps suggested by current trends. In this case separate scenarios that portray a specific trajectory in a variable or variables of interest may be used to evaluate empirical trends and, thus, the overall response of the system (e.g., Bureau of Reclamation 2012; Guan et al. 2019; Sampson et al. 2006; Sampson et al. 2007; Sampson et al. 2008). In some cases, scenarios may represent specific policy choices, examined to explore the separate and interacting influence of policy on system response (Gober et al. 2016; Withycombe Keeler et al. 2015). The goal, here, is to discover which policies and in what combinations would lead to the desired future (or a more desirable future). These analyses lead to a greater understanding of the specific policies that could be implemented, if politically and economically feasible, to achieve, say, a more sustainable future.

SCENARIOS FOR THE ANTHROPOCENE

Far too few know or acknowledge that human activities are directly and indirectly altering local, regional, and global environments resulting in dramatic changes to earth systems. Most of these changes are exacerbated by feedback (and feedforward) loops that accelerate the rate of change; global climate change represents one dominant focal challenge (e.g., clear felling and burning in the mid latitudinal tropical rainforests that impacts local and global precipitation patterns). If scenarios are to be of use for the challenges faced in the Anthropocene, they need to address both scale and context. For instance, local scenarios using cross-scale formulations and analyses are needed but they must be replicated across regions (and perhaps continents) so that they can provide context for pattern detection at global scales. Global scenarios need to be downscalable to regional feedbacks; "they need to consider the overall scale of ecological feedbacks, rather than continuing to categorize all ecosystem feedbacks as local" (Cumming and Peterson 2005, p. 50, box 3.2). Clearly, one central challenge moving forward is to accurately place system process and function within the context of scale with the inherent interconnections but, through collaboration, explore to detect (if present) emergent regional and global scale phenomena; if you will, a new look at the butterfly effect paradigm. That is, greater focus on studying weak signals (in context) as a means to understand potential patterns otherwise missed (discussed further below).

What Should be Known, and By Whom?

What needs to be known, and conveyed so that all understand, is that living systems operate in both simple and complex ways, and that complex systems can be understood (or at least mimicked). When first principals are known, robust numerical simulations of complex processes are possible (e.g., Farquhar et al. 1980; Landsberg and Waring 1997; Parton et al. 1988). Of course, arriving at first principals remains a challenge. Understanding first principals, however, and then creating algorithms to mimic them, illustrates near perfect understanding of a system. What also needs to be known is that perturbations to living systems often result in hidden consequences that go unnoticed and, when pushed far enough, cascading system failure ensues. I suspect that system failures are already underway; several phenomena, like the loss of honeybees (colony collapse disorder), bats (white-nose syndrome), and amphibians (the chytrid fungus *Batrachochytrium dendrobatidis*), and glacial ice worldwide, to name a few, may be harbingers of larger failures yet to be seen.

Now, some have been using the term "tipping" point to clarify biophysical positions on earth system function (e.g., Grainger 2017) or to highlight anthropogenic influences on the global climate system (e.g., Lovejoy and Nobre 2018; Winton 2006). First coined by Everett Rogers in 1962 in reference to the adoption of technologies, a tipping point may or may not be an inflection point that results in catastrophic loss of system stability (how we often think of them). It behooves scientists to remain cognitively indifferent so that (hopefully) the underlying principals marking a tipping point can be understood (or at minimum followed). While Al Gore and Malcolm Gladwell may have made the phrase "tipping points" a household word, the biophysical and ecophysiological reality is that we generally don't know what we don't know about the physical (and chemical) processes maintaining ecosystem homeostasis. Nor do we fully understand how ecosystems are connected across space and time and how suspected tipping points illustrate (or reflect) the actual, underlying processes that nominally keep systems intact.

Tipping points suggest an emerging issue; weak signals may serve as environmental (short-term) and horizon (long-term) scanning mechanisms for an emergent signal. Weak signals have been described as a valuable tool for use when anticipating future changes (Hiltunen 2008). Early work on weak signals was pioneered by Aguilar (1967) and Ansoff (1975) who explored adaptive management strategies for corporate organizations. Their work focused on managing "strategic surprise" by sensitive attention to weak signals. "Futurists" working in the corporate world, and those working in the social and biophysical sciences think of weak signals as indicators or, for some, "seemingly unimportant or unexceptional trends" that can have considerable impact on a system, but that must be interpreted and, after "correct" interpretation, become an early warning signal (Lesca and Lesca 2011). Cevolini (2016) argues that weak signals "should be" explained as the outcome of self-referential dynamics that finally leads to the paradox of knowing the unknown. I suggest that weak signals are not an indicator or early warning signal, per se, but rather they are the temporal embodiment of, or the continuity in, threaded cross-scale system processes as organized around the potential discontinuity of the system state. Anticipatory scenario analyses serve as one mechanism for identifying meaning in weak signals.

How Can This be Achieved?

Anticipatory scenarios are designed to be adaptive; to anticipate likely, but uncertain trajectories in the causal mechanisms through time that drive systems and, in the process, use statistical tools to look for patterns and to reduce complexity. One challenge of adaptation is to be prepared for an uncertain future; old approaches that would forecast, choose a desired trajectory, and then plan for it have been relegated as passé. New methods for planning under high uncertainty that: (1) explore a range of future possibilities, (2) help us think about these futures strategically and, (3) in real time, use these strategies to guide action as targeted events unfold are needed. One method, anticipatory scenario analysis, "recognizes that some aspects of the future are not knowable and that any forecast represents only one of many possible futures" (Quay 2010, p. 498). Under this approach hundreds, thousands, or tens of thousands of scenarios can be considered. Anticipatory scenario analyses, then, use statistical methods such as cluster analyses or principal component analyses to look for patterns in a response. Successful pattern detection can create strategic insight into the key variables and underlying mechanisms to address the questions at hand for the system of interest.

Let's look at one example. A colleague and I have a manuscript in review that focuses on the generally accepted notion that a metropolitan water utility could reduce overall water demand by promoting greater residential housing density. We created an agent-based optimization model for a large Western water utility to explore the interplay among (1) population ingrowth, (2) product type, for example, large single-family (SF) homes (LSF), typical SF homes (TSF), small SF homes (SSF), medium density multi-family housing (MMF), and so on, (3) housing density, and (4) irrigation efficiency on outdoor water demand. Briefly, we modeled the movement of new residents into building types of greater density than their originating type (i.e., the building type that they moved from). We defined short movements and long movements; a short movement would represent a shift from, say, a LSF into a TSF or a SSF while a long movement could be a shift from a LSF into MMF. A cluster analyses procedure (fastclus with k means in SAS™) of 5,040 scenarios detected four clusters in the response between residential density and outdoor water demand. The patterns reflected differences in the length of the movement of families into higher density building types with respect to their originating building type. These clusters yielded insight into the dynamics of product type, housing patterns and household movement, and residential density on water demand at the water utility scale. Anticipatory scenario analyses, in this case, provided one approach to address complex challenges that exhibit high uncertainty.

Another approach might use foresight scenarios to examine weak signals and emerging trends. Of course, one dilemma that Lesca and Lesca (2011) note is the accurate interpretation of a weak signal; organizational receptiveness remains one challenge of correctly interpreting weak signals (Rowe et al. 2017). In addition, either over-analyzing trivial "findings" or under-analyzing important ones remains another weak signal challenge (Schoemaker et al. 2013). So, how do we observe a weak signal? Or where do we find them? Hiltunen (2008) outlines a number of sources for weak signals from across multiple fields of study. For me, weak signals often present themselves through data mining and graphical exploration. That is, the repeated analyses of a data set viewed from different angles to ask different questions, and the use of sometimes unorthodox graphics, often yields insight.

Emergent phenomena exhibit weak signals. Emergent properties have a propensity to self-organize in ways that cannot be predicted from knowing the individual components; low-level rules of emergent properties give rise to higher-level complexity exhibiting new properties and behaviors that "emerge" without direction, with characteristics that are not discernable from low-level components. Although not predictable, they can leave evidence that allows the history of an emergent system to be reconstructed and quantitatively understood (Stern and Gerya 2020). Scenarios to explore emerging issues are challenging to create. One approach would be to focus on the apparent trajectory of the evidence, using surrogates for the underlying mechanisms. Scenarios created through the horizon scanning process, along with Intuitive Logics, may offer insight into emerging trends to advance foresight activities, to aid in strategy creation, and to enhance decision making (Rowe et al. 2017).

Once a future state has been envisioned, an analysis of the sustainability and/or resilience of the system would be ideal. Of course, this approach would require an independent analysis, separate from the processes and procedures used in the creation of the scenarios. Seldom, if ever attempted due to data and monetary constraints, this would be one approach to critically evaluate whether the envisioned future, and the policies and procedures created to reach that future, could come to fruition.

Regardless of the approach, scenarios for the Anthropocene will need to address five questions: (1) do we have sufficient, verifiable, knowledge of the system to even create plausible scenarios? If so, how far out into the future should we forecast? If not, what do we need to know to be able to move forward, (2) borrowing from atmospheric sciences, does the reference scenario exhibit sufficient known "teleconnections" among the scale-dependent processes to withstand random drift? (3) for those complex processes that we cannot seem to model accurately will a simple surrogate suffice? (4) what can we safely conclude from the results? And (5) can we frame the inherent uncertainty in ways that accurately, and effectively, communicate our findings? Addressing these questions increases the opportunity for effective science communication and, in the process, may allow for critical buy-in from interested (and disinterested) parties.

Of course, caution is always warranted when interpreting results from any foresight activity because either intended or unintended, conclusions based on model results are intricately linked to the assumptions and algorithms present, the data used to calibrate and verify the model, and the funding source. In all things, perceptions are inherently colored by legacy knowledge and priorities.

As such, a prudent observer would carefully and objectively evaluate numerical and statistical model results.

CONCLUSION AND FINAL THOUGHTS

Scenarios take on many forms, ranging from practical "normative" scenario planning used to articulate the values of a community or region by eliciting people's opinions about different possible visions of the future to visioning (exploratory) exercises that create policy timelines for imagined or perhaps desired futures. Scenarios may also be used to explore emergent phenomena and weak signals. We use scenarios because they provide real-time analyses of the temporal trajectory in the response of systems to processes and events that we could not otherwise evaluate. These scenarios may focus on: (1) practical decisions regarding the management of community resources in the preparation of its citizenry for growth or change (but also risk and resilience management), (2) end-point futures backcasted to create the trajectories of the policies and strategies needed to achieve a "desired" future or suite of possible futures, and (3) the ad hoc creation of variability in the known driving variables of a system to explore alternative trajectories in systems process and function towards some unknown future, but one molded by the storylines created. For the Anthropocene, adaptive anticipatory scenario development and analyses provides a modest, tractable approach to forecast and analyze the complex, inter-connected nature of local to global scale physical and biophysical process that are defining our quality of life and our wellbeing. Complex systems, often in the purview of wicked problems, may never be accurately modeled to our satisfaction but we must continue to try because life literally hangs in the balance.

ACKNOWLEDGMENTS

I thank Mrs Raphaële Bidault-Waddington for a review of the content of this chapter and the editors for the opportunity to contribute; thank-you. This work was funded, in part, by the National Science Foundation under Grant no. SES-1462086, DMUU: DCDC III: Transformational Solutions for Urban Water Sustainability Transitions in the Colorado River Basin.

REFERENCES

Aguilar, F. J. (1967). *Scanning the business environment*. Macmillian.
Ahmed, M. D., Sundaram, D. and Srinivasan, A. (2003). Scenario driven decision systems: concepts and implementation. *Proceedings of evaluation of modeling methods in systems analysis and design* (EMMSAD). Veldon, Austria. June.

Alexander, L. and Maiden, N. (2004). Scenarios, stories, and use cases: the modern basis for system development. *Computing and Control Engineering 15*, 24–29.

Ansoff, I. (1975). Managing strategic surprise by response to weak signals. *California Management Review, XVIII Winter* (2), 21–33.

Bankes, S. C. (1993). Exploratory modeling for policy analysis. *Operations Research 41*(3), 435–449.

Bankes, S. C., Lempert, R. J. and Popper, S. W. (2003). *Computer assisted reasoning for robust strategies. Scientific and technical: Final report*. Defense Advanced Research Projects Agency.

Berger, G. (1964). *Phénoménologies du temps et prospectives*. Presse Universitaires de France.

Bureau of Reclamation (2012). *Colorado River Basin Water Supply and Demand Study. Study report. December 2012*. Prepared by the Colorado River Basin Water Supply and Demand Study Team.

Burgess, M. G., Ritchie, J., Shapland, J. and Pielke, R., Jr. (2020, February 18). IPCC baseline scenarios over-project CO_2 emissions and economic growth. https://doi.org/10.31235/osf.io/ahsxw

Burke, J. and Ewan, J. (1999). Sonoran Preserve master plan: an open space plan for the Phoenix Sonoran Desert, Tempe, AZ: City of Phoenix Parks, Recreation and Library Department. Retrieved March 9, 2010, from http://phoenix.gov/PARKS/sonorcvr.pdf

Cevolini, A. (2016). The strongness of weak signals: self-reference and paradox in anticipatory systems. *Eur J Futures Res 4* (4). https://doi.org/10.1007/s40309-016-0085-1

Cumming, G. S. and Peterson, G. D. (2005). Ecology in global scenarios. In: A. Alonso, C. Field and R. Reid (eds.), *Scenarios*, Island Press, pp. 45–70.

Farquhar, G. D., von Caemmerer, S. and Berry, J. A. (1980). A biochemical model of photosynthetic CO_2 assimilation in leaves of C_3 species. *Planta 149*, 78–90. https://doi.org/10.1007/BF00386231

Gober, P., Sampson, D. A., Quay, R., White, D. Chow, D. and Winston, T.L. (2016). Urban adaptation to mega-drought: anticipatory water modeling, policy, and planning for the urban Southwest. *Sustainable Cities and Society 27*, 497–504. http://dx.doi.org/10.1016/j.scs.2016.05.001.

Grainger, A. (2017). The prospect of global environmental relativities after an Anthropocene tipping point. *Forest Policy and Economics 79*, 36–49.

Guan, X., Mascaro G., Sampson, D. and Maciejewski, R. (2019). A metropolitan scale water management analysis of the food-energy-water nexus. *Science of the Total Environment 701*, January 20. https://doi.org/10.1016/j.scitotenv.2019.134478

Hausfather, Z. and Peters, G. P. (2020). Emissions—the 'business as usual' story is misleading. *Nature 577*, 618–620.

Hiltunen, E. (2008). Good sources of weak signals: a global study of where Futurists look for weak signals. *Journal of Future Studies 12*(4), 21–44.

Holway, J., Gabbe, C. J., Hebbert, F., Lally, J., Matthews, R. and Quay R. (2012). *Opening access to scenario planning tools*. Policy Focus Report/Code PF031. The Lincoln Institute of Land Policy.

Hopkins, L. D. and Zapata, M. A. (eds.) (2007). *Engaging the future: forecasts, scenarios, plans, and projects*. Lincoln Institute of Land Policy.

Iwaniec, D. M., Childers, D. L., Vanlehn, K. and Wiek, A. (2014). Studying teaching and applying sustainability visions using systems modeling. *Sustainability 6*, 4452–4469. DOI 10.3390/su6074452.

Iwaniec, D. M., Cook, E., Davidson, M., Berbes-Blazquez, M., Georgescu, M., Krayenhoff, S., Middel, A., Sampson, D. and Grimm, N. (2020). The co-production of sustainable future scenarios. *Landscape and Urban Planning 197*, p. 2. DOI 10.1016/j.landurbplan.2020.103744.

Iwaniec, D. M. and Wiek, A. (2014). Advancing sustainability visioning practice in planning—the general plan update in Phoenix Arizona. *J Plann Practice and Res 29*, 543–568. DOI 10.1080/02674592014977004.

Kahn, H. (1965). *Thinking about the unthinkable*. Horizon Press.

Kahn, H. and Wiener, A. J. (1967). The next thirty-three years: a framework for speculation. *Daedalus, 96*(3), 705–732.

Landsberg, J. J. and Waring, R. H. (1997). A generalized model of forest productivity using simplified concepts of radiation-use efficiency, carbon balance, and partitioning. *For. Ecol. Manage 95*, 209–228.

Lempert, R. J., Popper, S. W. and Bankes, S. C. (2003). *Shaping the next one hundred years: New methods for quantitative, long-term policy analysis*. RAND Corp.

Lempert, R. J. and Schlesinger, M. E. (2000). Robust strategies for abating climate change. *Journal of Climatic Change 45*(3–4), 387–401.

Lesca, H. and Lesca, N. (2011). *Weak signals for strategic intelligence: anticipation tool for managers*. Wiley. http://dx.doi.org/10.1002/9781118602775.

Lovejoy, T. E. and Nobre, C. (2018). Amazon tipping point. *Science Advances 4*(2), 1. DOI: 10.1126/sciadv.aat2340.

Parton, W. J., Stewart, J. W. B. and Cole, C. V. (1988). Dynamics of C, N, P and S in grassland soils: a model. *Biogeochemistry 5*(1), 109–131.

Quay, R. (2010). Anticipatory governance: a tool for climate change adaptation. *Journal of the American Planning Association 76*(4), 496–511. DOI: 10.1080/01944363.2010.508428.

Ringland, G. (1998). *Scenario planning: managing for the future*. John Wiley & Sons.

Ringland, G. (2002). *Scenarios in public policy*. John Wiley & Sons.

Ringland, G. (2006). *Scenario planning*. John Wiley & Sons.

Rowe, E., Wright, G. and Derbyshire, J. (2017). Enhancing horizon scanning by utilizing pre-developed scenarios: analysis of current practice and specification of a process improvement to aid the identification of important 'weak signals.' *Technological Forecasting & Social Change 125*, 224–235.

Sampson, D. A., Cook, E. M., Davidson, M. J., Grimm, N. B. and Iwaniec, D. M. (2020). Simulating alternative sustainable water futures. *Sustainability Science 15*, 1199–1210. https://doi.org/10.1007/s1162 5-020-00820-y.

Sampson, D. A., Janssens, I. A., Curiel Yuste J. and Ceulemans, R. (2007). Basal rates of soil respiration are correlated with photosynthesis in a mixed temperate forest. *Global Change Biology 13*, 2008–2017.

Sampson, D. A., Quay, R. D. and White, D. (2016). Anticipatory modeling for water supply sustainability in Phoenix, Arizona. *Environmental Science and Policy 55*, 36–46. DOI: 10.1016/j/envsci.2015.08.014.

Sampson, D. A., Waring, R. H., Maier, C. A., Gough, D. M., Ducey, M. J. and Johnsen, K. H. (2006). Fertilization effects on forest carbon storage and exchange and net primary production: a new hybrid process model for stand management. *Forest Ecology and Management 221*, 91–109.

Sampson, D. A., Wynne, R. H. and Seiler, J. R. (2008). Edaphic and climatic effects on forest stand development, net primary production, and net ecosystem productivity simulated for Coastal Plain loblolly pine in Virginia. *Journal of Geophysical Research 113*, G01003, DOI: 10.1029/2006JG000270.

SAS (2013). *SAS/STAT user's guide*, Release 9.2 edition. SAS Institute, Cary, NC.

Schoemaker, P. J. H., Day, G. S. and Snyder, S. A. (2013). Integrating organizational networks, weak signals, strategic radars and scenario planning. *Technological Forecasting and Social Change, 80*(4), 815–824. https://doi.org/10.1016/j.techfore.2012.10.020.

Schwartz, P. (1991). *The art of the long view: planning for the future in an uncertain world*. Currency Doubleday.

Stern, R. J. and Gerya, T. (2020). Earth evolution, emergence, and uniformitarianism. *GSA Today 31*. doi.org/10.1130/GSATG479GW.1. CC-BY-NC.

van der Helm, R. (2007). Ten insolvable dilemmas of participation and why foresight has to deal with them. *Foresight, 9*(3), 3–17.

Weber, K. M. (2006). Foresight and adaptive planning as complementary elements in anticipatory policymaking: a conceptual and methodological approach. In: J. P. Voss, D. Bauknecht and R. Kemp (eds.), *Reflexive governance for sustainable development*. Edward Elgar Publishing, p. 198.

Winton, M. (2006). Do the artic sea ices have a tipping point? *Geophysical Letters 33* (23), 1–5. doi.org/10.1029/2006GL028017.

Withycombe Keeler, L., Wiek, L. A., White, D. D. and Sampson, D. A. (2015). Linking stakeholder survey, scenario analysis, and simulation modeling to explore the long-term impacts of regional water governance regimes. *Environmental Science and Policy 4*, 237–249.

27. The farthest we can see
Anthony Hodgson

INTRODUCTION

When we as individuals, organizations or whole societies such as cities, embark on projects to change the state of affairs, we have to include consideration of the conditions we believe we will face over the timespan of the project. We also try to assess the risks of the future not turning out as we expect. Otherwise, we are at risk of betting on the wrong future and being caught out. Cities have an ambiguous relationship to the future; they are subject to it and at the same time contribute to creating it. The Anthropocene has raised the stakes for anticipating the future.

What is going to happen next? The age of discontinuities has not passed (Drucker, 1969). When will significant events occur? What don't we know about in a time of impending synchronous failure? (Homer-Dixon, 2006). What can we make happen? These and similar questions pervade our thinking and our feeling for a variety of reasons, personal, technical and political (Kelly, 2006). Reflexively, they lead to a next level of questioning. Is it possible to *know* the future? *How* can we know the future? What are the *limits* of knowing the future? We can also go beyond knowledge in the sense of empirical information. Is it possible to *see* into the future? Can we *communicate* with the future? Indeed, is there a future *there* to be communicated with?

The Anthropocene demands of our society a huge step in futures literacy (Miller, 2018). How far we can see into the future depends on three main factors; first, how familiar we are with the domain we are looking into; second, the extent to which we realize that different types of futures thinking methods generate different results, so we need a portfolio of methods; third, recognition that it is individuals who look ahead, and who need to have developed a capacity we call *future consciousness* (Lombardo, 2006).

Underlying these factors, and rarely examined, are questions of the nature of time, the future, dimensionality and causality. Do we really know what we are dealing with in the matter of time? The Western dominated mindset assumes a sequential singular time, a clear distinction between past, present and future, and a limited view of causation which always proceeds from the past into the future. These linear assumptions need questioning and new insights brought to bear if we are to improve our capacity to see 'into the future'. As Roberto Poli (2019, p. 29) points out: 'An organisation unable to grasp new circumstances is an organism that behaves on the basis of stimuli that belong to the previous era.'

In contrast, there are other interpretations in which time is circular, recurrent and driven by patterns. For example, systems thinking asserts that effects can be causes. So, two guiding principles for addressing the question of this chapter emerge:

1. It is not only the farthest we can see but also the *deepest we can see*.
2. It is not just the time horizon we need to understand but the *richness of potentiality*.

FROM FORESIGHT TO THE PRESENT MOMENT

The practice of extending our experience of the present moment to include future consciousness requires relating our psychological states to different aspects of experience as shown in Figure 27.1 (Hodgson, 2019).

At the centre of our mind is the total set of immediate 'mental objects' that constitutes the conscious experience of the present moment. The horizontal dimension (x-axis) refers to the way the content of the present moment, in the form of traces, memories, and expectations and hopes, creates the experience of the flow of time. The span of this sets the distinction between a 'thin' or a 'thick' present moment. The thin present moment is the precise instant measured on the clock. The thick present moment extends into the chronological past and future and has a richer structure than just linear time. The vertical dimension (y-axis) represents latency in the form of potential patterns and appearance as passive forms. The diagonal dimension (z-axis in three dimensions) represents what we might call living commitments entering from the past but differently from causal time. It also represents, intriguingly, influences from choices or creative insights not yet made but held as possibilities in many minds.

The richness of a given present moment is a function of the extent and quality of these different types of influence. Each dimension is an influence entering and enriching the now. To visualize this, we need to suspend the convention that time flows only from left (past) to right (future) and consider the six influences converging on 'now' at the centre. The spatial axes x, y, and z enable us to visualize three time-like dimensions each influencing the present. In the figure all the arrows point *into* the sphere of the experienced present and

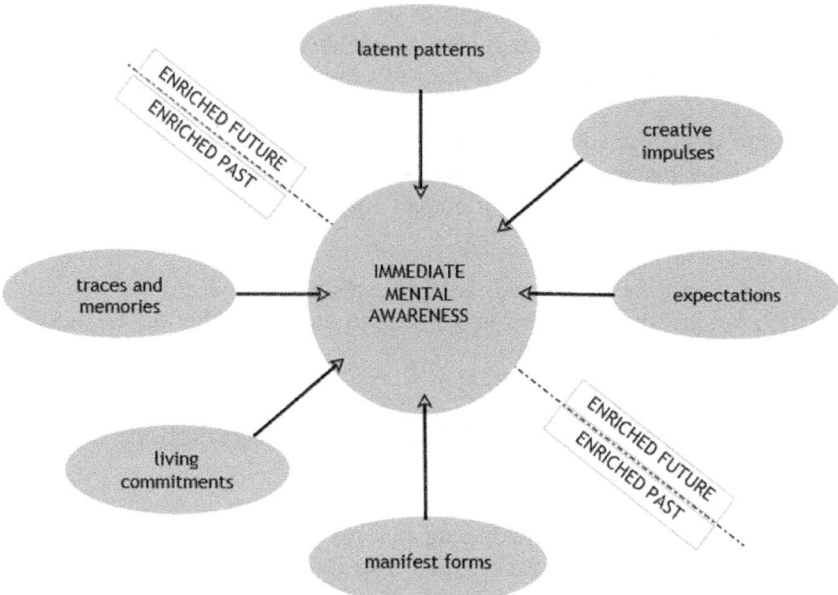

Source: Developed from Bennett (1966).

Figure 27.1 The multidimensionality of the present moment

impact the mental experiential content implying several forms of causation. This is the enrichment of future consciousness.

Conventional time is experienced through the traces and memories in the mind that imprint the present from the past. Retrocausality, whether imaginary or real, is experienced as expectations and hopes. The passive form is the constellation of relatively enduring forms. The active side of form is the vast superposition or latent multiple presence of potential patterns and states (Bohm, 2003). The additional degree of freedom of the third time-like dimension is the sense of meaningful interacting commitments that still prioritize in the present. Its future aspect is the region in which the present moment is open, creative and evokes choices and decisions. Thus, there are six sources of insight that can contribute to future consciousness. Conventional knowledge constrains to using extrapolation and continuing commitment. Without the additional influences of perspectives, extrapolation can leave us vulnerable to unanticipated trend breaks, and commitments can trap us in confirmation bias (Kahneman, 2012). Without openness we are trapped in memory and the creative potentialities are shut out.

SO HOW FAR?

Thinking differently about how we can know the future through the context of the present moment reshapes the way we attempt to anticipate how futures might unfold related to the conventional causal time scale. The proposition here is that in order to discover how far we can see, it is necessary to change our mental mode from analysis to synthesis, from dissection to wholeness, from certainty to openness. The key cognitive tool for this is the recognition of consciousness of the present moment as a developable faculty.

Four calibrations of the present moment are now described, each with a different approximate scale on the conventional time dimension. They are global geopolitical futures, technology futures, civilizational futures and epochal futures. They cover durations from roughly 20 years to 2,500 years. The four scales of present moment differ greatly in their multidimensional complexity. The unidimensional temporal measure gives the mind something to hold onto and clear some of the 'fog' obscuring seeing into the future. Very approximately:

1. 10–20 years – geopolitical futures
2. 10–50 years – technological futures
3. 50–250 years – civilizational futures
4. 250–2500 – epochal futures.

How we see and respond to the implications of the Anthropocene will differ qualitatively according to the present moment perspective adopted.

GEOPOLITICAL FUTURES

In the complexity of how geopolitical forces play out and shape the future we see the competitive play of ideologies and paradigms (Hardt and Negri, 2000). We can identify the main contrasts or opposites between the different ideologies. It is then a matter of identifying the variety of how they can be juxtaposed and from that anticipate different images or scenarios of the future. In classical scenario planning the struggles are relatively short term and tend to a resolution where one scenario takes over and becomes predominant; for example, in a national election where one party gains a substantial majority or in religion where one belief system becomes dominant in a society pretty much to the exclusion of any others.

However, at a deeper level of human values, aspirations and beliefs, the tensions and conflict may endure for several generations so that we see a continuing drama where, on a larger time scale, there is an oscillation of dominance from which events unfold from the state of play of the different tensions analogous to the interaction of characters in a drama. The scenario analysis is placed into the context of a larger present moment.

At the current stage of the Anthropocene, we are in the midst of a huge complexity of factors, forces and complex systems with emergent properties (Steffen et al., 2015). We can get an overview of this by juxtaposing two especially important dimensions of deeper human value. One of them is how far people adhere to the egocentric values of 'I', or centre in the participative values of 'we'. The other is how far people adhere to tribal isolation in their cities or centre in a one world orientation. Interestingly, although these dimensions interact strongly, they are distinct and interdependent in different ways. How they interact generates four distinct scenarios which we will name *Ethical Trading*, *Paradigm Wars*, *Global/Anti-global* and *Techno-assimilation*.

Instead of a crossroads after which only one or another scenario will prevail, we consider the more subtle dynamic where the impact of defining events switches a current predominant scenario to one of the others dominating for a period. In this futures model[1] (ISRG & IFF) all remain present and playing out in some region of global society. Thus, we see the future spun out of the present in a complex dance of structures and influences. We entered this century with the drive to *Ethical Trading*. The 9-11 event switched to *Paradigm Wars* between major cultures. Now the build-up of the superstorms and pandemic are switching towards *Global/Anti-global*. However, increasing pervasiveness of digital technology points to a further possible switch to *Techno-assimilation* where society becomes dominated by exploiters of artificial intelligence. At the moment *Ethical Trading* now has a back seat but rising concerns over the human ecological footprint and the acceleration of urbanization, question the whole basis of global societal life in the Anthropocene. So, we can see this 'dance of the scenarios' going on in four-to-seven-year cycles for at least three decades until a new, as yet unforeseen, collective ideology might enter to compete in the geopolitical arena.

TECHNOLOGY FUTURES

Attempts are made by government and businesses to foresee the technological future beyond geopolitics and short-term business. For example, the UK Government Foresight group in 2006 commissioned a 'technology forward look' to consider the next 50 years of intelligent infrastructure (Sharpe and Hodgson, 2006). The method of three horizons (Sharpe, 2013) was applied as a way of stretching the timespan. This look ahead of 50 years, takes us well beyond the 15 to 20-year outlook of most major technology road-mapping exercises and scenario exercises. Each horizon provides a distinct way of thinking about future technology.

Horizon 1 – Invent, Develop, Deploy

This horizon, about a decade into the future, covers the roll out of new, but understood, technical capabilities to solve problems and address opportunities within the dominant socio-technical systems of the present day. It may still involve significant levels of new invention, but of the sort that experts in the field consider to be well within known art or its extension. A lot of the 'smaller, cheaper, stronger, faster, smarter' types of invention lie here.

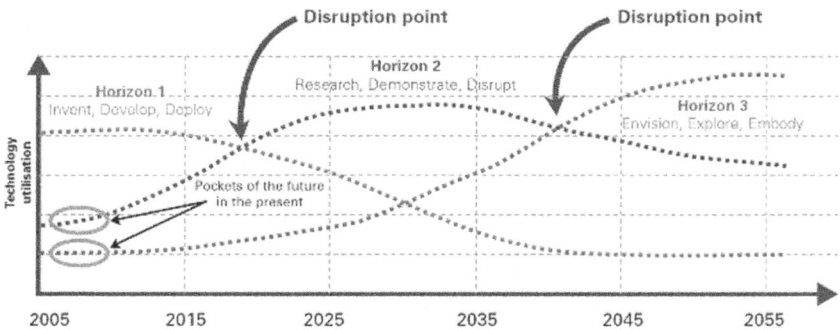

Figure 27.2   Taking a long view in three horizons

Horizon 3 – Envision, Explore, Embody

Looking beyond the horizon of known science and beyond many years of unpredictable invention, what can be considered valid?

First, if something does not contravene known natural laws, and we really want it, we can try to accomplish it. The effect of such a vision is that it configures the use of resources in the present moment in making the initial steps based on a likely switch of what is valued. Such an exercise can legitimately include anything that can be imagined and that would have high impact, with particular significance to those where large resources are already committed to the goal by researchers and their backers.

Second, there is the intent is to bring the future into the present moment in such a way that it draws research and invention around it in pursuit of new goals – such a vision puts us on a different path. A decision to switch totally to net zero carbon, to living in an urban community without cars, to achieve organic vertical agriculture in cities, are examples relevant to the Anthropocene. These are pursued as societal goals that influence the direction of technology development.

The term 'pockets of future in the present' is used for these exemplary innovations. As attributed to William Gibson: 'The future is already here, it's just not widely distributed yet.'

Horizon 2 – Research, Demonstrate, Disrupt

Having set up the contrast and tension between H1 and H3 we can examine the transition phase of Horizon 2 with more discrimination. This time horizon is the domain of research, invention and market disruption as well as the navigation of multiple dilemmas.

Some innovations, designated H2+, will facilitate the emergence of H3 systems in a transformative way. Others, designated H2-, will be absorbed back into the H1 systems and contribute only to marginal or incremental change. Whether the H2 innovations become subsumed or controlled by existing patterns and actors or go on to generate transformative outcomes, is an area of major uncertainty, and will depend on an array of social, political, and technological factors, including the extent of the ability of different actors to influence regulatory demands.

CIVILIZATIONAL FUTURES

There are numerous studies of the life cycle of civilizations interpreted in various developmental stages[2] (Toynbee, 1946). The key point about these models for seeing into the future on this scale is that, to the extent that the position of a civilization (or large country) can be diagnosed today, a guess can be made as to where it is in its life cycle. A second layer of guessing is how long it might endure. This whole inquiry can be usefully represented by the three horizons framework to look at the rise and fall followed by displacement as one civilization takes over from another.

Horizon One is the immediate landscape which is formed by the manifestation of the current worldview or power paradigm that holds the civilization together. *Horizon Three* is in the distance and is shaped by a new emerging worldview or power paradigm. It has very different characteristics from Horizon One. *Horizon Two* represents the wave of turbulent change that enables the deconstruction of the old world and reconstruction of the new world.

Underlying this picture of displacement are recurring systemic patterns underlying patterns of the three horizons at this scale of present moment. These patterns can help us diagnose the likely viability of large cultural and ethnic complex societies.

EPOCHAL FUTURES

Beyond the rise and fall of civilizations a much deeper and large-scale sequence of development is taking place at the present time, to what is referred to as the Global Age (Albrow, 1997). Humanity is very young indeed in the evolutionary scale and can be seen as having many lessons of such a nature that they take two or three thousand years to learn, and then imperfectly. The scale of the whole of humanity now has to be considered since the Anthropocene includes all peoples and, being global, is beyond the scale of civilizations and highly determined by environmental conditions (Fernández-Armesto, 2000).

The following account draws on the thinking of J.G. Bennett in his account of historical cycles (Bennett, 1966). He saw the evolution of humanity taking place in cycles of about 2.5 thousand years which he termed *epochs*. He described the pattern of an epoch as fundamentally shaped by a Master Idea of that specific evolutionary phase. We are currently in a transition present moment of perhaps 500 years in which a turbulent transition between epochs is taking place. The last 2.5 thousand years he referred to as the Megalanthropic Epoch, implying the development from a 'divine right' of exploitation of humans by an elite to an emerging recognition of, and respect for, the individuality and conscience of individuals. The new emerging cycle he characterized as the Synergic Epoch of global cooperation. A sketch of these Master Ideas is summarized in Figure 27.3.

From a three horizons perspective we can see the arc from 500 BCE to 2250 CE as Horizon One. Horizon Three is seeded around 1750 (these chronological times are fuzzy rather than precise) and takes us to somewhere around 4500 CE. Of course, some time before then according to this hypothesis a new epochal transition will have begun to yet another Master Idea, as yet unidentified. The present moment that can embrace the entire cycle of displacement, in time terms, is around 5000 years. This greater present moment (GPM) spans the transformation from fragmented humanity to global humanity. The large circle of this span represents the present moment capacity needed for a full transitional intelligence. Two smaller circles represent the two overlapping epochs of 2500 years, each being the present moment of that Master Idea. A smaller present moment of around 500 years still represents quite a stretch for our comprehension and is the recognition that a major paradigm shift beyond civilizational displacement is taking place.

Each Epoch has a negative or shadow side. The current positive transition is from *Individual Realization* to *Collective Wisdom*. This does not leave individual realization behind but builds on it as an essential component. *Megalomaniac domination* is the aberration, or the shadow side of the current period and the emerging Synergic Epoch could fall into its shadow side of

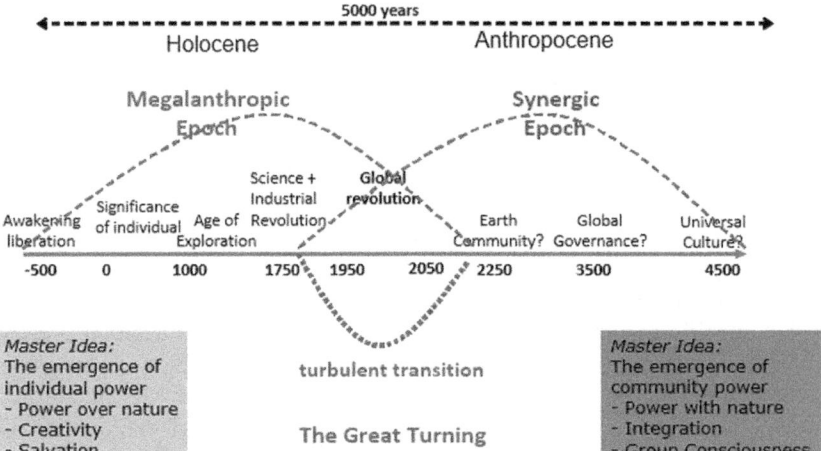

Figure 27.3   The current displacement between old and new epochs

'*cyborg assimilation*' through misapplication of brain technology, artificial intelligence, genetic manipulation and mass conditioning.

The future is open as to whether humanity will 'make it' into and through the Synergic Epoch. It is not a foregone conclusion. Perhaps synergy is also an essential precondition for humanity as a whole to participate in the greater community of the universe beyond simply technological capability.

CONCLUDING REFLECTIONS

This conceptual and methodological speculation has now brought us to the point where we can begin to reframe and re-perceive pretty much everything around the question of seeing into the future and what it may hold for us. The value of this speculation does not lie in some new predictive ability, which itself would still be locked into limited determinism, but in whether it is able to help generate new questions, new values, new insights, new understandings (Sharpe and Hodgson, 2017) and new viable actions for the transformational epoch we are characterizing as the Anthropocene. The key to this step is an enriched understanding of time and the present moment (Hodgson, 2013, 2019). In short, 'seeing the future' requires future consciousness and, despite all the advances in technology and big data, the prime instrument is the human mind. The challenge for knowledge cities is whether their sub-communities can together cultivate a shared future consciousness related to global, bioregional and city scales that dynamically sustain human and biospheric wellbeing through the unavoidable shocks and changes of the larger planetary system.

I am confident our human nature has more capacity to see into and to anticipate the future than is yet practiced in our current cultures (Poli, 2019). Future consciousness is seen as the capacity to identify the future in the present moment and then articulate that into possible pathways of unfolding, including the dynamics of how deeper world views, ideologies and belief systems both enable and prevent clear sight in their interactions. How far we can see into the future also requires how clearly we can see into the past. There is a historical legacy which drives us but only partially determines the future. There is a constantly churning dynamics of the tribal nature of humans which operates in tension on many scales. There is the inheritance of the forms of settlements, power and exploitation that cannot easily be shaken off. In the richer future accessing the potential is sensed in many ways in different cultures which we simplify as the 'desire for a better world'. In the stretch of calendar time there is the vision of what might actually be achieved. But with all the difficulties of seeing the future the most difficult and, paradoxically, the most promising trend is the emergence of shared creativity (Hodgson, 2020) that alone is likely to generate the new kind of knowledge we need to flourish in harmony with mother earth in the Anthropocene.

NOTES

1. Based on an unpublished study of the International Strategy Research Group and the International Futures Forum conducted over the period 2000 to 2004.
2. For example, Toynbee described the cycle as genesis, growth, breakdown and disintegration.

REFERENCES

Albrow, M. (1997). *The Global Age*. Stanford University Press.
Bennett, J.G. (1966). *The Dramatic Universe*. Vol. 4. Hodder and Stoughton.
Bohm, D. (2003). The Enfolding-Unfolding Universe and Consciousness, in L. Nichol (ed.), *The Essential David Bohm*, pp. 78–120. Routledge.
Drucker, Peter (1969). *The Age of Discontinuity*. Elsevier Publishing Company.
Fernández-Armesto, F. (2000). *Civilizations*. Macmillan.
Hardt, M. and Negri, A. (2000). *Empire*. Harvard University Press.
Hodgson, A. (2013). Towards an Ontology of the Present Moment. *On the Horizon* *21*(1) 24–38.
Hodgson, A. (2019). Foresight and the Seven Dimensions of Experience: A Transdisciplinary Transcultural Approach. *World Futures 75*, 3113–3134.

Hodgson, A. (2020). *Systems Thinking for a Turbulent World: A Search for New Perspectives*. Routledge.

Homer-Dixon, T. (2006). *The Upside of Down – Catastrophe, Creativity and the Renewal of Civilization*. Random House.

Kahneman, D. (2012). *Thinking Fast and Slow*. Penguin Books.

Kelly, E. (2006). *Powerful Times – Rising to the Challenge of Our Uncertain World*. Wharton School Publishing.

Lombardo, T. (2006). *The Evolution of Future Consciousness*. AuthorHouse.

Miller, R. (2018). *Transforming the Future: Anticipation in the 21st Century*. UNESCO.

Poli, R. (2019). *Working with the Future: Ideas and Tools to Govern Uncertainty*. Bocconi University Press.

Sharpe, B. (2013). *Three Horizons: The Patterning of Hope*. Triarchy Press.

Sharpe, B. and Hodgson, A. (2006). *Towards a Cyber-Urban Ecology: Intelligent Infrastructure Futures Technology Forward Look*. Office of Science and Technology.

Sharpe, B. and Hodgson, A. (2017). Anticipation in Three Horizons, in R. Poli (ed.), *Handbook of Anticipation*, pp. 1–18. Springer.

Steffen, W., Richardson, K., Rockstrom, J., Cornell, S.E., Fetzer, I., Bennett, E.I., Biggs, R. et al. (2015). Planetary Boundaries: Guiding Human Development on a Changing Planet. *Science 347*, 6223, 736–747.

Toynbee, A. (1946). *A Study of History*. Oxford University Press.

28. Knowledge for the Anthropocene: an agenda

Francisco Javier Carrillo

INTRODUCTION

This chapter carries on from Chapter 18, where major fields of institutionalized science applied to the study of knowledge were reviewed. Setting out from the current understanding arising from the most established sciences of knowledge, this chapter now looks at a wider balance of knowns and unknowns, including meta-scientific knowledge. It starts by looking at philosophies of knowledge and the reflexivity involved in our knowledge being constrained by the way we understand knowing, its processes and outcomes. After reviewing empirical epistemology, other areas of knowledge studies beyond the established sciences of knowledge from Chapter 18 are reviewed. It then turns to some areas of meta-scientific knowledge studies looking for a broader picture of relevant knowledge. Finally, it concludes sketching out some elements for a knowledge for Anthropocene agenda that can be drawn from both chapters.

KNOWLEDGE KNOWNS AND UNKNOWNS

Given its foundational nature, one would expect knowledge itself to be a well-understood phenomenon. Trendy expressions such as 'Knowledge Economy', 'Knowledge Society', and so on, seem to take knowledge for granted. However, a closer look at such concepts reveals the lack of proper terms and definitions, the variety of perspectives and the pending foundational work. Certainly, progress has occurred, as in many other fields. In the following sections, advances in the understanding of knowledge beyond the sciences of knowledge are summarized, pointing towards the challenges ahead.

Philosophies of Knowledge

A broader arena for tackling the unknowns of human understanding is to be found in Philosophy. Two branches of philosophical inquiry deal closely with a theory of knowledge. Epistemology is directly concerned with the nature, methods, validity, scope and justification of knowledge, particularly the distinction between justified belief and opinion. The field is as old as philosophical practice, and a wealth of epistemological approaches populate the history of philosophy. While the Philosophy of Science galvanized epistemological efforts in Western philosophies, a recent turn is re-evaluating not only philosophical perspectives from other civilizations, but also previously disregarded forms of knowledge captured by native, feminist, pragmatic, minorities, post-colonial, southern, and more-than-human experiences, hence widening the realm of Epistemology to different forms, means and ends of knowing. The second branch of philosophy that bears a close interdependence with Epistemology is Ontology, or the study of being, including existence, reality, becoming, and the categories and relations of beings. Whereas there is also a long tradition of ontologies defining the scope of human knowledge, and particularly of scientific explanation, recent turns in philosophy have brought such interdependence to a first plane. Reliance of Epistemology upon other branches of philosophy, such as Axiology (Ethic and Aesthetic values) and Metaphysics have also become more widely recognized over recent years. Chapter 1 provides a current appraisal of some prominent philosophical developments concerned with the Anthropocene and by extension, to knowledge in this new epoch.

Realizations of contemporary philosophy regarding Knowledge for the Anthropocene include: (1) the overcoming of the dichotomy between the sensory and symbolic realms of experience, acknowledging the material base of a more-than-human world; (2) the demotion of human agency from a self-assigned central stage on time, meaning and value; (3) a 'flat ontology' giving equal existential status to all beings; (4) the de-colonization of areas where some form of supremacism was taken for granted, either openly or surreptitiously, such as in gender, race and geography; (5) the search for new terms of understanding between humans and Earth; (6) the acknowledgement of Anthropocene as a hyper-complex reality escaping the sense-making of individual intellects; (7) the need to integrate action and value through a pragmatist perspective that keeps a balance between critical analysis and bounded rationality and realism.

Knowledge Reflexivity

Knowledge Studies have gained momentum through the application of scientific explanation to knowledge as a natural phenomenon: knowledge on knowledge. Such reflexive move has generated a feedback process within modern philosophy that has in turn provided wider epistemic frameworks. The reflexive turn in science and knowledge at large has taken several shapes, including *Scientometrics* (measurement and analysis of scientific production), *Second-Order Science* (extrapolation of Second-Order Cybernetics into designing scientific systems) and *Metascience* (self-awareness movement about the constraints of scientific research and the means to overcome those). We will now focus on two major reflexive movements: on the one hand, Empirical Epistemology and the Science of Science movement and, on the other, Science and Technology Studies.

Empirical Epistemology and the Science of Science

Besides the wider interpretative framework provided by Philosophy of Science thus regarded as the *Science of Sciences* (Flint, 2015, Part I) or *Primary Epistemics* (Goldman, 1986, p. 378) a number of scientific disciplines have ventured into specific aspects of knowledge phenomena, complementing the former with a *Sciences of Science* perspective. In the 1960s, a movement known as *Science of Science*, building on the progressive social perspectives of John D. Bernal undertook a critical reflection of science as a public matter of social concern (Bernal, 1967). Figures such as Derek de Solla Price, Charles P. Snow, and Maurice Goldsmith advocated "the examination of the phenomenon of 'science' by the methods of science itself" (Goldsmith, 1965, p. 10). The Science of Science movement laid the foundations for the Social Responsibility of Science movement, as well as the emergence of the sciences of knowledge (Chapter 18). Empirical or Naturalistic Epistemology complements the flow of philosophical analysis with the factual inputs and theoretical insights from the scientific disciplines that have gained ground into aspects formerly restricted to philosophical analysis. Rather than displacing the latter, empirical epistemology has contributed to advance the understanding of knowledge by furthering philosophical inquiry with new realizations and perspectives.

In Chapter 18, the most prominent Sciences of Knowledge were reviewed, showing how some areas of empirical epistemology evolved into full-fledged disciplines within the major realm of the Knowledge Studies. It was also pointed out how most of these fields of Knowledge Studies partially subsumed within the contemporary field of 'Science and Technology Studies' or 'Science, Technology and Society Studies'.

Science and Technology Studies

Over recent decades, *internal* or *realist* and *external* or *postmodernist* perspectives of science (Carrillo, 1983, p. 19), have become highly interdependent, often through a hate-love relationship (also referred to as the 'science wars') at the borderline field of 'Science and Technology Studies' or 'Science,

Technology and Society' (STS) (Felt et al., 2017). Some core STS stands include the strong conviction that scientific and technological knowledge is *constructed*, that scientific issues are also *translated* between research and policy and that identity categories such as gender and ethnicity become diluted by the received scientific ethos (Howell, 2017, p. 50). The tension between the internalist effort to affirm the social value of science before the public eye and the externalist efforts of STS to critically expose the vulnerability of scientific practices to ideology and power, may prove finally valuable as animosities reside and a common challenge emerges. Nowhere have the bitter disputes become so irrelevant and marginal as in the coming to terms of science and humanities with the Anthropocene. While constructivists are now keen to stress the role of scientific evidence before climate change, scientific approaches to the Anthropocene have opened up to richer perspectives on agency, techno-determinism, gender, colonialism, and so on. The common ground of materiality, ecology and nature, where the socio-technical and the natural-material no longer see themselves as two cultures but as the single irreducible base of the climate crisis, has brought several STS scholars to the frontline of the philosophical framing of the Anthropocene, such as Donna Haraway (e.g., 2016) and Bruno Latour (e.g., 2018).

Other Studies of Knowledge

Besides the Sciences of Knowledge reviewed in Chapter 18, and the fields of knowledge studies covered above, there are other important contributions from the sciences that provide significant inputs to a general theory of knowledge despite not being established as independent disciplines. Without being exhaustive, the following deserve attention.

1. Knowledge in Physics. Regarding information as an elementary physical variable or relating knowledge as a fundamental process, there have been a number of relevant developments. This includes John Wheeler's (1990) ontological attribution by information ("it from bit") and Seth Lloyd's (2006) attempts to calculate the Universe's computing capacity and look at it as a monumental quantum computer. Building on Ludwig Boltzmann's views on entropy, Cesar Hidalgo (2015) at MIT has undertaken a universal theory of information, from atoms to societies.
2. Information Theory and Cybernetics. A significant precedent of the former, Information Theory is a mathematical rendering of information as a natural phenomenon originally proposed by Claude Shannon, with wide applications. Emphasis on information processes brings this field close to Knowledge Management, particularly through the work of Doede Nauta and Winfried Nöth, who draw on Pierce's ideas to develop a semiotic

information theory (Nauta, 2019). Also closely related is Management Cybernetics, a field introduced by Stafford Beer. He developed the Viable System Model, of special significance to Knowledge in the Anthropocene. An exploration of such significance is offered by Chapter 12 in the companion volume *City Preparedness for the Climate Crisis*.

3. System dynamics. Another development that resonates at both the organizational and urban levels is this mathematical modeling visualization method to tackle complex systems originated by Jay Forrester. *The Limits to Growth*, the 1972 Club of Rome study – a precursor of global environmental consciousness, is perhaps its best-known application, still relevant today (Meadows et al., 1992). Complexity and cybernetics are part of the major field of Systems Science, covering other aspects relevant to Knowledge in the Anthropocene, such as chaos theory, systemic interaction, emergence and adaptation, under the common perspective of Systems Thinking (Hodgson, 2019).

4. Computer Science & Artificial Intelligence (AI). Insofar as the core question of computer science is 'What can be automated?' (Denning, 2000), it is of relevance to the understanding of knowledge. Can knowledge be automated? The Dartmouth Proposal (a call for the 1955 presumably founding event of AI) opened with a bold conjecture: "that every aspect of learning or any other feature of intelligence can in principle be so precisely described that a machine can be made to simulate it" (McCarthy et al., 2006, p. 12). Since then, the field of AI has evolved with varying fortunes in its attempts to emulate human capacities such as reasoning, learning, planning, knowledge representation, natural language processing and perception. Today, its applications are as impressive and as ubiquitous as worrisome are its ethical, economic and political implications (Renda, 2019; Zuboff, 2019).

5. Consciousness and time studies. Of all knowledge objects, the closest to our attention is also the most difficult to grasp: ourselves. 'Know thyself' was the first of three aphorisms said to be inscribed at the forecourt of the Temple of Apollo at Delphi. Ever since, it remains the most challenging task to human understanding. In its contemporary form, consciousness studies deal with 'the hard problem' of explaining that which we experience as we experience it, such as our feelings. The self-referential character of looking at consciousness makes it logically slippery. However, a race is on to establish neurological correlates of experience and other empirical grounds for its qualities (Damasio, 2021). Pending on that search are potential reformulations of other phenomena such as identity, thinking, time, and along with it, major implications on anything from physical to behavioral sciences.

Indigenous Knowledge Systems

The studies of Indigenous Knowledge (Kanu and Ndubisi, 2020), Traditional Knowledge (ICSU, 2002) and Cultural Heritage (Sullivan, 2016) tend to be integrated under the label of Indigenous Knowledge Systems (IKS). This now consolidated field had been long neglected despite calls for vindication such as *vernacular knowledge* from Ivan Illich (Illich, 1981) and *lifeworld* from Jürgen Habermas (Fairtlough, 1991) both reinterpreted as *experiential knowledge* by André Gorz (2010, p. 30), that go back to the concept of *phronesis* (practical wisdom) in classical Greece (Nonaka and Toyama, 2007). But contemporary IKS also carry a distinctive agenda, notably driven by post-colonial epistemologies and politics. The full vindication of IKS is at the core of epistemic justice (Ndlovu-Gatsheni, 2020).

In Chapter 18, reference was made to Berger and Luckmann's (1996) consolidation of social constructionism. Of particular relevance here is the concept of *Social Stock of Knowledge*. This acknowledges the existence and importance of modes of knowing beyond scientific and theoretical knowledge that are just as significant for everyday life, including values, beliefs, proverbs, myths, institutions, routines, customs, conventions, roles, division of labor, practices, procedures, and so forth. This also includes forms of knowledge transmission such as oral tradition and rituals, as well as knowledge contexts such as herbalism, midwifing, and more recent ones such as home economics and do-it-yourself (DIY). In a broader sense, most knowledge systems are at once localized and permeable, endogenous and exogenous. Each of these attributes exhibits a continuous distribution. Also, all knowledge systems occur in specific spatiotemporal coordinates and under particular contexts of significance. Therefore, all knowledge systems can be said to be 'indigenous' in a wider sense (Marsden, 2004; Carrillo, 2015, p. 2). Hence, IKS invites to understand and regulate the multidimensional and multidirectional knowledge transfer processes in a fully globalized world (Batra et al., 2013; Polónia et al., 2018; Galan et al., 2020). Such understanding seems particularly relevant to a knowledge agenda for the Anthropocene (Tom et al., 2019; Breidlid and Krøvel, 2020). Rather than demarcated and insulated, all these globalized and diachronic knowledges (Castree, 2017) can be integrated in creative new ways (Kimmerer, 2013; Braidotti, 2017) and further projected by the leverage of artistic imagination (Buckland et al., 2017).

In this sense, IKS are just part of a wider de-colonialization process from geopolitical, gender and racial constraints, amongst others, regarding the space of possibilities for Knowledge in the Anthropocene (de Souza Santos, 2018). As Ulloa (2017) suggests, rather than a "green colonialism" seeking to homogenize "diverse ways of thinking about the environment into a single knowledge system" (p. 112), we should embrace a "willingness to reimagine contemporary change discourse, to allow for the emergence of other knowledges" (p. 117). Such de-colonization is a prerequisite for Knowledge in the Anthropocene: "What would it mean, then, for the imperative to de-colonize an overly Westernized world if we were to ask, 'what kind of planet is this that enables multiple forms of human collective, that makes many-world worlds possible, that proliferates telluric spirits and earth beings'?" (Clark and Szerszynski, 2021, p. 167).

DESIGN ELEMENTS FOR A KNOWLEDGE IN THE ANTHROPOCENE

One can only feel humble before the huge amount and variety of disciplinary and non-disciplinary sources for the current understanding of knowledge. However, the urging questions remain: How can we integrate the best of human understanding to make sense of this unprecedented global emergency? What is the most critical knowledge for coping with the likely impacts of the Climate Crisis? How can that be achieved?

Before drafting some preliminary guidelines, here are some considerations consistent with Chapter 18 and the sections above in this chapter:

1. UNEP's GEO6 Report (UNEP, 2019, p. 66) poses three questions regarding data and knowledge for human-environment interactions: Who pays for data and knowledge and for what sorts of data and knowledge? Whose interests do the existing data and knowledge serve? And whose data and knowledge counts and why? Hence, we need to account for all relevant agents, their value systems and how they interrelate.
2. At the Lund Centre for the History of Knowledge website, Maria Simonsen and Laura Skouvig are credited with the following idea: "rather than try to define knowledge, there is a need for a pragmatic conceptualization." This invites the concept of knowledge systems discussed in Chapter 18.
3. Three principles of producing actionable knowledge have been outlined as follows: (1) engaging in substantive interactions between producers and users of knowledge, (2) ensuring equitable relationships among parties engaged in the process, and (3) producing knowledge that is usable by decision-makers (Mach et al., 2020, p. 35).
4. A core idea in redesigning received knowledge systems is knowledge co-production. Current perspectives emphasize "the importance of (re)politicizing co-production by allowing for pluralism and for the contestation of knowledge" (Turnhout et al., 2020, p. 15). The following 'areas of improvement of co-production practice' have been suggested: (1) Support longer-term co-production and diversify the types of engagement approaches used; (2) Use co-production to ask fundamental questions in

addition to science translation or value-addition of existing science; (3) Increase collaborative framing and designing of projects; (4) Increase evaluation of outcomes (Jagannathan et al., 2020, pp. 26–27).
5. The RESCUE Report issues the following recommendations: (1) build an institutional framework for an open knowledge system; (2) re-organize research so disciplines share knowledge and practices, and, from the onset, work together with each other and with stakeholders; (3) initiate long-term integrated demonstration projects; (4) develop sustainability education and learning that builds capacity for knowledge sharing and research across disciplines in an innovative open knowledge system; (5) respond to the challenges and opportunities created by the internet for an open knowledge system ready for transitions towards sustainability; and (6) create a dynamic, adaptive and integrated information and decision-support system on global change issues (Jäger et al., 2011).

AN AGENDA FOR KNOWLEDGE IN THE ANTHROPOCENE

Rather than a grand narrative manifesto, what seems urgent is a series of consensual criteria and policies, distributed performances and ongoing re-examination. Whether humanity can evolve into a global consciousness of the type envisaged by Pierre Teilhard de Chardin and Vladimir Vernardsky (the Noosphere), remains highly speculative. Hopefully the very circumstances of the Anthropocene might precipitate a critical mass of global consciousness and action for establishing the bases of a new epistemic paradigm in consonance with a more-than-human world. Conceivably, the willingness to collaborate will be directly proportional to the pain felt by the most affluent and powerful societies (Hulme et al., 2020).

For the time being, there are concrete pathways that might eventually lead into that direction. This humble start aims at bringing these issues into public debate and formal research programs. Based on the former discussions, the following seven criteria provide a preliminary roadmap:

1. An Integrated Theory of Knowledge and a Global Open Knowledge Base on the Anthropocene

Braiding together knowledge bases from different cultural contexts demands a specific know-how. From sorting out the epistemic challenges of not getting lost in translation (commensurability management), to integrating technical capacities through appropriate technology and deep innovation, to facilitating a critical pedagogy and a self-directed learning that promotes epistemic empowerment (Carrillo, 2001). Some advancements in that direction are provided by *open disciplinarity, convergence, bridging* and *actionable knowledge*.

Open disciplinarity entails a deliberate attempt to overcome the constraints of disciplinary chauvinism by exploring the possibilities of intersection and complementarity. Marilyn Stember has proposed the following modalities: intradisciplinary, cross-disciplinary, multidisciplinary, interdisciplinary and transdisciplinary (Stember, 1991).

Besides taking down walls, bridges must be built. 'What to' and 'how to' become relevant. Managing convergence could catalyze cross-disciplinary mirroring (National Research Council, 2014). Angeler, Allen and Carnaval (2020) examine the conditions for a convergence science. Knowledge bridging involves specific dispositions and competencies (Mistry and Berardi, 2016), as well as widely studied knowledge processes such as transfer, gatekeeping, interfacing, sharing and brokering.

Intradisciplinary advances towards an integrated theory of knowledge have been produced within several disciplines. Further progress should be reached within each disciplinary field and across disciplines. Besides, and most important, an encompassing framework to unify contributions from the knowledge sciences within a coherent epistemological ground would be required. Chapters 18 and 28 aim at drawing a coherent perspective from Pragmatism and Critical Theory, to intersect with Speculative Realism, New Materialism and Object-Oriented Ontology.

The knowledge area to set an example should be precisely the emerging one of Knowledge for the Anthropocene. How to articulate the best understanding of the realities of the Anthropocene from the natural and social sciences and the humanities so as to lead to an effective global activism and collaboration is a major knowledge management undertaking (Grundmann, 2007; Jamison, 2010; Sarewitz, 2011; Clark and Szerszynski, 2021).

2. Anthropocene Literacy

While there is wide scientific consensus about the anthropogenic footprint on the geo-stratigraphical record, as well as overwhelming evidence on the associated existential risks, the level of scientific analphabetism on the Anthropocene is disheartening. There are several organizations doing commendable work in educating on the Anthropocene (see Chapter 6), but they have limited reach comparative to the global extent of ignorance, denial and disregard. Part of that situation is due to the deliberate and systematic promotion of disinformation through social media and the debasement of international cooperation.

A major effort is required from leading international agencies such as UNESCO as well as academic networks, to counter the latter source of confusion and apathy with a strong program of Anthropocene literacy. There is little point is saving human culture if there are no new generations to further it. This includes the individual development of a socially interdependent critical framework to make sense of the Anthropocene world (Adams, 2020). A group of academic activists and students have developed an agenda for Higher Education Institutions before the Climate Emergency (Vargas Roncancio et al., 2019).

A Pedagogy for the Anthropocene is not independent from the narrative from the hard sciences providing the empirical basis to document the climate emergency. The integration of empirical knowledge and the communication exercise in conveying convincing and motivating narratives are both themes to be advanced conceptually and pragmatically.

3. Human Response to the Anthropocene

Why is there such a huge gap between the level of risk posed by the Anthropocene global existential risks and the utterly disproportionate global organized response? There is substantial background on human response before catastrophe anticipation, misfortune denial, biased decision-making, and functional stupidity (see Chapter 3). Behavior Analysis, Evolutionary and Cognitive Psychology, Neurology and Behavioral Economics are contributing to unravel this enigma (Wong-Parodi and Feygina, 2020).

An international taskforce should urgently be set to address this critical vacuum. Unless effective leads are established to understand and counter the cognitive, emotional, behavioral and cultural constraints that are preventing the wider public from becoming aware of and actively engaging in climate mitigation and adaptation, there is little hope that high-level policies can be properly implemented.

4. Radical Science and Education Overhaul

What is the use of an educational system that drove us here? There is an urgent need to re-invent collective learning. The COVID-19 pandemic is providing a great opportunity for de-schooling society and recovering the meaning of learning experiences. The Global Commission on Adaptation (GCA) report calls for "a revolution in understanding" (2019, pp. 4–5). Wark and Jandrić (2016) revalue a critical pedagogy for the Anthropocene.

Also, the scientific establishment is in need of a major re-design. The Alliance of World Scientists is issuing an influential series of 'warnings' on aspects of the climate emergency. A warning on the urgent transformation of science itself in generating the appropriate conscience and action should be prominent (Cuhra, 2019; Fazey et al., 2020). Critical accounts of some of the inner aspects of scientific practice that urgently require a redesign, such as the prevailing system of incentives, are becoming commonplace (Priem et al., 2010; Bladek, 2014; McCain, 2016; Moosa, 2018). In reflecting on how science could contribute to an Earth System governmentality, Lövbrand, Stripple and Wiman (2009) urge clarification of strategies in terms of who would be governed, tasks distribution between authorities, and guiding governance principles.

Likewise, Technology Transfer and Innovation chains must be re-invented. The Globelics report "Research on Innovation and Development in the Anthropocene" concludes that "learning, competence-building and development of innovation capabilities may be placed at the centre of the sustainability discourse instead of being a side-track as now" (Johnson et al., 2017, p. 17). In substitution of market-validated 'innovation' often of negative social value, we should aim at a *Responsible Innovation* that provides true social value while being accountable for its environmental costs as well as a *Deep Innovation* that brings up the best in human creativity for the public good.

5. International Monitoring Plan

As the environment continues to deteriorate and climate impacts become increasingly hazardous, the need to count on a universally available real-time state of the planet dashboard is imperative. This should include prominent indicators on who is doing what in terms of Nationally Determined Contributions (NDCs) commitments to subsequent Conferences of the Parties (COPs). It should also enable public global accountability of other major agents such as oil companies and other environmentally liable corporations, financial institutions, national governments and environmentally regressive movements.

The widespread availability of this information would empower citizens and communities to take appropriate action, from the local to the national to the international level. Eventually, this global mirror of climate responsibilities would enable legal actions not only at local jurisdictions, but also at the International Court of Justice for crimes against humanity and ecocides on related grounds.

6. Community Actionable Knowledge Management

The purpose of referring the experience of the Center for Knowledge Systems (Chapter 18), was to exemplify the singular benefit for the study of knowledge systems provided by an environment that is not only conducive to research and reflection, but also intensely driven by expected outcomes from applied projects (Antonacopoulou, 2008, p. 14; Lambe, 2011). One of the most promising and underdeveloped lines of action is to merge Actionable Knowledge and Knowledge Management (KM). While a proper KM strategy starts with a participatory definition of the value coordinates of a community, Actionable Knowledge addresses urgent needs, building transdisciplinary collaboration and developing participative knowledge generation and application. It might be time that we capitalize on the design of knowledge systems (Carrillo, 1997; Bennet and Bennet, 2014). Basically, these involve operationalizing the attributes of knowledge objects, agents and contexts, so as to achieve the maximum possible alignment between those attributes (Carrillo et al., 2019). It will become increasingly important to integrate all relevant forms of knowledge, particularly those that will be required by most local communities. Rather than mere procedural knowledge we are looking for the set of contextualizable, actionable knowledge that might be best fit for the foreseeably demanding and unpredictable scenarios of the Anthropocene (Kirchhoff et al., 2013; Mach et al., 2020). From this standpoint, actionable knowledge and knowledge systems seem to be fairly consistent with the recent concept of adaptive intelligence that integrates creative, analytical and practical skills with "wisdom-based" ones, or "skills that help to ensure that our ideas contribute towards achieving a common good both in the short term and the long term balancing our own, others' and higher-level interests" (Sternberg, 2021, p. 4). According to Sternberg, adaptive intelligence enables individuals to either "(1) change themselves so they better fit their environment, (2) shape the environment to better fit their and others' needs or desires, or (3) find a new environment that is a better fit than the one presently inhabited" (Sternberg, 2020, p. 38). Such concept is of obvious relevance to Knowledge for the Anthropocene.

Developing community actionable knowledge involves a dialogue between the bottom end of community life and the upper end of disciplinary science. It also involves community development as a core area of (urban) preparedness, as emphasized in the companion volume to this book *City Preparedness for the Climate Crisis* (Carrillo and Garner, 2021). Know-what, know-why and know-how become critical capacities. Capacities encapsulated in the history of ideas by *Phronesis*: "practical wisdom that has been derived from learning and evidence of practical things. Phronesis leads to breakthrough thinking and creativity and enables the individual to discern and make good judgements about what is the right thing to do in a situation" (The Oxford Review Encyclopedia of Terms, n/d).

7. Global Common Knowledge Markets

How could the global community be best served by knowledge systems aimed at providing adequate support in likely times of extreme disruption, failed states, collapsed economies and deep scarcity? This is a question that deserves the best of human creativity and innovation now that it might still be possible to get prepared for those scenarios. A UCLA's report on Knowledge

Infrastructures covers a number of key issues (including 'Knowledge infrastructures in the Anthropocene'), issuing a set of rules to develop and sustain knowledge infrastructures (Borgman et al., 2020, pp. 11–13).

As the Climate Crisis unfolds, an open common market of knowledge is at the core of human possibilities. A major challenge is to revert the privatization and monetization of knowledge that has so far impaired its developmental potential (Johnson and Ludvall, 2020). While the emergence of Surveillance Capitalism seems to achieve the ultimate accumulation of knowledge and power (Zuboff, 2019), there are three major developments that show the viability of knowledge systems as a common good. First, are *Knowledge Commons* as production systems, clearly documented by Nobel laureate Elinor Ostrom. Building on empirical evidence, she identified design criteria for effective commons governance (Hess and Ostrom, 2011). A second major development is *Open Knowledge*, a successful non-for-profit collective knowledge generation and distribution scheme including Open Source in software development, Open Access in scientific research and Open Educational Resources (García-Peñalvo et al., 2010). Third, is the wider concept of Open Knowledge Market, that encompasses the former two and a whole host of other value-exchange schemes that have proven able to operate outside the canonical constraints of free-market capitalism (Carrillo, 2016). Many of these schemes are now scalable thanks to Smart Contract technology, thus dispensing any intermediation. Chapter 14 in the twin volume on city preparedness presents a taxonomy of minimum viable transaction regimes for the Anthropocene, combining physical and intangible assets.

Finally, we must think of a universal knowledge base that should be available to every human being as a natural right, the equivalent of a Universal Basic Income. As these lines are written, the 20th anniversary of Wikipedia is celebrated. Despite its shortcomings (Greenstein and Devereux, 2017), Wikipedia remains a formidable achievement in global collaboration and volunteer engagement, providing one of the most powerful knowledge sharing platforms in history (Mesgari et al., 2015; Proffit, 2018). However, Wikipedia is largely a repository of formal knowledge as opposed to the *lifeworld*, *phronesis* or *vernacular knowledge* concepts mentioned above (*connaissance* vs. *savoir* in Gorz's distinction).

Anticipating scenarios of high increasing disruption and volatility, a Universal procedural knowledge e-Handbook seems pertinent, either online or as a portable device (Dartnell, 2015). Internet *How To* portals such as *1000 Life Hacks* (http://1000lifehacks.com) and *WikiHow* (https://www.wikihow.com/Main-Page) are precursors to such resource. In terms of hardware, it could learn from the One Laptop Per Child initiative of 2005. A global *How-To* encyclopedia on a sturdy, self-contained, self-powered, multi-lingual and multi-cultural digital device that operates on- and off-line and is up- and down-scalable in instrumental support conditions is conceivable. For example, setting the initial conditions of tools and inputs available, energy supply, and level of understanding, the replies to how-to questions would become the most effective in a situated context. In order to be effective, it should include all domains of expertise, from mechanical through biological, to behavioral. Besides procedural knowledge, this could also serve as a last memory frontier storing the best of human knowledge in arts and the sciences: a portable Alexandria (Radley, 2021). It need not be a universal canon, as it could be customized to an extent. In Bradbury's dystopian novel *Fahrenheit 451*, a few dissidents memorize segments of books to preserve them (Bradbury, 2012). Not a completely inadequate metaphor for a world where a temperature threshold threatens with burning out civilization as we know it.

CONCLUSION

To sum up, the elements here identified for designing a global Knowledge for the Anthropocene agenda are:

1. An independent international scientific network that compiles and integrates state-of-the-art knowledge on the natural, social and cultural aspects of the Anthropocene and is open to coexisting alternative discourses.
2. A parallel international association of progressive education, communication and media agencies that serves the public with the best available Anthropocene literacy resources as open knowledge commons.
3. An international taskforce charged with investigating the cultural and behavioral drives of the Anthropocene so as to contribute to redesign the terms of relationship between humans and the Earth System.
4. A cluster of alternative science, technology and innovation management experiences designed to optimize global knowledge capital and overcome the restrictions of the prevailing rewards system for scientific practice.
5. A global emergency plan that follows up on the COP agreements and feeds back the former three agenda items with performance indicators paired with an international federation of climate civic associations that promotes active local communities' engagement and drives political will for effective action.
6. A global shared infrastructure for the development of community actionable knowledge management capacities able to quickly adapt to unpredictable and fast-changing scenarios.
7. An open global knowledge market that could encompass the outcomes of all the previous agenda items and facilitate exchanges throughout the world.

REFERENCES

Adams, M. (2020). *Anthropocene psychology: Being human in a more-than-human world.* Routledge.
Angeler, D. G., Allen, C. R., & Carnaval, A. (2020). Convergence science in the Anthropocene: Navigating the known and unknown. *People and Nature, 2*(1), 96–102.
Antonacopoulou, E. P. (2008). Actionable knowledge. In Clegg, S. and Bailey, J. (Eds.), *International Encyclopedia of Organization Studies,* 14–17. Sage.
Batra, S., Payal, R., & Carrillo, F. J. (2013). Knowledge village capital framework in the Indian context. *International Journal of Knowledge-Based Development, 4*(3), 222–244.
Bennett, A., & Bennett, D. (2014). Knowledge, theory and practice in knowledge management: Between associative patterning and context-rich action. *Journal of Entrepreneurship, Management and Innovation, 10*(4), 7.
Berger, P., & Luckmann, Y. (1966). *The social construction of reality.* Anchor.
Bernal, J. D. (1967). *The social function of science, 1939.* Routledge.
Bladek, M. (2014). DORA: San Francisco declaration on research assessment (May 2013). *College & Research Libraries News, 75*(4), 191–196.
Borgman, C. L., Darch, P. T., Pasquetto, I. V., & Wofford, M. F. (2020). *Our knowledge of knowledge infrastructures: Lessons learned and future directions.* UCLA.
Bradbury, R. (2012, reprint). *Fahrenheit 451.* Simon & Schuster.
Braidotti, R. (2017). Critical posthuman knowledges. *South Atlantic Quarterly, 116*(1), 83–96.
Breidlid, A., & Krøvel, R. (Eds.) (2020). *Indigenous knowledges and the sustainable development agenda.* Routledge.
Buckland, D., Gray, O., & Wood, L. (2017). The cultural challenge of climate change. *South Atlantic Quarterly, 116*(1), 97–109.
Carrillo, F. J. (2016). Knowledge markets: a typology and an overview. *International Journal of Knowledge-Based Development, 7*(3), 264–289.
Carrillo, F. J. (2015). Knowledge-based development as a new economic culture. *Journal of Open Innovation: Technology, Market, and Complexity, 1*(2), 15.
Carrillo, F. J. (2001). Meta-KM: a program and a plea. *Knowledge and Innovation: Journal of the KMCI, 1*(2), 27–54.
Carrillo, F. J. (1983). *El Comportamiento Científico.* Limusa-Wiley.
Carrillo, F. J., & Garner, C. (Eds.) (2021, forthcoming,). *City preparedness for the climate crisis.* Edward Elgar Publishing.
Carrillo, F. J., Edvardsson, B., Reynoso, J., & Maravillo, E. (2019). Alignment of resources, actors and contexts for value creation: bringing knowledge management into service-dominant logic. *International Journal of Quality and Service Sciences, 11*(3), 1–31.
Carrillo, J. (1997). Managing knowledge-based value systems. *Journal of Knowledge Management, 1*(4), 280–286.
Castree, N. (2017). Global change research and the "people disciplines": Toward a new dispensation. *South Atlantic Quarterly, 116*(1), 55–67.
Clark, N., & Szerszynski, B. (2021). *Planetary social thought: the Anthropocene challenge to the social sciences.* Polity.
Cuhra, M. (2019). The scientist: creator and destroyer—"scientists' warning to humanity" is a wake-up call for researchers. *Challenges, 10*(2), 33.

Damasio, A. (2021). *Feeling and knowing: A manifesto on consciousness.* Pantheon.

Dartnell, L. (2015). *The knowledge: How to rebuild civilization in the aftermath of a cataclysm.* Penguin.

De Chardin, P. T. (2018). *The phenomenon of man.* Lulu Press.

Denning, P. J. (2000). Computer science: The discipline. *Encyclopedia of Computer Science, 32*(1), 9–23.

de Sousa Santos, B. (2018). *The end of the cognitive empire: The coming of age of epistemologies of the South.* Duke University Press.

Fairtlough, G. H. (1991). Habermas' concept of "Lifeworld". *Systems Practice, 4*(6), 547–563.

Fazey, I., Schäpke, N., Caniglia, G., Hodgson, A., Kendrick, I., Lyon, C., ... & Verveen, S. (2020). Transforming knowledge systems for life on Earth: Visions of future systems and how to get there. *Energy Research & Social Science, 70*, 101724.

Felt, U., Fouché, R., Miller, C. A., & Smith-Doerr, L. (Eds.) (2017). *The handbook of science and technology studies.* MIT Press.

Flint, R. (2015) (reprint). *Philosophy as scientia scientiarum: And, a history of classifications of the sciences.* Palala Press.

Galan, J., Bourgeau, F., & Pedroli, B. (2020). A multidimensional model for the vernacular: Linking disciplines and connecting the vernacular landscape to sustainability challenges. *Sustainability, 12*(16), 6347.

García-Peñalvo, F. J., De Figuerola, C. G., & Merlo, J. A., (2010). Open knowledge: Challenges and facts. *Online Information Review, 34*(4), 520–539.

Global Commission on Adaptation (2019). *Adapt now: A global call for leadership on climate resilience.* Global Center on Adaptation and World Resources Institute.

Goldman, A. I. (1986). *Epistemology and cognition.* Harvard University.

Goldsmith, M. (1965). The science of science foundation. *Nature 205*, 10.

Gorz, A. (2010). *The immaterial.* Translated by Chris Turner. Seagull.

Greenstein, S., & Devereux, M. (2017). *Wikipedia in the spotlight.* Kellogg School of Management Cases.

Grundmann, R. (2007). Climate change and knowledge politics. *Environmental Politics, 16*(3), 414–432.

Haraway, D. (2016). *Staying with the trouble: Making kin in the Chthulucene* (Experimental Futures). Duke University Press.

Hess, C., & Ostrom, E. (Eds) (2011). *Understanding knowledge as a commons.* MIT Press.

Hidalgo, C. (2015). *Why information grows: The evolution of order, from atoms to economies.* Basic Books.

Hodgson, A. (2019). *Systems thinking for a turbulent world: A search for new perspectives.* Routledge.

Howell, J. P. (2017). Space for STS: An overview of science and technology studies. In Warf, B. (Ed.), *Handbook on geographies of technology*, 50–64. Edward Elgar Publishing.

Hulme, M., Lidskog, R., White, J. M., & Standring, A. (2020). Social scientific knowledge in times of crisis: What climate change can learn from coronavirus (and vice versa). *Wiley Interdisciplinary Reviews. Climate Change.*

Illich, I. (1981). Vernacular values. In Illich. I., *Shadow work*, Ch. 2. Marion Boyars Publishers.

International Council for Science (ICSU) (2002). *Science and traditional knowledge.* ICSU.

Jagannathan, K., Arnott, J. C., Wyborn, C., Klenk, N., Mach, K. J., Moss, R. H., & Sjostrom, K. D. (2020). Great expectations? Reconciling the aspiration, outcome, and possibility of co-production. *Current Opinion in Environmental Sustainability*, *42*, 22–29.

Jäger, J., Pálsson, G., Goodsite, M., Pahl-Wostl, C., O'Brien, K., Hordijk, L., & Avril, B. (2011). Responses to environmental and societal challenges for our unstable earth (RESCUE), ESF forward look–ESF-COST 'Frontier of science' joint initiative. *RESCUE report*.

Jamison, A. (2010). Climate change knowledge and social movement theory. *Wiley Interdisciplinary Reviews: Climate Change*, *1*(6), 811–823.

Johnson, B., & Lundvall, B. Å. (2020). Possible socialisms and the challenges of the globalizing learning economy in the Anthropocene age. In Brundenius, C. (Ed.), *Reflections on socialism in the twenty-first century* , 17–45. Springer.

Johnson, B., Lema, R., & Villumsen, G. (2017). *Research on innovation and development in the Anthropocene* (No. 2017-01). Globelics-Global network for economics of learning, innovation, and competence building systems, Aalborg University.

Kanu, I. A., & E. Ndubisi (Eds.) (2020). *African indigenous knowledge systems. Problems and perspectives*. APAS.

Kimmerer, R. W. (2013). *Braiding sweetgrass: Indigenous wisdom, scientific knowledge and the teachings of plants*. Milkweed.

Kirchhoff, C. J., Carmen Lemos, M., & Dessai, S. (2013). Actionable knowledge for environmental decision making: broadening the usability of climate science. *Annual Review of Environment and Resources*, *38*, 393–414.

Lambe, P. (Ed.) (2011). Knowledge management special issue: connecting theory and practice. *Journal of Entrepreneurship, Management and Innovation*, *10*(1), 1–147.

Latour, B. (2018). *Down to earth: Politics in the new climatic regime*. Polity.

Lloyd, S. (2006). *Programming the universe: A quantum computer scientist takes on the cosmos*. Vintage.

Lövbrand, E., Stripple, J., & Wiman, B. (2009). Earth system governmentality: Reflections on science in the Anthropocene. *Global Environmental Change*, *19*(1), 7–13.

Mach, K. J., Lemos, M. C., Meadow, A. M., Wyborn, C., Klenk, N., Arnott, J. C., … & Stults, M. (2020). Actionable knowledge and the art of engagement. *Current Opinion in Environmental Sustainability*, *42*, 30–37.

Marsden, D. (2004). Indigenous management and the management of indigenous knowledge. In Wright, S. (Ed.), *Anthropology of Organizations*, 39–53. Routledge.

McCain, K. (2016). *The nature of scientific knowledge*. Springer.

McCarthy, J., Minsky, M. L., Rochester, N., & Shannon, C. E. (2006). A proposal for the Dartmouth summer research project on artificial intelligence, August 31, 1955. *AI Magazine*, *27*(4), 12.

Meadows, D. H., Meadows, D. L., & Randers, J. (1992). *Beyond the Limits: Confronting Global Collapse, Envisioning a Sustainable Future*. Chelsea Green.

Mesgari, M., Okoli, C., Mehdi, M., Nielsen, F. Å., & Lanamäki, A. (2015). "The sum of all human knowledge": A systematic review of scholarly research on the content of Wikipedia. *Journal of the Association for Information Science and Technology*, *66*(2), 219–245.

Mistry, J., & Berardi, A. (2016). Bridging indigenous and scientific knowledge. *Science*, *352*(6291), 1274–1275.

Moosa, I. A. (2018). *Publish or perish: Perceived benefits versus unintended consequences*. Edward Elgar Publishing.

National Research Council. (2014). *Convergence: Facilitating transdisciplinary integration of life sciences, physical sciences, engineering, and beyond*. National Academies.
Nauta, D. (2019) (reprint). *The meaning of information*. De Gruyter Mouton.
Ndlovu-Gatsheni, S. J. 2020. *Decolonization, development and knowledge in Africa: Turning over a leaf*. Routledge.
Nonaka, I., & Toyama, R. (2007). Strategic management as distributed practical wisdom (phronesis). *Industrial and Corporate Change, 16*(3), 371–394.
Polónia, A., Bracht, F., Conceição, G. C., & Palma, M. (2018). *Cross-cultural exchange and the circulation of knowledge in the first global age*. Cambridge Scholars.
Priem, J., Taraborelli, D., Groth, P., & Neylon, C. (2010), Altmetrics: A manifesto, October 26, 2010. http://altmetrics.org/manifesto.
Proffitt, M. (Ed.) (2018). *Leveraging Wikipedia: Connecting communities of knowledge*. American Library Association.
Radley, A. (2021). The universal knowledge machine. In Cipolla-Ficarra, V. (Ed.), *Handbook of research on software quality innovation in interactive systems*, 102–132. IGI Global.
Renda, A. (2019). *Artificial intelligence. Ethics, governance and policy challenges*. Centre for European Policy Studies.
Sarewitz, D. (2011). Does climate change knowledge really matter?. *Wiley Interdisciplinary Reviews: Climate Change, 2*(4), 475–481.
Stember, M. (1991). Advancing the social sciences through the interdisciplinary enterprise. *The Social Science Journal, 28*(1), 1–14.
Sternberg, R. J. (2021). We've got intelligence all wrong – and that's endangering our future. *New Scientist, 3317* (16 January), 1–10.
Sternberg, R. J. (2020). Rethinking what we mean by intelligence. *Phi Delta Kappan, 102*(3), 36–41.
Sullivan, A. M. (2016). Cultural heritage & new media: a future for the past. *J. Marshall Rev. Intell. Prop. L., 15*, 604.
The Oxford Review Encyclopedia of Terms (n/d). *Phronesis: Definition and meaning*. https://www.oxford-review.com/oxford-review-encyclopaedia-terms/phronesis-definition-meaning/.
Tom, M. N., Huaman, E. S., & McCarty, T. L. (2019). Indigenous knowledges as vital contributions to sustainability. *Int Rev Educ, 65*, 1–18.
Turnhout, E., Metze, T., Wyborn, C., Klenk, N., & Louder, E. (2020). The politics of co-production: Participation, power, and transformation. *Current Opinion in Environmental Sustainability, 42*, 15–21.
Ulloa, A. (2017). The geopolitics of carbonized nature and the zero-carbon citizen. *South Atlantic Quarterly, 116*(1), 111–120.
United Nations Environment Programme (UNEP) (2019). *Global environment outlook – Geo-6: Healthy planet, healthy people*. UNEP.
Vargas Roncancio, I., Temper, L., Sterlin, J., Smolyar, N. L., Sellers, S., Moore, M., ... & Egler, M. (2019). From the Anthropocene to mutual thriving: an agenda for higher education in the Ecozoic. *Sustainability, 11*(12), 3312.
Wark, M., & Jandrić, P. (2016). New knowledge for a new planet: Critical pedagogy for the Anthropocene. *Open Review of Educational Research, 3*(1), 148–178.
Wheeler, J. A. (1990). Information, physics, quantum: The search for links. In Zurek, W. (Ed.), *Complexity, Entropy, and the Physics of Information*, 3–29. Addison-Wesley.

Wong-Parodi, G., & Feygina, I. (2020). Understanding and countering the motivated roots of climate change denial. *Current Opinion in Environmental Sustainability, 42*, 60–64.

Zuboff, S. (2019). *The age of surveillance capitalism: The fight for a human future at the new frontier of power*. Public Affairs.

Conclusion to *Knowledge For The Anthropocene*

Günter Koch

The World Capital Institute's (WCI) decision to shift from its narrower dedication to the theme of knowledge cities and regions since its inception to the larger question of the future of the planet in a new Earth Age, namely the Anthropocene, laid the foundation for an ambitious new programme of work for the Institute. The title of this book addresses two dimensions, each of which encompasses an entire scientific agenda, and which would have to find its reflection in an extensive scientific infrastructure. Both "disciplines", that of knowledge science and that of the science of the Anthropocene, already fill entire libraries. For our book project, the difficulty was to make the right choice of "spotlights" on the range of topics here, a decision which was excellently mastered by its main editor.

It would be hubris if I, as co-editor, were now to venture a summary or even a consistent conclusion from the many – altogether 28 – high-quality and substantial contributions. What can be said, however, is that these many exciting and intellectually excellent chapters, result in a jigsaw scenario that must undoubtedly remain incomplete, but which nevertheless allows us to see in outline where the discussion stands today and where the scientific journey will go. What I can do is to work out what the major areas of agreement are, where there are still clear gaps in the spectrum of topics and what the "lessons" from these contributions can be.

The clear consensus is the recognition that we have entered an age in which our planet's ability to survive is already being shaped today and will be shaped in the future by the actions of people and thus by man-made political decisions. The major political signal for this was set in 2015 by the Paris Climate Agreement. In this book, at best harmless doubts are raised as to whether a separate term such as the Anthropocene is needed for this, or whether we are no further along than the final stage of the Holocene. What is striking, however, is that in many contributions a variety of new terms appear, all of which attempt to condense new phenomena and facts into a single word. It is a characteristic of science that an object is considered to be "understood" and sufficiently elaborated when a separate particular term can be found for it that summarizes the new concept in one word.

On this point I can report a second conspicuous feature: the frequency of emergence of new conceptualizations has increased exponentially over the last few decades. I would even go so far as to claim that this phenomenon can be taken as a measure of both the accelerated speed of the acquisition of new knowledge and the increased complexity of the subject of discourse. The fact that we are already in a jungle of new terms was also the reason for dedicating a separate chapter in this book to a glossary as well as to the digital sources for researching the many (new) topics.

This brings us to the topic of complexity. It is clearly the case that the intersection of questions about the Anthropocene and the provision of knowledge to find one's way scientifically in the Anthropocene is complex. The emergent facts about a new world and the necessary requirements for action is not only complex in itself, but also requires complex action. In other words, it goes beyond the ability of a single person or a small group to cope with the complexity of the world we live in.

Personally, everyone knows what complexity is because everyone has a subjective or a professional interpretation and a way of coping with it. For example, a mathematician aims to find formulas describing phenomena that have proven to be difficult to capture, such as changes in weather or financial markets. The interest of a technician is to learn how to master a huge and virtually impossible understanding of new machinery or facility. A medical doctor is hardly in a position to consider all factors which are relevant with respect to a disease which is new or not yet curable. A psychologist or a cognitive scientist may wonder how people take decisions in situations in which they do not have all data and facts needed and may be confused. At the "lower end" of obfuscation, computer scientists do research on the complexity of algorithms, and, at the "upper end" they are confronted in understanding the complexity of large systems, the largest for them being the internet. Finally, sociologists seem to have surrendered in interpreting how society in its rapid development can be explained.

There is no question that in such a situation, where large parts of society no longer understand in a logical sense what is happening in politics, economy and science, there is a tendency to adopt esoteric models of explanations communicated by some self-appointed gurus. Such individuals want to make us believe that complexity can be reduced, as if factual complexity could be spirited away. In the best case we may admit that there exist "recipes for thinking" about how such hydra can be fought. At a minimum, we may acquire the courage to fight against it.

On the other side there exist options in terms of research institutes and researchers on a world scale who take up the challenge to study and to resolve complexity. For example, the Santa Fé Institute in New Mexico, the IIASA Institute near Vienna, Austria, devoted to "Grand Challenges Research", or the department for foresight research at the Austrian Institute of Technology (AIT) or the complexity research group of Stefan Thurner in Vienna – to name a few as examples from my neighbourhood. Their intellectual foundation can be found in the constructivist school of thinking well-anchored in Vienna. All these experts are quite engaged. However, within the larger community of scientists they are represented only in subcritical numbers. (One of those colleagues in the domain of complexity research told me that for his scientific career he has to work during the day in a classical discipline – say, physics – and in the evenings he invests in the still little-respected discipline of complexity science.)

As a veteran of many years in science management it is clear to me that a full generation and several failed existences of scientists are needed until a new direction becomes accepted, the benefit of which cannot be immediately recognized. Nevertheless, I cannot understand that a discipline that might be called "complexity science", and the resulting body of knowledge about complexity and its mastering, still has not yet been widely introduced in universities and in their curricula, given that everybody perceives that the big problems at hand (i.e., the Grand Challenges raised by the Anthropocenic changes) by their nature are complex. This holds especially for all the different aspects of securing our future living conditions, for example, climate, demography, diseases, finance, to name but a few. It is even worse: the ignorance of methodologies for coping with complexity seduces one into taking decisions too often with catastrophic consequences, which, in retrospect, we must wonder how naive and with how little analytical competence the authorized decision makers have acted. There exist many such examples – think of Chernobyl, of the catastrophes in the financial sphere a decade ago and now the pandemic – which seem to hit us like unpredicted meteors.

More and more citizens are starting to see that those owning the power and who are sitting in the governments evidently are not as much in the know as they make us believe. Polls made, for example, in the Austrian Parliament a few years ago on finding out about competences of the MPs in the field of financial politics disclosed conclusive results. And this is not a singular case. We must recognize that even highly respected personalities in world politics lack the know-how in the kinds of complex matters they are engaged to resolve. The more we become aware of such deficits, the more we are requested to find solutions on how to build a stable future architecture of political decision making. To defend and further develop democracy needs ideas on how to benefit from the social and political intellectual capital available for developing new and innovative ways of decision making in regulations – one subject discussed in this book. And even regulatory framework constructions need to be reconsidered in order for them to keep pace.

In general, I claim that the reason for the inability to understand this need for change has its roots in the still predominant paradigm in the Western Hemisphere of the past which makes people believe that mankind still has to consider itself to be challenged "to subdue the earth" and that this is doable somehow with the deterministic thinking that has made us successful up to now. Albert Einstein, who himself was convinced of the causal rational and logical foundations of the world, was intelligent enough to state that problems cannot be solved with the methods which have produced them. In a metaphoric (!) sense, this conforms to Kurt Goedel's discovery, which tells us that within a (formal) system of mathematics there may exist mathematical problems which cannot be decided if they can be solved within that system. Thus, the paradigm which limits our potential in understanding the complexity of our world of today results from the ways of thinking re-enforced in the enlightenment period 300 years ago. To date, its "laws" were helpful and produced the whole body of knowledge on which our modern society and technologies rely. However, the example of not understanding complexity demonstrates that we need new foundations of thinking. This will be achieved in science – one of the best examples is quantum physics – but is certainly not yet acquired in everyday life.

In my own environment, so far, I can identify one direction in founding a new paradigmatic basis that is helpful in dealing with complexity. The "ideological" precondition, which makes us humble in our expectations, comes from the discussion on constructivism, following the famous dictum of Wittgenstein, that "that which one cannot speak of, one must remain silent about". This insight tells us that in the end we only can master what we can understand, that is, what we can explain, being founded on scientifically sound insights. We may not believe that beyond our own competences and capacities there exist absolute and objective truths, which at the end of time will solve all our problems anyway. The only way to cope with and (if you want) to "master" complexity is to behave in a competent and educated way versus complex situations, which means not to lose one's head or even to panic. The future, which is always a complex beast, cannot be foreseen, but what we can do is to participate actively in its design, and that is mainly what constructivism tells us.

What is the consequence for our discussion through this book? Besides the formal accreditation of "complexity research" as an official meta- and trans-disciplinary art, its methods and insights have to become part of any programme in research and education.

One of the lessons from the well-conceived contributions of this book is that there exist sufficiently many scientists who are able to establish and to grow research groups in new areas such as those mentioned above. There is an urgent need for results from "complexity science" in politics and management, mainly in domains of environmental sciences, financial business, new regulations, infrastructure development, social organizations to name but a few.

The obviously still-unresolved question of how the new complexity can be mastered scientifically, intellectually, perceptively, cognitively and emotionally brings us exactly to the edge that we want to outline more clearly than before with this book. At first it may seem that with the 28 contributions we have enlarged the forest so much that one can no longer see the wood for the trees. At the same time, however, this clarifies the requirement, to be discussed later, that we are on the threshold of new paradigms and methods that have not yet been written and that construct a new programme for science. If in 1899 the President of the American Patent Office, Charles H. Duell thought that his agency could be closed because all inventions had been made, this proved to be one of the most fundamental misunderstandings of ingenious human history. With a different intention and for entirely different motives, Francis Fukuyama declared in his famous work *The End of History* (1992, Free Press), a conclusion without a continuation, as an end-point of mankind's' ideological evolution and universalization of liberal democracy after the Western model as an ultimate blueprint of human government, just as politicians proclaimed the final victory of capitalism over socialism with the collapse of the Berlin Wall. The contributions to the discourse developed in the chapters of this book point in a completely different direction, namely that now, more than ever, the reinvention of society and with it, inevitably interwoven, the future history of the planet must be rewritten. Science is being reset to a new beginning.

Now one could sit back and think about how such a new beginning could be approached on the basis of what has been argued so far. If one uses the well-known permutations of knowing versus not knowing, that is, knowing the known, not knowing the known, knowing the unknown, and not knowing the unknown one is trapped in the fact that it is precisely the fourth case that prevents us from being able to plan the future: The unknown unknown. We know a lot about the major developments on our globe, but we are always hit by surprises, like the current coronavirus pandemic. We have to acquire the ability of intellectual resilience in order to constantly reinvent the world in which we live. But that is what makes our lives so exciting.

Until now, human history has been easy for humanity to grasp, as it has been dominated by Western culture for the last hundreds of years. The Christian religion gave it carte blanche to do whatever it wanted. The disaster began with the Genesis, the story of creation as reported in the Bible, namely the divine command to Adam and Eve, known as the injunction "Dominium terrae" with the words "Be fruitful and increase in number; fill the earth and subdue it. Rule over the fish in the sea and the birds in the sky."

The next most important "wrongdoer" in Western history was Aristotle, who taught us the method and programme of scientifically logical and stringent work, which in turn found its heyday in the phase of the Enlightenment. Francis Bacon (1561–1626) declared that "knowledge is power" and meant by this that the mastery of nature and planetary conditions was the main task of science to be accomplished with human intellect. So far, this has worked quite well. The first concerns that this programme was no longer working properly came to the natural scientist Alexander von Humboldt who, around the year 1800, by studying the lowering of the water level of a lake in Ecuador, discovered that the uncontrolled deforestation of the neighbouring forest was the cause. It was to take almost 200 years for this kind of insight on ecological balancing to reach us.

What I found most interesting in our book were the contributions of authors from the non-western and non-Christian societies such as from the indigenous peoples of Australia or Africa, who, from their cultural background, provide a new approach to the observation of and interaction with nature. I have found at least as exciting the approaches of those authors who methodically open up a new perspective to people who are formal scientists like me, *nota bene* the colleagues from the field of artistic disciplines excellently represented in this book.

If I promised at the beginning to also point out gaps in the collection of contributions, as a computer scientist I am struck by the fact that the topic of a virtual Anthropocene complementary to the real Anthropocene has not been dealt with much. As these times are emerging, the big boss of Alphabet, the parent company of Google, made the statement that the time had come to reach an agreement on the future of the internet complementary to the Paris climate agreement. The inventor of the World Wide Web, Tim Berners-Lee, already stated in the mid-teens that "The internet is broken", by which he meant to say that the internet has developed in a direction that was not wanted nor foreseen by its protagonists and that can be seen as synonymous with the environmental catastrophe.

If one looks for a common thread in the contributions, a momentum emerges that is relevant to the question of what future research on the topics raised should look like. Taking the discussion on "new economics" as an example, it can be stated that the current economic system, which is considered to be co-responsible for current undesirable developments, needs a profound change. Quite a few of our authors address the question of how radical such a renewal step must be: evolutionary or revolutionary. How difficult this question is to answer can be demonstrated by the example of the discussion on the Gross National Product (GNP; as well as on GDP with D for Domestic): Although the national economic ratio has been considered useless and misleading for years, no alternative proposal has yet prevailed. It seems that the theoretical framework of economics that has evolved over the last hundred years continues to provide a stable framework within which economic measures are developed or refined. The question that arises here is whether this edifice should not be redesigned from the ground up, because it is becoming increasingly apparent that the axiomatic nature of the models no longer corresponds to a description of the new reality of an Anthropocene economy. The question is whether, in the prevailing economistic worldview, it is sufficient to state that costs of environmental burdens may no longer be externalized and thus outsourced to the detriment of us all, that is, that here no more than relatively simple accounting calculations are applied and that it is ultimately only a matter of politically answerable questions of distribution and imputability. In view of the claim of economists to justify their models formally, that is, mathematically, just as natural scientists do, the question now becomes exciting whether and how economics, like physics already today, will be able to open up a new and completely different dimension in the sense of mindboggling quantum philosophy. In the field of finance, such upheavals are on the horizon, as is evidently already being experienced today in practice in the still simple example of virtual money – the keyword is Bitcoin or Blockchain, its technological basis.

In order to generalize these kinds of thoughts, the question arises at the end of this conclusive review whether the programmes of science should not be rethought from the ground up. In any case, this thought suggests itself when one reads the following overview of the agglomeration of philosophies, methods and schools of thought from the first chapter of this book written by Carlos Jesús García Meza stating in summary:

> Latest philosophical schools: new materialisms and posthumanism, actor-network theory, assemblage theory, agential realism, object oriented ontology and the theory of hyperobjects – represent a promising philosophical shift in terms of a different trifecta of ontology, epistemology and ethics that instead of holding itself to a logic of ontological separation and superiority (and the associated violence to other beings and entities and to ourselves), may recognize the interdependence of the human with the rest of the world. Let us call this reconfiguration a "more-than-human" world, an ontological shift that may well allow us to acknowledge our entanglement(s) in more-than-human realities that can relocate us in new, symbiotic, biophilic relationships with the planet, instead of the usual ones of separation, parasitism and destruction.

I could not think of a better conclusion, which for this book also serves as an introduction.

Index

abrupt climate change 54
abyssal thinking 167
accelerated tautological vision 290
The Accursed Share (Bataille) 299
actant 17, 54–5
actionable knowledge 345, 349–50
actor-network theory 17
adaptation 55
    deep adaptation 145–54
    inequality and climate justice 183
    transformational 150
adaptive capacity 55
adaptive intelligence 350
adaptive knowledge 107–15
    actionable 109–10, 113
    cultural learning 113–14
    deep adaptation 110, 111
    glocalism 114
    human security 108
    pedagogical 109, 113–15
    perceptual 108, 114, 115
    positive deviant 113–14
    strategic 108–9, 112, 115
    Sustainable Development Goals 110–11
    see also knowledge
aerosol 55
afforestation 55
agenda for knowledge 346–52
    Anthropocene literacy 347–8
    community actionable knowledge management 349–50
    global common knowledge markets 350–52
    global open knowledge base 346–7
    human response 348
    integrated theory of knowledge 346–7
    international monitoring plan 349
    radical science and education overhaul 348–9

The Age of Surveillance Capitalism (Zuboff) 228
agrarian state economies 122–3
agriculture, climate change impacts 249–50
AI Superpowers: China, Silicon Valley, and the New World Order (Lee) 215
Ake, Claude 168
albedo 55
Albrecht, Glenn 20
Alliance of World Scientists (AWS) 77
alterity 292
Amin, Samir 173
Anders, Günther 27
Anthropocene and Digital Technologies 77
Anthropocene Arts 77
Anthropocene Curriculum 77
Anthropocene Knowledge 78
anthropocentrism 11–12
anthropogenic climate change 56
anthropogenic emissions 56
anthropomorphism 285
anthroposphere
    ecological standards 135–6
    economy reorientation of 135–6
    genuine human sphere 134–5
    im-material spatial entity 135
    self-constituting space of culture 134
artificial intelligence 216–17, 343
Asante, Molefi Kete 169
assemblages 19
atmosphere 56
atmospheric aerosols 56
Australia, bushfires in 250–52
Australian Bureau of Meteorology 250

Bacon, Francis 363
Ball, Philip 214

Barad, Karen 13, 19
Barth, Fredrik 225
Barton, Calabrese 102
baseline for cuts 56
Bataillean perspective 300
Bataille, George 299
Beck, Ulrich 27
Becoming Human by Design (Fry) 270
Beer, Stafford 343
Belief in a Just World (BJW) 33
Bendell, Jem 110
Bennett, Jane 13, 16
Bennett, J.G. 335
Berardi, Franco 288, 289
Berger, Peter 231
Bernal, John D. 226, 341
Bernal, Martin 169
Berners-Lee, Tim 363
Bhambra, Guminder K. 170
biocapacity 56
biofuel 56
biogeochemical cycle 57
biosensors 217–18
biosphere 57
Black Athena: The Afroasiatic Roots of Classical Civilization (Bernal) 169
Black Marxism: The Making of the Black Radical Tradition (Robinson) 169
Boff, Leonardo 277
Boltzmann, Ludwig 342
Book of Imaginary Beings (Borges) 293
Borges, Jorge Luis 293
Braidotti, Rosi 13
Broken Nature: Design Takes on Human Survival 78
brotherhood, French Revolution 208
bushfires in Australia 250–52
Butler, Judith 20

cap and trade 57
Cape Farewell 78
capitalism
　communicative 293
　plastic 297–304
　semiocapitalism 289, 294
　surveillance 351
Capitalocene 57
capital systems 3, 4, 228, 229
Capra, Frithjof 138

carbon capture and storage 57
carbon cycle 57
carbon dioxide 57
carbon footprint 58
carbon inequality 179–80, 183
carbon neutral 58
carbon offsetting 58
carbon sequestration 58
carbon sink 58
carrying capacity 58
Cartesian-based science 187
Center for Climate and Energy Solutions (C2ES) 78
Center for Knowledge Systems 3, 349
Centre for Humans & Nature 78
Centre for International Environmental Law (CIEL) 79
Chaosmosis (Guattari) 281
Chardin, Pierre Teilhard de 2, 207
chlorofluorocarbons (CFCs) 58
cities of refuge 59
civilizational futures 334
Clean Air Act 198
climate
　definition of 59
　doublethink 148
　feedback 59
　injustice 34
　mainstreaming 148–50
　model 59
　refugee 59
　sensitivity 60
　system 60
Climate Advocacy Lab 79
Climate and Clean Air Coalition 79
climate change
　adaptation and traditional knowledge 252
　debate in media 43–4
　deep adaptation 147–8
　definition of 59
　human-made 245
　impact on agriculture 249–50
　scientific knowledge 219–20
climate change crisis 44–9
　bias on corporations and organizations 46–8
　bias on digital platforms 45–6
　bias on individuals 48–9
climate change denial 30–36

climate engagement 35–6
motivated cognition 35
personality tendencies 33
political orientation and ideology 34
social identity process 31–2
system justification 32–3
Climate Change Library 79
Climate Clock 80
climate crisis
definition of 59
as socio-economic and political
crisis 190
Climate Emergency EU 80
climate engagement, psychological
insights 35–6
climate justice
carbon inequality 179–80
between countries 194–5
within countries 194
dimensions of 181–2
inequality and 182–5
intersecting individual
inequalities 184–5
mitigation and adaptation
action 183
national inequalities 183–4
macroeconomic cost-benefit
analysis 195
principles 181
tax-and-bonds transfer strategy
196–7
types of 182
climate science 186–7
Climate Science Special Report (CSSR)
79
Coderre, Denis 102
cognitive bias 48
Colebrook, Claire 291
collapsology 150–51
see also deep adaptation
Collyer, Fran 172
Colomina, Beatriz 269
Colonialism 171
coloniality of being 167
commons 60
Common Sense Education (CSE) 80
communication 32, 34, 36
communicative capitalism 293
communion 280

community actionable knowledge
management 349–50
complexity science 360, 362
comprehensive wealth 159
computational chemistry 216–17
confirmation bias 48–9
Connect4Climate 80
Connell, Raewyn 172
consciousness and time studies 343
consilience 27
consumerism 125
consumption tax 194, 196
contaminant 60
Convention on Biological Diversity
(CBD) 253
convergence science 347
Conway, Erik M. 26
Copernicus Europe's Eyes on Earth 80
coral bleaching 60
corrective justice 182
cosmogenesis 280
cosmology 264–5
see also Toward Alien Cosmologies
Covering Climate Now (CCNow) 80
COVID-19 pandemic 15, 146, 154, 175,
202, 213, 219, 221, 232, 263–4,
270, 273, 307–11, 348
critical posthumanism 60
Crutzen, Paul J. 133
Cuarón, Alfonso 290
cubism 284–5, 287
cultural learning 113–14
cultural schizophrenia 171
cultural superstructures 266, 273–4
culture, definition of 112
cybernetics 342–3
cyborg assimilation 336

Daes, Erica-Irene 254
Dartmouth Proposal 343
Darwin, Charles 277
data-images transformation 287–90
decolonial critique 61
*Decolonizing Methodologies: Research
and Indigenous Peoples* 168
*Decolonizing the Mind: The Politics of
Language in African Literature*
(Ngugi wa Thiong'o) 168
*Decolonizing Theory: Thinking Across
Traditions* (Nigam) 173

deep adaptation 145–54
   adaptive knowledge 110, 111
   awareness/emptiness/action 148–50
   climate change 147–8
   climate mainstreaming 148–50
   collapsology 150–51
   on continuum of responses 145–6
   definition of 150
   degrowth details 152–4
   limitations 147–8
   vulnerability responses 151–2
deforestation 61
degrowth 152–4
DeLanda, Manuel 13
Deleuze, Gilles 281
deliberative democracy 313–14
Delillo, Don 289
de-mobility 273
democracy 307–14
   cascading systemic failures 310
   climate change 309
   COVID-19 pandemic 307–10
   crisis learning 308–9
   deliberative 313–14
   ecological design 312
   economic growth 312–13
   Fukushima disaster 308
   globalization 309–10
   nuclear power plants 308
   political economy 311
   progress traps 312
   thinking in systems 311
Descartes, René 277
desertification 61
Dezzeen 81
diagonal dimension 329
differentiation 280
digital economies 46
digital platforms 45–6
digital resources 76–95
Dion-Routhier, Justine 102
Diop, Cheikh Anta 169
distributional justice 182
Drucker, Peter 229, 240
Du Bois, William E. B. 169
Duell, Charles H. 362
Dunne, Gerry 168
Durkheim, Emile 231
Dussel, Enrique 168

Earth 61
Earth Law 81
Earth System 2, 20, 22–5, 28, 60, 246, 307
ecocide 61
ecological economics 136–7
ecological fiscal reform 201
ecological footprint 61
ecological spirituality 280
ecological standards 135–6
economics for Anthropocene 132–43
   alternative sources/economics 137–8
   anthroposphere
      ecological standards 135–6
      economy reorientation of 135–6
      genuine human sphere 134–5
      im-material spatial entity 135
      self-constituting space of culture 134
   ecological economics 136–7
   genuine savings 157–63
      future climate change 160–61
      gross domestic product 157–9
      in long run 161–3
      natural capital 159, 160
      produced capital 159
      sustainability approach 160
      weak and strong sustainability 159
   metrics, measurability and evaluations 141–2
   negative externalities 138–41
      active communicative participation 141
      entrepreneurial design elements 139–40
      socio-economic re-cultivation 140
      strategic reorientation 140
   neoliberalistic economic theory and practice 133
   sustainability 143
economies as complex ecosystems 213–18
   artificial intelligence 216–17
   biosensors 217–18
   computational chemistry 216–17
   exponential *vs.* linear growth 215–16
Ecosia 81

ecosophy 281–2
ecosystem
  definition of 61
  multi-dimensional 272–3
  see also economies as complex ecosystems; societies as complex ecosystems
Ecowarriors.app 81
educating Anthropocene
  principles for 99–101
  sociopolitical actions 99–101
  teaching strategies 101–3
    court hearings 103
    documentaries 102–3
    environmental organization representatives 103
    graphic novels 101
    opinion-building and action-research processes 102
Einstein, Albert 361
Elizondo, Virgil 277
Elton, Charles 278
empirical epistemology 341
The End of History (Fukuyama) 204, 362
The *End of the Cognitive Empire: The Coming of Age of Epistemologies of the South* (Santos) 170
Engelke, Peter 125
environmental ecology 279, 282
environmentalism 62
environmental organizations list 84
environmental pricing reform 201
Environmental Protection Agency (EPA) 81
epistemic cultures 243–4
epistemic freedom 174–5
epistemicides 167
epistemic injustice 167–75
  black radical tradition 169
  cognitive empire and commissions 170–73
    colonial turn 170
    cultural bomb 171–2
    cultural schizophrenia 171
    empire of mind 170
    intimate enemy 170
    metaphysical empire 170
  counter-cartography of the history of human life 173

current conjuncture 174–5
Euromodernity 167, 168, 170
non-synchronous synchronicities 173
resolution of 173–4
sociogeny 173
struggles for epistemic freedom 174–5
Subaltern studies 169
Epistemic Injustice: Power and the Ethics of Knowing (Fricker) 168
epistemic justice
  climate justice 185–6
    epistemic problems 186–8
    knowledge complementarity 186
    local knowledges 187–8
  entrenching effects 185–6
Epistemology 11, 25–6, 340–41
epochal futures 335–6
equality, French Revolution 208
ergosphere 247
Eurocentrism 169, 170–71, 173
Euromodernity 167, 168, 170
European Renaissance 168
existential risks 62
exosphere 62
experiential knowledge 344
exploratory scenario planning 317
externality 128
extinction 62
Extinction Rebellion 81, 190

Fanon, Frantz 173
feedback loop 62
feedback mechanisms 62
Ferrando, Francesca 264
finitude 15, 19
flat ontology 14, 16, 17
fluorinated gases 62
Food and Agricultural Organization of the United Nations (FAO) 249
forcing mechanism 63
Forrester, Jay 343
fossil economy 63
fossil fuel economies 124–5
fossil-fueled industrialism 124
Foucault, Michel 231
fragile world hypothesis 152
Frances, Pope 312

freedom, French Revolution 208
Fresco, Jacques 208, 209, 210
Fricker, Miranda 168, 169
Fridays for Futures 81
Friends of the Earth 82
Fry, Tony 270
Fukuyama, Francis 204, 362
Future Earth 82

Gaïa Hypothesis 27
García Meza, Carlos Jesús 364
Geertz, Clifford 112
general circulation model (GCM) 63
general ecology 63
generalized ecology 281–2
genuine savings 157–63
    future climate change 160–61
    gross domestic product 157–9
    in long run 161–3
    natural capital 159, 160
    produced capital 159
    sustainability approach 160
    weak and strong sustainability 159
geological epoch 63
geological sequestration 63
geological time scale 63–4
"Geology of mankind" (Crutzen) 133
geopolitical futures 331–2
geopolitics 64
George, Henry 128
Gershenfeld, Neil 221
Giacometti, Alberto 292
Gibson, William 334
gift economies 120
Gildea, Robert 170
Giussani, Bruno 215
Gladney, Jack 289
Gladwell, Malcolm 321
Global Commission on Adaptation (GCA) report 348
global common knowledge markets 350–52
Global Diversity Foundation (GDF) 82
global ecology 27
global energy budget 64
global environmental governance 194–202
    climate justice
        between countries 194–5
        within countries 194
    macroeconomic cost-benefit analysis 195
    tax-and-bonds transfer strategy 196–7
    Green New Deal 197–202
        economic foundation 198–9
        framework 198
        historical foundation 197
        implementation 199–202
            behavioral insights 201
            ecological fiscal reform 201
            environmental pricing reform 201
            financialization strategy 199
            future research endeavors 202
            monetary and credit policies 199
            sustainable tourism 201–2
Global Footprint Network 82
global heating 64
global knowledge 240–47
    as capital 245–6
    competition and contestation 244–5
    definition of 241, 244–5
    depiction of 242
    epistemic cultures 243–4
    inequalities 243–4
    information and data 241–3
    knowledge distribution 243–4
    structures of 244–5
    see also knowledge
global open knowledge base 346–7
global resilience 272
global warming 64
glocalism 114
Godard, Jean-Luc 292
Goedel, Kurt 361
Goldsmith, Maurice 341
Goody, Jack 168
Gore, Al 321
Gorz, André 344
grains 122
Gramsci, Antonio 169
Great Acceleration 64, 108, 109, 125, 126
Great Depression 124
Great Divergence 65
Great Transition Initiative 82
green colonialism 344–5
Green Dreamer 82–3

greenhouse effect 65
greenhouse gases (GHGs) 65
Green New Deal (GND) 197–202, 272
    economic foundation 198–9
    framework 198
    historical foundation 197
Grosfoguel, Ramon 172
gross domestic product (GDP) 157–9
Gross Domestic Production (GDP) 195
Guattari, Félix 281

Habermas, Jürgen 344
Haeckel, Ernst 277
Haldane, J.B.S. 222
Haldane, Richard Burdon 220
Half-Earth Project 83
Harari, Yuval 214, 270
Haraway, Donna 14, 16, 287, 342
Harbisson, Neil 270
Harman, Graham 13
Haus der Kulturen der Welt (HKW) 83
Havel, Vaclav 314
Hawkins, Gay 301
heat island 65
Hellenocentrism 170
heuristics 48
Hidalgo, Cesar 342
Hockfield, Susan 219
Hodgson, Anthony 228
Holocene 1, 2, 20, 22, 25, 27, 54, 55, 65, 107, 120–30, 134, 312, 358
Holocene economies
    alternative energy sources 129–30
    ecological bottleneck 129–30
    economic thinking evolution 125–8
        externality 128
        labor, land, and capital 127
        reproducible capital 127
    hunter-gatherer economies 120–25
        agrarian state economies 122–3
        economic inequality 121
        fossil fuel economies 124–5
        gift economies 120
        horticultural economies 121–2
Holton, Gerald 226
horizontal dimension 329
horticultural economies 121–2
Hountondji, Paulin J. 172–3
human capital 159, 160
human exceptionalism 11, 12

Humanities, Arts, Science, and Technology Alliance and Collaboratory (HASTAC) 83
human security 108
human, social animal 42–3
hunter-gatherer economies 120–25
    agrarian state economies 122–3
    economic inequality 121
    fossil fuel economies 124–5
    gift economies 120
    horticultural economies 121–2
hydrocarbons 66
hydrologic cycle 66
hyperobjects 15, 27

Illich, Ivan 344
immanence 19
im-material spatial entity 135
impressionism 284
inclusive wealth 159
indigenous knowledge 66
indigenous knowledge systems (IKS) 344–5
inequality and climate justice 182–5
    intersecting individual inequalities 184–5
    mitigation and adaptation action 183
    national inequalities 183–4
information theory 342–3
Inhabiting the Anthropocene 83
inheritance tax 194, 197
Institute for Public Policy Research (IPPR) 83
integral ecology 276–82
    arcadian approach 277
    cosmogenesis 280
    ecological spirituality 280
    ecosophy 281–2
    environmental approach 279, 282
    generalized ecology 281–2
    integrative discipline 278
    mental approach 279, 281
    principle of natural selection 278
    social approach 279, 281–2
integrated knowledge 26
integrated theory of knowledge 346–7
intellectual capital management 233–4
intellectual property 255–7
intergenerational justice 181

Intergovernmental Panel on Climate
    Change (IPCC) 66, 69, 84, 185
International Environmental
    Organizations 84
International Geosphere Biosphere
    Programme (IGBP) 23, 24
international justice 181–2
International Organization for Migration
    147
International Panel on Climate Change
    249
intersectionality 169
intranational justice 182
ionosphere 66
Irigaray, Luce 276

Jamieson, Dale 26
Jordan, Chris 303

Kahneman, Daniel 227
Keil, Frank 220
Keynes, John Maynard 198
kin 66
Kjellén, Rudolf 64
Klausen, Soren 225
Knorr-Cetina, Karin 243
knowledge
    adaptive 107–15
        actionable 109–10, 113
        cultural learning 113–14
        deep adaptation 110, 111
        glocalism 114
        human security 108
        pedagogical 109, 113–15
        perceptual 108, 114, 115
        positive deviant 113–14
        strategic 108–9, 112, 115
        Sustainable Development Goals
            (SDGs) 110–11
    agenda for 346–52
        Anthropocene literacy 347–8
        community actionable
            knowledge management
            349–50
        global common knowledge
            markets 350–52
        global open knowledge base
            346–7
        human response 348

        integrated theory of knowledge
            346–7
        international monitoring plan
            349
        radical science and education
            overhaul 348–9
    for Anthropocene 22–3, 246–7,
        339–52
    anthropology of 225
    appropriation 27
    biology and 226–7
    in cybernetics 342–3
    definition of 240
    design elements 345–6
        actionable knowledge 345
        knowledge co-production
            345–6
        RESCUE Report 346
        UNEP's GEO6 Report 345
    experiential 344
    future of 24–7
    global see global knowledge
    history of 226
    implementation 26
    indigenous 66
    indigenous knowledge systems
        344–5
    in information theory 342–3
    integrated theory of 346–7
    integration 25–6
    intellectual capital management
        233–4
    other studies of 342–3
    philosophies of 339–40
    in physics 342
    political science 232–3
    psychology of 227–8
    reflexivity 340–41
    science and technology studies
        341–2
    sciences of 225–35
    scientific see scientific knowledge
    sociology of 231–2
    Technocene 24–5
    traditional see traditional knowledge
        (TK)
    traditional ecological 252
    vernacular 344
    as vital economic resource 240
    as world capital 240–47

Knowledge and Global Power: Making New Sciences in the South (Collyer, Connell, Maia and Morrell) 172
Knowledge as Culture (McCarthy) 231
knowledge-based attributes 230
knowledge-based culture 2
knowledge-based development (KBD) 229, 233
knowledge complementarity 186
knowledge economy 229–31
    disruptive view 229–30
    incremental view 229
    instrumental view 229
knowledge markets 350–51
Knowledge Societies 229
knowledge system (KS) 3, 227–8
Krauss, Rosalind 286
Kurzweil, Ray 215, 273
Kyoto Protocol 67

La Casas, Bartolome de 171
Lai, Jessica 255
Land Use, Land-Use Change, and Forestry (LULUCF) 67
La Région centrale (Snow) 285
Latour, Bruno 13, 342
Law, John 17
Lee, Kai Fu 215
Lentzos, Filippa 215
Lloyd, Seth 342
Local Indicators of Climate Change Impacts (LICCI) 186
Lovelock, James 273
Luckmann, Thomas 231
Lund University Centre for Sustainability Studies (LUCSUS) 84

machine learning 215
Machlup, Fritz 229
macroeconomic cost-benefit analysis 195
Maia, Joan 172
Malabou, Catherine 227, 270
Malthus, Thomas Robert 126
"Man and Nature" (Marsh) 22
Mannheim, Karl 231, 233
Marder, Michael 301
Marginson, Simon 244
Markey, Edward 198
Marsh, George Perkins 22
Masood, Ehsan 221
Mataatua Declaration 255–6
Material Practices: Earth in the Making 84
Max Planck Institute for the History of Science (MPIWG) 85
Mazrui, Ali A. 171
Mbembe, Achille 170, 171
McCarthy, Doyle 231
McKibben, Bill 219
McNeill, J. R. 125
Meadows, Donella 311
media accountability 42–9
    climate change crisis 44–9
        bias on corporations and organizations 46–8
        bias on digital platforms 45–6
        bias on individuals 48–9
    climate change debate 43–4
    human, social animal 42–3
megacities 67
megalomaniac domination 335
Melancholia (von Trier) 284
mental ecology 279, 281
Mercator Research Institute on Global Commons and Climate Change (MCC) 85
Merkel, Angela 137
Merton, Robert 231
mesosphere 67
meta-narratives 273–4
methane 67
Miéville, Anne-Marie 292
Mignolo, Walter D. 168
mitigation 183
Modern Monetary Theory 197
Monet, Claude 276
Morrell, Robert 172
Morton, Timothy 13, 15, 27
motivated cognition 35
motivated reasoning 35
multi-dimensional ecosystems 272–3
mysterious machines 269–70

Nancy, Jean-Luc 304
Nandy, Ashis 170
NatGeo 85
National Academies Press (NAP) 85

National Aeronautics and Space
    Administration (NASA) 85
natural capital 159, 160
naturalistic epistemology 341
Natural Resources Defence Council
    (NRDC) 86
natural source 67
Nature 86, 133, 214
natureculture 14
Nauta, Doede 342
neoliberalistic economic theory and
    practice 133
neomaterialist flat ontology 18
new materialisms
    agency 16–18
    materiality 13–16
    ontological turn 12–13
    relationality 18–19
Newton, Isaac 126
Ngugi wa Thiong'o 168, 170, 171
Nigam, Aditya 173
nitrogen cycle 67
nonhuman agency 67
nonlinearity 147–8
noosphere 205, 207–8, 210
Nordhaus, William 127
normative scenario planning 317
Nöth, Winfried 342
Nymphaeaceae 276

object-oriented ontology (OOO) 15–16,
    68
Ocasio-Cortez, Alexandria 198
Odum, Eugene 278
OLCreate: Climate Change Resources 86
The Old Place (Miéville) 292, 293
open disciplinarity 347
Open Educational Resources (OER)
    Commons 86
Oreskes, Naomi 26
Organisation for Economic Co-operation
    and Development (OECD) 86
Organizations Relating to Climate 87
Ostrom, Elinor 351
overshoot 68
ozone 68

Pareto-optimal strategy 196
Paris Agreement 178, 184
    Article 2.2 of 182
Parkin, Sara 113
particulate matter 68
parts per million (ppm) 69
parts per million by volume (ppmv) 68
Payne, Marise 255
pedagogical adaptive knowledge 109,
    113–15
perceptual adaptive knowledge 108, 114,
    115
Periodization 171
permaculture 68
permafrost 68
Perrow, Charles 308
personality tendencies 33
phronesis 234, 344, 350
Piaget, Jean 226
piece-meal social engineering 207
Pierce, Franklin 136
Piketty, Thomas 126
planetary boundaries 69
Planet Earth 71
planet equivalent 69
plastic capitalism 297–304
    aesthesis of 302–4
    economic expression 302–4
    as reingestion of energy 301–2
    as restricted energy economy
        297–301
    waste art 297
Polanyi, Karl 228
Poli, Roberto 329
political injustice
    breaking through impasse 190–91
    consumption economy and
        extractive industries 189
    description of 188
    entrenching effects 185–6
    Lancet Commission Report 188–9
    political capture 190
    shareholder-first business model
        189–90
political orientation and ideology 34
Pomeranz, Kenneth 65
Popper, Karl 204, 205, 206
positive deviant 113–14
Post Carbon Institute 87
posthuman agency 18
post-human Anthropocene futures
    263–74

challenges 263–4
critical paradigm 263
foresight methodology 264–7
   cosmology 264–5
   cultural superstructures 266
   multi-scalar and multi-perspectives canvas 265
   use- and human-centric pragmatism 265–6
   weak-signals research 265
LIID Future Lab experiences and conceptual tools 267
Toward Alien Cosmologies 265, 267–74
   art of future 273–4
   being and ontologies 269–70
   bodily and cognitive diversity 269–70
   cultural superstructures 273–4
   lifestyle design and cultural reset 271–2
   mapping planetary impact chains 272–3
   meta-narratives 273–4
   multi-dimensional ecosystems 272–3
   mysterious machines 269–70
   relational, collaborative and spatial diversity 271–2
posthumanism 60, 69
present moment 329–36
   civilizational futures 334
   diagonal dimension 329
   epochal futures 335–6
   geopolitical futures 331–2
   horizontal dimension 329
   mental objects 329
   multidimensionality of 329–30
   retrocausality 330
   technological futures 332–4
      horizon 1 – invent, develop, deploy 333
      horizon 2 – research, demonstrate, disrupt 334
      horizon 3 – envision, explore, embody 333–4
   thin and thick 329
   vertical dimension 329
Price, Derek de Solla 341

principle of symmetry 17
Problems of a Sociology of Knowledge (Scheler) 231
procedural justice 182
Proceedings of the National Academy of Sciences (PNAS) 87
produced capital 159
progressive tax 194, 196
Promethean gap 27
Promethean worldview 69
psychoterratic emotions 20

quadruple injustice 184
Quijano, Anibal 168

radiative forcing 69–70
Rand, Ayn 308
Rapid Transition Alliance 87
recognition justice 182
reflectivity 70
relative sea level rise 70
relinquishment 110
Renaissance perspectives 285
renewable energy 70
Renn, Jürgen 25, 247
reproducible capital 127, 160
resilience 70, 110
Resilience Alliance (RA) 87
resource-based civilization model 208
Responses to Environmental and Societal Challenges for our Unstable Earth (RESCUE) Program 5
restoration 110
Restoration for Sustainable Coastal Ecosystems (RESCuE) 87
restricted energy economy 297–301
retrocausality 330
rewilding 70
Ribas, Moon 270
Ricardo, David 126
Right Wing Authoritarianism (RWA) 33
Robinson, Cedric 169
Rogers, Everett 321
Romer, Paul 229
Roosevelt, Franklin D. 197
Roy, Édouard Le 2
Rozenblit, Leonid 220

Sachs, Jeffrey 199

Saffo, Paul 222
Sakaiya, Taichi 229
Salk, Jonas 223
Santos, Boaventura de Sousa 167
Savulescu, Julian 220
scenario(s) 316–24
    actionable perspective 316
    for Anthropocene 320–24
        anticipatory scenario analyses 321–2
        first principals 320
        questions 323
        tipping points 321
        weak signals 322–3
    definition of 316
    planning perspectives 317–18
        exploratory scenario planning 317
        normative scenario planning 317
    real-time analysis 324
    reasons for using 317–20
    specific policy choices 319–20
    storylines 316–17
    urban sustainability perspectives 318–19
Scheler, Max 231
Schellnhuber, Hans Joachim 25, 145, 152
Schiffman, Richard 307
Science as a Vocation (Weber) 231
ScienceDirect 88
Science of Science movement 341
sciences of knowledge 225–35
scientific knowledge 213–23
    climate change 219–20
    definition of 244
    economies/societies as complex ecosystems 213–18
        artificial intelligence 216–17
        biosensors 217–18
        computational chemistry 216–17
        exponential *vs.* linear growth 215–16
    future of science 223
    Haldane principle 221
    knowledge/trust 220–21
    limitations 222–3
    machine learning 215
    traditional knowledge and 253
    see also knowledge
Scientists Warning Foundation 88
Searle, John 107
Secularism 171
self-determination 254–5
self-referential characters 343
semiocapitalism 289, 294
Sepulveda, Gines De 171
Seydel, Rutherford 91
Shakur, Hassan 98, 104
Shannon, Claude 342
Shaviro, Stephen 288
Shaw, Jim 297, 300
Simonsen, Maria 345
sink 70
sixth extinction 71
Skouvig, Laura 345
Smith, Adam 126, 127
Smith, Linda T. 168
Snow, Charles P. 341
Snow, Michael 285
soalr radiation 71
social animal (human) 42–3
social capital 240
social constructionism 231–2
The Social Construction of Reality: A Treatise in the Sociology of Knowledge (Berger and Luckmann) 231
Social Dominance Orientation (SDO) 33
social ecology 279, 281–2
social engineers 43
social identity process 31–2
social psychological drivers 30–34
    personality tendencies 33
    political orientation and ideology 34
    social identity process 31–2
    system justification 32–3
Social Responsibility of Science movement 341
Social Science as Imperialism: The Theory of Political Development (Ake) 168
societies as complex ecosystems 213–18
    artificial intelligence 216–17
    biosensors 217–18
    computational chemistry 216–17
    exponential *vs.* linear growth 215–16

Society of Environmental Journalists (SEJ) 88
sociogeny 173
sociology of knowledge 231–2
Solow, Robert 127
Sontag, Susan 287
spheres 71
Steffen, Will 145
Stehr, Nico 232
Stember, Marilyn 347
Stiglitz, Joseph 198
Stockholm Resilience Centre (SRC) 88
Stoekl, Allen 300
strategic adaptive knowledge 108–9, 112, 115
stratosphere 71
subjectivity 280
subsiding/subsidence 71
Sunrise Movement 88
surveillance capitalism 351
survival 71
sustainability
    ecological economics 136–7
    economy creation 143
    in genuine savings 159, 160
    measurability and evaluations 141–2
Sustainability Zeroline 136–8, 141–3
sustainable development 71
Sustainable Development Goals (SDGs) 110–11
sustainable tourism 201–2
Swimme, Brian 280
Swisher, Kara 219
system dynamics 343
Systemiq Earth 88–9
system justification 32–3

Tansley, Arthur 278
tax-and-bonds transfer strategy 196–7
Technocene 24–5
technodiversity 72
technofossils 72
technological futures 332–4
    horizon 1 – invent, develop, deploy 333
    horizon 2 – research, demonstrate, disrupt 334
    horizon 3 – envision, explore, embody 333–4
technosoil 72

technosphere 72, 89
10 in 10 76
10000 changes 77
TerraBrasilis 89
terrestrial 72
Thatcher, Margaret 137
The Anthropocene Project 89
The Breakthrough Institute 89
The Climate Mobilization (TCM) 89
The Dark Mountain Project 90
The Earth Institute 90
The Ecologist 90
The Theft of History (Goody) 168
The Green New Deal 90
The Guardian 90
The Hartwell Paper 91
The Nation 91
The New Weather Institute (NWI) 91
The Oxygen Project 91
thermosphere 72
The Royal Canadian Geographic Society (RCGS) 91
The UK Labour Party 92
The UN Climate Change Learning Partnership 92
The Washington Post 92
The World Bank (WBG) 92
The World Economic Forum (WEF) 92
The World Future Council (WFC) 93
The World Weather Attribution (WWA) 93
thick present moment 329
Think Tank Networks 93
thin present moment 329
Third Industrial Revolution 43
350 parts per million 77
tipping point 72
Tobin, James 127
tourism 287
    see also sustainable tourism
Toward Alien Cosmologies 265, 267–74
    art of future 273–4
    being and ontologies 269–70
    bodily and cognitive diversity 269–70
    cultural superstructures 273–4
    lifestyle design and cultural reset 271–2
    mapping planetary impact chains 272–3

meta-narratives 273–4
multi-dimensional ecosystems 272–3
mysterious machines 269–70
relational, collaborative and spatial diversity 271–2
traditional cultural expressions (TCEs) 257
traditional ecological knowledge (TEK) 252
traditional knowledge (TK) 249–57
   Aboriginal people 251–2, 256
   bushfires in Australia 250–52
   climate change adaptation 252
   definition of 253
   intellectual property and 255–7
   scientific knowledge and 253
   self-determination 254–5
   see also knowledge
transformational adaptation 150
transition agendas 204–9
   French Revolution, freedom, equality and brotherhood 208
   Noosphere 205, 207–8, 210
   piece-meal social engineering 207
   resource-based civilization model 208
   trial-and-error basis 206
transition paths 205–9
trial-and-error basis 206
troposphere 72
Tversky, Amos 227
Tyndall Center for Climate Change Research 93

UK Student Climate Network 93
ultraviolet radiation 72
Union of Concerned Scientists (UCS) 93–4
Union of International Associations (UIA) 87
United Nations Declaration on the Rights of Indigenous Peoples (UNDRIP) 254–5
United Nations Environment Programme (UNEP) 198
United Nations Framework Convention Climate Change (UNFCCC) 94, 249
United Planet 94

Universal Declaration on Cultural Diversity 256
UN's Sustainable Development Goals (UN SDGs) 135

Van Gogh, Vincent 276
veil of ignorance 194–5
vernacular knowledge 344
Vernadsky, Vladimir 2, 71, 207, 346
vertical dimension 329
vibrant matter 73
visuality conditions 284–95
   accelerated tautological vision 290
   alterity 292
   anthropomorphism 285
   communicative capitalism 293
   cubism 284–5, 287
   data-images transformation 287–90
   experimental film 284–6
   images as cognition 290–91
   impressionism 284
   Renaissance perspective 285
   semiocapitalism 289, 294
vitality 16
von Humboldt, Alexander 137, 363
von Trier, Lars 284
VOX 94
vulnerability 151–2
vulnerable world hypothesis 152

Warhol, Andy 285, 293
The War of the Worlds (Wells) 42
waste art 297
Weber, Max 231
Welles, Orson 42
Wells, George 42
Westbroek, Peter 27
Westernization 170
Wheeler, John 342
Whitehead, Alfred North 23
White Noise (Delillo) 289
Wigley, Mark 269
Wilson, Edward O. 27
withdrawal 15–16
Wolin, Sheldon 313
World Capital Institute (WCI) 94, 358
World Economic Forum 272
World Intellectual Property Organization (WIPO) 253, 255

World Meteorological Organization (WMO) 59, 94
World Wide Web 363
World Wildlife Fund (WWF) 94–5
Worster, Donald 277

Wright, Ronald 312
Wynter, Sylvia 173

Zuboff, Shoshana 228